# Algorithms for Intelligent Systems

This book series publishes research on the analysis and development of algorithms for intelligent systems with their applications to various real world problems. It covers research related to autonomous agents, multi-agent systems, behavioral modeling, reinforcement learning, game theory, mechanism design, machine learning, meta-heuristic search, optimization, planning and scheduling, artificial neural networks, evolutionary computation, swarm intelligence and other algorithms for intelligent systems.

The book series includes recent advancements, modification and applications of the artificial neural networks, evolutionary computation, swarm intelligence, artificial immune systems, fuzzy system, autonomous and multi agent systems, machine learning and other intelligent systems related areas. The material will be beneficial for the graduate students, post-graduate students as well as the researchers who want a broader view of advances in algorithms for intelligent systems. The contents will also be useful to the researchers from other fields who have no knowledge of the power of intelligent systems, e.g. the researchers in the field of bioinformatics, biochemists, mechanical and chemical engineers, economists, musicians and medical practitioners.

The series publishes monographs, edited volumes, advanced textbooks and selected proceedings.

More information about this series at http://www.springer.com/series/16171

Dipti Singh · Amit K. Awasthi · Ivan Zelinka ·
Kusum Deep
Editors

# Proceedings of International Conference on Scientific and Natural Computing

## Proceedings of SNC 2021

Springer

*Editors*
Dipti Singh
Department of Applied Mathematics
Gautam Buddha University
Greater Noida, India

Amit K. Awasthi
Head, Department of Applied Mathematics
Gautam Buddha University
Greater Noida, India

Ivan Zelinka
Faculty of Electrical Engineering
and Computer Science
Technical University of Ostrava
Ostrava, Czech Republic

Kusum Deep
Department of Mathematics
Indian Institute of Technology Roorkee
Roorkee, India

ISSN 2524-7565 ISSN 2524-7573 (electronic)
Algorithms for Intelligent Systems
ISBN 978-981-16-1527-6 ISBN 978-981-16-1528-3 (eBook)
https://doi.org/10.1007/978-981-16-1528-3



This Springer imprint is published by the registered company Springer Nature Singapore Pte Ltd.
The registered company address is: 152 Beach Road, #21-01/04 Gateway East, Singapore 189721, Singapore

# Preface

This book comprises the outcomes of the International Conference on Scientific and Natural Computing, SNC 2021, which provides a platform for researchers, academicians, engineers, and practitioners to present and discuss their innovative, interesting ideas in order to stimulate and expand the horizon of their research. The book presents the latest developments and challenges in the area of soft computing like Genetic Algorithm, Self-Organizing Migrating Algorithm, Particle Swarm Optimization, Biogeography-based Evolutionary Algorithms, Artificial Neural Network, Grey Wolf Optimization, Chaotic Atom Search Optimization, etc. and its applications in the various interdisciplinary areas including weather forecasting, image registration, order reduction, power point tracking, disaster management, etc.


Greater Noida, India — Dipti Singh
Greater Noida, India — Amit K. Awasthi
Ostrava, Czech Republic — Ivan Zelinka
Roorkee, India — Kusum Deep

# Contents

# About the Editors

**Dr. Dipti Singh** is currently working as an Assistant Professor at Department of Applied Mathematics, Gautam Buddha University, Greater Noida. She received her M.Sc(2002) and Ph.d(2007) degrees from Indian Institute of Technology, Roorkee. She has more than 30 papers in reffered journals of high impact factor/International conferences. She has contributed chapters in many books and also has one Edited book "Problem Solving and Uncertainty Modeling through Optimization and Soft Computing Applications", DOI:10.4018/978-1-4666-9885-7, to her credit. She has supervised 2 Ph.D thesis, 7 M.sc and 9 M.tech Dissertations. She is life member of Operations Research Society of India, Soft Computing Research Society of India and also Member of International Research Group Unconventional Algorithms and Computing, Czech (NAVY), University of Ostrava. She is the advisory board member of series of International Conference on Soft Computing for Problems Solving (SocProS). Her research interests include Optimization Techniques, Nature Inspired Algorithms, Computational Intelligence, and their application to real life engineering problems.

**Dr. Amit K. Awasthi** is working as an Assistant Professor at Gautam Buddha University. He achieved Gold Medal in M. Sc. in Mathematics (1999) from M. J. P. Rohilkhand University, Bareilly. He received his Ph.D. in 2007 from Dr. B.R. Ambedkar University, Agra, in Cryptology. His main areas of research interests include Cryptography, Security Protocols, and Coding Theory. He has published more than 30 papers in international journal/conference/Cryptography Archives. He co-edited a volume of LNICST from Springer. He got the certificate for the most cited international research paper from (ESI). He was also appraised by By Elsevier Publishing (New York) for Significant contribution to the scientific community. Where he has served as associate editor in computer and electrical engineering. Currently, he is also serving for the Elsevier journal - computer standard and Interface. He is a member of Indian Mathematical Society, Group for Cryptographic Research, Cryptography Research Society of India, and Computer Society of India.

**Ivan Zelinka** is currently working at the Technical University of Ostrava (VSB-TU), Faculty of Electrical Engineering and Computer Science. He graduated consequently at Technical University in Brno (1995 – MSc.), UTB in Zlin (2001 – PhD) and again at Technical University in Brno (2004 – assoc. prof.) and VSB-TU (2010 - professor). Before academic career, he was an employed like TELECOM technician, computer specialist (HW+SW) and Commercial Bank (computer and LAN supervisor). During his career at UTB, he proposed and opened more than 10 different lectures series. He also has been invited for lectures at numerous universities in different EU countries as well as keynote and/or tutorial speaker. The field of his expertise if mainly on AI, unconventional algorithms and cybersecurity. He is and was responsible supervisor of numerous grant of fundamental research of Czech grant agency GAČR, co-supervisor of grant FRVŠ - Laboratory of parallel computing. He was also working on numerous grants and two EU project like a member of the team (FP5 - RESTORM) and supervisor (FP7 - PROMOEVO) of the Czech team and supervisor of international (security of mobile devices, Czech - Vietnam) and national applied research (founded by TACR agency). Currently, he is a professor at the Department of Computer Science and in total, he has been the supervisor of more than 50 MSc. and 25 Bc. diploma thesis. Ivan Zelinka is also supervisor of doctoral students including students from the abroad and a guarantor of magister study programme (Cybersecurity) and doctoral programme (Computer sciences). He was awarded by Siemens Award for his PhD thesis, as well as by journal Software news for his book about artificial intelligence. Ivan Zelinka is a member of British Computer Society, Editor in chief of Springer book series: Emergence, Complexity and Computation, Editorial board of Saint Petersburg State University Studies in Mathematics, a few international program committees of various conferences and international journals. He is the author of journal articles as well as of books in Czech and English language and one of three founders of TC IEEE on big data. He is also head of research group NAVY.

**Dr. Kusum Deep** is a full Professor, with the Department of Mathematics, Indian Institute of Technology Roorkee, India and Visiting Professor, Liverpool Hope University, UK and University of Technology Sydney, Australia. With B.Sc Hons & M.Sc Hons. School from Centre for Advanced Studies, Panjab University, Chandigarh, she is an M.Phil Gold Medalist. She earned her PhD from UOR (now IIT Roorkee) in 1988. She has been a national scholarship holder and a Post Doctoral from Loughborough University, UK assisted by International Bursary funded by Commission of European Communities, Brussels. She has won numerous awards like Khosla Research Award, UGC Career Award, Starred Performer of IITR Faculty, best paper awards by Railway Bulletin of Indian Railways, special facilitation in memory of late Prof. M. C. Puri, AIAP Excellence Award. She has authored two books, supervised 20 PhDs, and published 125 research papers. She is a Senior Member of ORSI, CSI, IMS and ISIM. She is the Executive Editor of International Journal of Swarm Intelligence, Inderscience. She is Associate Editor of Swarm and Evolutionary Algorithms, Elsevier and is on the editorial board of many journals. She is the Founder President of Soft Computing Research Society, India. She is the General

Chair of series of International Conference on Soft Computing for Problems Solving (SocProS). Her research interests are Evolutionary Algorithms, Swarm Intelligence and nature inspired optimization techniques and their applications.

# Chapter 1
# Salp Swarm Algorithm for Multimodal Image Registration

**Sanjeev Saxena and Mausumi Pohit**

## 1 Introduction

Registration of different modality images can provide very useful information about a scene. For example, infrared (IR) image possesses distinct reflective properties and hence can be used for crop analysis from satellite images as healthy vegetation reflects more energy in IR region than that in visible region and hence appears brighter whereas, water body appears darker in infrared image [1]. In medical field, the magnetic resonance (MR) images and computer tomography (CT) images provide different information for the same body part, which is very helpful for patient diagnosis. There are numerous applications of IR and visible image registration like vehicular traffic movement assessment, texture classification, security surveillance, remote-sensing application, land covers study, crop/vegetation study, urban area planning, etc.

Image registration techniques are fundamentally based on an optimization process, which searches the n-dimensional space of geometric transformations. The solution is an n-element vector having the parameter values of the registration transformation. The search is guided by a similarity metric, a function that measures the degree of resemblance between the input images after the alignment [2, 3]. The presence of noise, discretization of images, illumination difference makes it difficult to apply the traditional numerical method approach to optimization. However,

S. Saxena (✉)
Department of Electronics and Communication Engineering, Amity University, Noida, Uttar Pradesh, India
e-mail: sanjeevnsaxena22@gmail.com

M. Pohit
School of Vocational Studies & Applied Sciences, Gautam Buddha University, Greater Noida, India

 
D. Singh et al. (eds.), *Proceedings of International Conference on Scientific and Natural Computing*, Algorithms for Intelligent Systems,
https://doi.org/10.1007/978-981-16-1528-3_1

a number of metaheuristic algorithms have been reported in the literature to solve complex real-world problems of image processing [4].

Swarm-based algorithms are a class of metaheuristic algorithms, which have become very popular due to their simplicity and ease of implementation. These algorithms basically mimic the social behavior of the species present in nature. Many new algorithms have been reported in the last few years. Since a single algorithm cannot be the best algorithm for all practical applications [5], the improvement in the existing algorithms and the proposal of new algorithms are a new emerging area of research. Salp Swarm Algorithm (SSA) is a recently proposed algorithm [6] for the optimization of single and multi-objective functions. SSA is based on the swarming behavior of salps in the ocean. Since its inception, SSA has been successfully applied in various real-world optimization problems [7–10] but according to the information available, it has not been used yet in multimodal image registration problems.

The paper is organized into the following sections. Section 2 explains the concept of mutual information function as applicable to image registration. Section 3 described the proposed SSA algorithm for image registration. In Sect. 4, the results of the experiment are presented. Section 5 discusses the conclusion of our study.

## 2 Mutual Information-Based Image Registration

Image registration is a process in which an image captured from a sensor is aligned with a reference image in the database [1]. The alignment between two images is specified as a spatial transformation, mapping the content of one image to the corresponding area of the other. If T and R are the images used for image registration, then the registration problem can be described by the following equation:

$$T(X, Y) = R\left(t\left(X', Y'\right)\right) \tag{1}$$

where $(X, Y)$ and $(X', Y')$ are the corresponding coordinates in T and R and t is the geometric transformation function. Transformation function can be different for different types of misalignment. In the present study, only two types of transformations are considered, viz., translation and rotation. The transformation function for the mutually translated and rotated images is given as,

$$t(X, Y, \theta) = \begin{bmatrix} cos\theta & sin\theta & t_X \\ -sin\theta & cos\theta & t_Y \\ 0 & 0 & 1 \end{bmatrix} \tag{2}$$

where $t_X$ and $t_Y$ are the translational misalignment parameters in X and Y directions, respectively, and $\theta$ is the rotational misalignment parameter. Precise information of the parameters is required for correct image registration.

Intensity-based image registration is usually based on two separate measures, based on either Correlation or Mutual Information (MI) [11]. For multimodal image registration, where correlation techniques cannot be used, MI provides superior results [12]. MI is a measure of similarity between two images. If two images to be registered have no common part between them, value of MI is zero; whereas, if a substantial part is common between the two images, value of MI is high. Mutual information between two images T and R depends on the individual and joint probabilities of the two images to be registered and is defined as,

$$M(T, R) = H(T) + H(R) - H(T, R) \tag{3}$$

where H(T) and H(R) are the entropy of the individual images and is given by,

$$H(T) = -\Sigma_i p_T(i) \log p_T(i) \tag{4a}$$

and,

$$H(R) = -\Sigma_j p_R(j) \log p_R(j) \tag{4b}$$

$p_T(i)$ and $p_R(j)$ being the measure of the occurrence of the ith and jth grey levels respectively. For G discrete grey values, H(T) and H(R) are the sum of individual grey level probabilities. H (T, R) is defined as the joint entropy for images T and R and can be obtained by the following equation:

$$H(T, R) = -\Sigma_{i,j} p_{T,R}(i, j) \log p_{T,R}(i, j) \tag{5}$$

P is defined as the joint probability distribution of the two images. In the context of image registration problem, M (T, R) can be defined as a measure of information common to both images. The problem of multimodal image registration can be summarized as the maximization of the mutual information between two images. The Salp Swarm algorithm (SSA) uses MI as the optimization function, which should be maximized between two images. The next section explains this process.

## 3 Salp Swarm Algorithm for Image Registration

Salp Swarm Algorithm (SSA) is a nature-inspired algorithm based on swarm intelligence originally proposed by Mirjalili et al. [6]. SSA is based on the unique swarming behavior of Salps where they form a chain-like structure. The chain of salps has one leader and the rest are followers whose movement is guided by the leader. The goal of the swarm is to search for food, which exists somewhere in the search space. The leader is the one who is nearest to the food and leads the rest of the swarm towards it. The ultimate goal of the swarm is to converge on the food.

Individual salp in the population can be denoted as $x_j^i$, where j $= 1, 2, \ldots$, D, and D is the dimension of the solution space of the ith salp. Like any other optimization algorithm, SSA starts with the random generation of the initial population.

In the context of the image registration problem, the image in the database is referred to as the reference image (R) and the target image is called as the Test image (T). To generate an initial population of size N, N is the number of transformations that is applied on T defined by a D-dimensional vector with random coordinates, where each dimension corresponds to a single transformation. The population set of N images is equivalent to each salp in the swarm. To choose the leader among the N candidates, the mutual information $M(R, T_i)$ from Eq. (3) is determined for each candidate image, where $T_i$ is the $i^{th}$ random test image. The leader is chosen as the one having the highest value of $M(R, T_i)$.

To update the position of the leader at the $l^{th}$ iteration, the following equation is used:

$$ {}_l x_j^1 = \begin{cases} {}_l F_j + c_1\big((ub_j - lb_j)c_2 + lb_j\big)c_3 \geq 0.5 \\ {}_l F_j - c_1\big((ub_j - lb_j)c_2 + lb_j\big)c_3 < 0.5 \end{cases} \quad (6) $$

Here $x_j^1$ is the first salp position in the $j^{th}$ dimension and Fj is the food source position in the respective dimension. ub and lb are the upper and lower bounds in the respective dimensions, l varying from 1 to L, where L is the maximum limit of iterations. To maintain the balance between exploration and exploitation behavior of the swarm during search process, coefficient $c_1$ is used, which is updated after every iteration. It is defined as

$$ c_1 = 2e^{-\left(\frac{4l}{L}\right)^2} \quad (7) $$

where $c_2$ and $c_3$ are two random numbers drawn from the interval [0, 1], generated randomly to control the optimization process.

At each iteration, the ith follower salp's position in the $j^{th}$ dimension is updated by the position of itself and the preceding salp in the previous iteration. The updated equation is given by,

$$ {}_l x_j^i = \frac{1}{2}\left({}_{l-1}x_j^i + {}_{l-1}x_j^{i-1}\right), \ \ where\ i \geq 2 \quad (8) $$

Once a new population set is generated at the $l^{th}$ iteration, the fitness function is again generated from Eq. (3). The highest value of MI determines the leader for the $(l+1)^{th}$ iteration. Its coordinates are assigned to the food position ${}_{l+1}F_j$ for the next iteration. From Eqs. (6), (7) and (8), it is clear that (i) the SSA updates the position of its leader with the help of the best food source position and (ii) the position of the followers is updated with respect to the leader in each iteration. As the global optimum is not known initially, the best solution up to current iteration is considered as the global optimum at the beginning of each iteration. To register an image with only translational and rotational distortion, each salp represents a source image in the 3D space with a different value of (x, y, θ). The algorithm searches the solution space for the position of the candidate image having maximum value for the MI function.

## 4 Experimental Work and Results

For the experiment, three image sets of infrared and visible images are used. These image sets are created from the Image dataset of infrared and visible images [13] with randomly chosen translational and rotational misalignment. Image size for the first set is taken as 180 × 180 while the other two image sets consist of images with sizes 150 × 150. As explained in Sect. 2, Mutual Information (MI) is used as an objective function to register the images. The number of iterations for SSA algorithm is fixed at 500 and the number of candidate solutions is taken as 30. The experiment was conducted with Matlab software (version 2017a). Figure 1a shows the three visible images used in our experiment and Fig. 1b shows the corresponding set of infra-red images. Figure 1c shows the registration result for each case. Figure 1d shows the

(a) (b) (c) (d)

**Fig. 1** Image sets used and registration results. **a** Visible image. **b** Infrared image. **c** Registration result. **d** Difference image

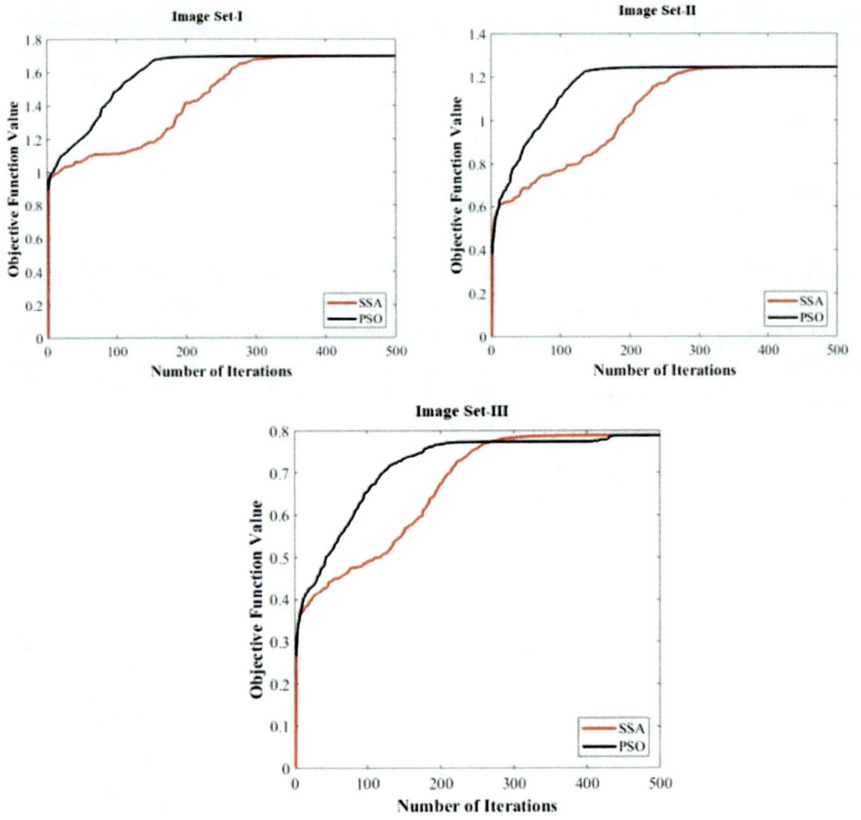


**Fig. 2** SSA versus PSO convergence behavior

difference between the two registered images emphasizing the part, which is common to both the images.

## 4.1 Comparison with Particle Swarm Optimization Algorithm

In the second part of our experiment, the performance and convergence behavior of this algorithm is assessed with respect to Particle Swarm Optimization (PSO) algorithm, which is another well-known swarm-based algorithm used in the multimodal image registration. PSO, a frequently used multimodal image registration algorithm [14–17] was originally proposed by Kennedy and Eberhart in 1995 [18]. The PSO algorithm is based on the swarming behavior of the birds. It is one of the most popular swarm-based algorithms, which has been used in various real-world engineering optimization problems successfully. In PSO, particles update their positions and velocities by the following equations:

$$V_d^i \leftarrow \omega * V_d^i + c_1 * rand1_i^d * \left(pbest_i^d - X_i^d\right) + c_2 * rand2_i^d * \left(gbest^d - X_i^d\right) \quad (9)$$

$$X_i^d \leftarrow X_i^d + V_i^d \quad (10)$$

where, $X_i^d$'s and $V_i^d$'s are the ith particle position and velocity in $d^{th}$ dimension, respectively. $pbest_i^d$ is the best position obtained by the $i^{th}$ particle till this iteration and $gbest_i$ is the best-updated position achieved by all the particles. Here, ω is the inertia weight parameter, which controls the behavior of the swarm by maintaining the balance between exploration and exploitation. $c_1$ and $c_2$ are the acceleration coefficients. In PSO, self-experience of the swarm in earlier iterations and the overall swarm confidence for the global search are decided by the values of $c_1$and $c_2$. rand represents the random numbers.

For the comparative study, standard PSO parameters are used. The value of the coefficients $c_1$ and $c_2$ is taken as 1.49. Standard linearly decreasing inertia weight (from 0.9 to 0.4) in PSO is used for the study. The number of iterations is 500 and the swarm particles are taken as 30 as used in SSA algorithm. Standard approach is used to study the convergence behavior in which each algorithm is run 30 times on each image set and parameter values are compared. The average best value in 30 runs is plotted to study the convergence behavior of both the algorithm as shown in Fig. 2. Comparative results are presented in Table 1.

Although the convergence behavior shows that PSO algorithm is converging faster than the SSA for each image sets and hence can register the images quickly in comparison to the SSA but the data presented in the table show that the SSA can achieve the maximum value of objective function (MI) in each case. The average value of objective function in 30 runs is higher in case of SSA. The standard deviation is also lower in each case for SSA. SSA although converges slower than PSO algorithm but eventually reaches the maximum value of MI and hence can provide more accurate registration results. Hence the proposed SSA algorithm is expected to provide better registration results and can be used in the registration of infrared and visible images.

**Table 1** Comparative results between SSA and PSO algorithm after 30 independent runs

| Image set | Algorithm | Average objective function value | Standard deviation | Maximum objective function value | Minimum objective function value |
|---|---|---|---|---|---|
| Set I | SSA | *1.6971* | *0.0007* | *1.6978* | 1.6937 |
| | PSO | 1.6967 | 0.0008 | 1.6976 | ***1.6936*** |
| Set II | SSA | *1.2445* | *0.0010* | *1.2455* | 1.2427 |
| | PSO | 1.2440 | 0.0011 | 1.2455 | ***1.2414*** |
| Set III | SSA | *0.7893* | *0.0007* | *0.7903* | 0.7875 |
| | PSO | 0.7889 | 0.0011 | 0.7902 | ***0.7856*** |

## 5 Conclusion

Swarm-based algorithms have become very popular due to their simplicity and ease of implementation leading to the development of many new algorithms in recent years. For any swarm algorithm, there are two phases, initial one is the exploration phase in which swarm particles vigorously search in the search space for best solutions, followed by the exploitation phase in which the fine-tuning of all best solutions is conducted. For the success of any swarm-based algorithm, it is imperative that the balance between exploration and exploitation process is maintained; otherwise the solution of the problem may be trapped in local optima.

Multimodal image registration is often necessary to extract useful information about a scene. Due to different reflective properties of various objects in infrared and visible region, this registration can be very useful in real-world practical applications. This work explores the possibility of SSA for infrared and visible image registration. The performance of SSA is compared with the PSO algorithm, which has been extensively used for multimodal image registration problems. For the comparison, standard method is used in which each algorithm is run 30 times for each image set. Results show that both the SSA and PSO algorithms provide equally accurate registration results. However, PSO converges faster in each case but SSA optimizes the objective function more effectively at a slower pace increasing the chances of exploration. Therefore, SSA has less chance to be trapped in a local minimum. Thus, we can conclude that SSA is quite promising to solve difficult image registration problems.

## References

1. Le Moigne J, Netanyahu NS (eds) (2011) Image registration for remote sensing. R.D. Eastman, Cambridge University Press
2. Registration I (2012) Principles tools and methods. A A Goshtasby, Springer
3. Zitova BJ, Flusser J (2003) Image registration methods: a survey. Image Vis Comput 21(11):977–1000. https://doi.org/10.1016/S0262-8856(03)00137-9
4. Yang X (2010) Nature-inspired metaheuristic algorithms (2nd ed). In: Conference held in Granada, Spain. Luniver Press, United Kingdom, Springer, London Ltd, London
5. Wolpert DH, Macready WG (1997) No free lunch theorems for optimization. IEEE Trans Evol Comput 1(1):67–82
6. Mirjalili S, Gandomi AH, Mirjalili SZ, Saremi S, Faris H, Mirjalili SM (2017) Salp swarm algorithm: a bio-inspired optimizer for engineering design problems. Adv Eng Softw 114:163–191
7. Faris H, Mafarja MM, Heidari AA, Aljarah I, Ala'M AZ, Mirjalili S, Fujita H (2018) An efficient binary salp swarm algorithm with crossover scheme for feature selection problems. Knowl -Based Syst 154:43–67
8. Ibrahim RA, Ewees AA, Oliva D, Abd Elaziz M, Lu S (2019) Improved salp swarm algorithm based on particle swarm optimization for feature selection. J Ambient Intell Hum Comput 10(8):3155–3169

9. Abbassi R, Abbassi A, Heidari AA, Mirjalili S (2019) An efficient salp swarm-inspired algorithm for parameters identification of photovoltaic cell models. Energy Convers Manage 179:362–372
10. Sayed GI, Khoriba G, Haggag MH (2018) A novel chaotic salp swarm algorithm for global optimization and feature selection. Appl Intell 48(10):3462–3481
11. Pluim JP, Maintz JA, Viergever MA (2003) Mutual-information-based registration of medical images: a survey. IEEE Trans Med Imaging 22(8):986–1004
12. Maes F, Collignon A, Vandermeulen D, Marchal G, Suetens P (1997) Multimodality image registration by maximization of mutual information. IEEE Trans Med Imaging 16(2):187–198
13. Brown M, Süsstrunk S (2011) Multi-spectral SIFT for scene category recognition. In: CVPR 2011. IEEE, pp 177–184
14. Wachowiak MP, Smolíková R, Zheng Y, Zurada JM, Elmaghraby AS (2004) An approach to multimodal biomedical image registration utilizing particle swarm optimization. IEEE Trans Evol Comput 8(3):289–301
15. Li Q, Sato I (2007) Multimodality image registration by particle swarm optimization of mutual information. In: International conference on intelligent computing. Springer, Berlin, Heidelberg, pp 1120–1130
16. Zhuang Y, Gao K, Miu X, Han L, Gong X (2016) Infrared and visual image registration based on mutual information with a combined particle swarm optimization–powell search algorithm. Optik 127(1):188–191
17. Liang JJ, Qin AK, Suganthan PN, Baskar S (2006) Comprehensive learning particle swarm optimizer for global optimization of multimodal functions. IEEE Trans Evol Comput 10(3):281–295
18. Kennedy J, Eberhart R (1995) Particle swarm optimization. In: Proceedings of ICNN'95-International conference on neural networks. IEEE, vol 4, pp 1942–1948

# Chapter 2
# Multi-objective Chaotic Atom Search Optimization for Epistasis Detection in Genome-Wide Association Studies

**S. Priya and R. Manavalan**

## 1 Introduction

In medical research, identifying the relationship between genes and diseases is a key step towards disease prevention, diagnosis, and treatment. Genetic and environmental risk factors serve as a major contributor to human diseases. Genome-wide association studies (GWAS) researchers aim to detect genotype variants of interest for several diseases such as diabetes, cancer, hypertension, bipolar disorder, rheumatoid arthritis, chronic illness, cardiovascular disease, diabetes, psoriasis, etc. [1]. In Mendelian's disease, the mutation occurs in only one gene. The complex diseases are influenced by multiple genes and interactions among them increase the disease susceptibility [2]. In 1909, Bateson used the term Epistasis. Genetic interactions occur in multiple genes and significantly increase pathogenicity [3]. Single nucleotide polymorphism (SNP) is a common indicator of genetic differences, which plays a pivot role in many complex traits of the disease [4]. The SNP is a difference in a sequence of DNA dependent on the four nucleotides Cytosine (C), Thyamin (T), Adenine (A), and Guanine (G), and these SNPs change the amino acid sequence [5]. Every SNP is correlated with a characteristic that identifies the disease's genetic predisposition by observing the gene regulatory pathways [6]. The GWAS use genotype data from a wide range of SNPs to identify the single gene effects and genetic interactions in human diseases like diabetes, arthritis, hypertension, and autism, etc. [7].

S. Priya (✉) · R. Manavalan
Department of Computer Science, Arignar Anna Government Arts College, Villupuram, Tamilnadu, India
e-mail: priyasri.ash@gmail.com

R. Manavalan
e-mail: manavalan_r@rediffmail.com

 
D. Singh et al. (eds.), *Proceedings of International Conference on Scientific and Natural Computing*, Algorithms for Intelligent Systems,
https://doi.org/10.1007/978-981-16-1528-3_2

This research's main goal is to develop an efficient multi-objective ASO that accelerates the identification of disease-related SNP–SNP interactions from hundreds of SNPs. The primary objectives of this research are given below.

(1) Proposed a multi-objective ASO for epistasis (MASO-Epi) strategy and multi-objective Chaotic ASO (MCASO-Epi) to classify disease-related SNPs. In MCASO-Epi approach, the initial population of Atoms is systematically determined using chaotic maps.
(2) Employ a two-stage selection process to reduce the computation overhead. In the screening stage, MASO-Epi and MCASO-Epi are used to select candidate solutions (two-way interactions) that have a stronger correlation with disease status. The G-test, a statistical strategy is applied in the clean phase to assess important associations between candidate SNPs and disease status.
(3) Investigate the power of the proposed approaches such as MASO-Epi and MCASO-Epi algorithm with MACOED using simulated disease models with marginal effects (DMEs) and disease models with no marginal effects (DNMEs).

## 2 Related Work

Researchers have typically used diverse search strategies like stochastic search algorithms, exhaustive search, and evolutionary optimization algorithms to infer genetic associations of complex diseases. Recently, evolutionary optimization algorithms attract much interest due to its ability to efficiently solve NP-hard issues in polynomial time in developing the epistasis model for reducing computational burden [8].

A multi-objective ant colony optimization technique (MACOED) was implemented to detect epistasis. ACO is applied in the screening stage to filter the SNPs and chi-squared strategy was adapted in the clean stage to find significant SNP combinations [9]. Epistasis based on ACO (epiACO) is introduced to recognize SNP interactions. The different strategies for path selection, and a memory-based approach are used to improve epiACO [10]. An Epistatic Interaction Multi-Objective Artificial Bee Colony Algorithm Based on Decomposition (EIMOABC/D) model is proposed for epistasis interaction detection. Two objective functions such as the Bayesian network score and the Gini index score are used as a measure in this algorithm to characterize different epistatic models [11]. The multi-objective bat optimization algorithm (epiBat) is suggested for genetic interaction. The Gini score and K2 score were used as a fitness function. Finally, G-test evaluates the significant SNP pairs [12].

The primary problem of the existing epistasis detection algorithms always incurs huge computational cost and minimal detection power. The proposed approach aims to identify the disease correlated SNPs with high detection power compared to the existing approaches.

# 3 Multi-objective Atom Search Optimization (ASO) for Genetic Interactions

The two versions of ASO such as MASO-Epi and MCASO-Epi are proposed. The objective functions such as AIC and K2 scores are implemented in epistasis detection model to examine the interaction between the SNP combination and disease status. The difference between MASO-Epi and MCASO-Epi lies in the population initialization of Atoms. In MASO-Epi, the population is initialized randomly, whereas MCASO-Epi initializes the population using ChebyShev chaotic map. The proposed epistasis model consists of two stages such as screen and clean stage. The general architecture of the proposed system is expressed in Fig. 1.

## *3.1 Multi-objective Chaotic Atom Search Optimization for Epistasis Detection*

Zhao et al. [13] proposed an Atom Search optimization (ASO) algorithm, a meta-heuristic search technique. ASO quantitatively simulates and mimics the behaviour of atomic motion process, where molecules interact with one another by interaction forces, leading to Lennard–Jones and bond length constraints. The ASO usually begins the optimization process by creating random populations of molecules. The atoms change their locations and velocities in each iteration, and each iteration updates the position of the better atom found so far. In MASO-Epi, each atom is assigned a position (x) based on the random combinations of SNPs. Atom's velocity (v) and positions (x) are assigned in the two-dimensional space, which are computed using Eqs. (1) and (2).

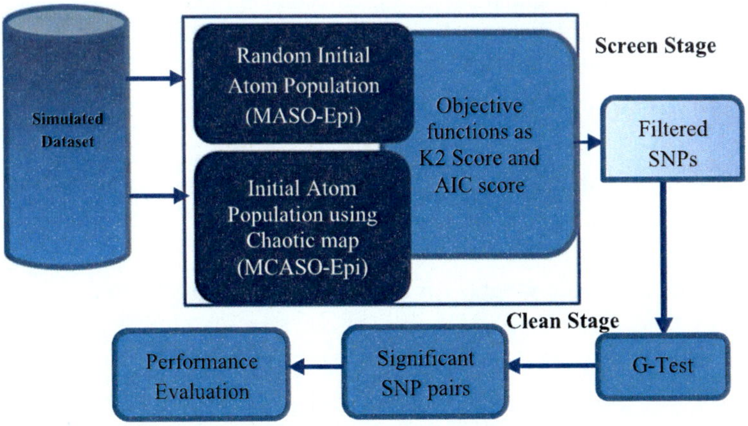


**Fig. 1** Architecture of proposed system

$$x_i(k+1) = x_i(k) + v_i(k+1) \quad (1)$$

Here, $x_i(k)$ represents the initial locations of the atoms, the position of the atom indicates the SNP number, $x_i(k+1)$ indicates the position of the molecules in the (k + 1)th iteration. The velocity of the atoms is represented as $v_i$, the velocity at (k + 1)th iteration is updated as

$$v_i(k+1) = rand \times v_i(t) + a_i(k) \quad (2)$$

where $a_i$ represents the acceleration of the ith atom in kth iteration, which is calculated using mass of the molecule (m), interaction force (F), and constraint force (G) of the atomic molecules.

In MCASO-Epi, each atom is assigned a position (X) based on ChebyShev chaotic map using the Eq. (3).

$$x_i(k+1) = \cos\left(k\cos^{-1}(x_k)\right) \quad (3)$$

The simplified version of acceleration is calculated as

$$a_i(k) = -\eta(k) \sum_{j \epsilon Kbest} \frac{rand\left[2 \times \left(h_{ij}(k)^{13}\right) - \left(h_{ij}{}^{7}\right)\right]}{m_i(k)} \times \frac{x_i(k) - x_i(k)}{x_i(k).x_i(k)_2} + \beta e^{-\frac{20T}{T}} \frac{s_{best}(k) - x_i(k)}{m_i(k)} \quad (4)$$

where

Kbest—subset of best interacting SNPs.
T—maximum number of iterations.
$s_{best}$—best-interacting SNPs in the current iteration.
$\beta$—multiplier weight.
$\eta$—depth function to regulate the atoms attraction or repulsion.
$h_{ij}(k)$—a feature that distinguishes the distance between two molecules.
$m_i(k)$—mass of the atom in kth iteration.

$\eta, h_{ij}(k) and m_i(k)$ are calculated using the Eqs. (5), (6), and (7).

$$\eta(k) = \alpha \times \left(1 - \frac{t-1}{T}\right)^3 xe^{-\frac{20T}{T}} \quad (5)$$

where, $\alpha$ represents depth weight, T is the maximum number of iterations.

$$h_{ij}(k) = \begin{cases} h_{min} \frac{r_{ij}(k)}{\sigma(k)} < h_{min} \\ \frac{r_{ij}(k)}{\sigma(k)} h_{min} \le \frac{r_{ij}(k)}{\sigma(k)} \le h_{max} \\ h_{max} \frac{r_{ij}(k)}{\sigma(k)} > h_{max} \end{cases} \quad (6)$$

where, $h_{min}$ and $h_{max}$ indicate the upper and lower boundary of h, $r_{ij}(k)$ identifies the distance between the two atoms i and j at the kth iteration, $\sigma(k)$ represents the scale length, $h_{min}$ and $h_{max}$, The mass M is calculated as follows,

$$M_i(k) = e^{-\frac{Fit_i(t)-Fit_{best}(t)}{Fit_{worst}(t)-Fit_{best}(t)}} \tag{7}$$

The best scoring values among the atom populations are considered as best global solution. The algorithm is stopped once the maximum iteration is reached.

In contrast to a single objective function, multi-objective (MO) function stores several optimal solutions. The multiple optimal solutions are known as Pareto solution. MO optimization methods produce a collection of appropriate solutions that allow designers to explore wider options and also assist in arriving the best optimal solution. The pseudo-code of MASO-Epi and MCASO-Epi for selecting the best combinations of disease-related SNPs is exposed in Fig. 2.

#### 3.1.1 Scoring Functions

The objective functions such as Akaike information criterion (AIC), K2 scores, are chosen as multi-objective functions for MASO-Epi and MCASO-Epi approach in stage I. In stage II, the G-test score is used to find significant diseased SNP pairs. The detailed description of the AIC and K2 score is presented in [9]. The G-test score is available in [8].

#### 3.1.2 Pareto Optimal Approach

The proposed methods such as MASO-Epi and MCASO-Epi employ AIC and K2 scores, as objective functions. It is essential to choose the optimal SNP subset that fulfils both objective functions. The non-dominated SNP subset can be chosen based on the results of two fitness functions using pareto optimal approach [9].

## 4 Experimental Results and Discussion

The proposed epistasis models are implemented using MATLAB R2018(b) software. Sections 4.1 and 4.2 provide a description of the simulated dataset and performance measure, respectively. Section 4.3 exposes the experimental outcome of epistasis disease models.

```
Multi-objective Atom Search Optimization for Epistasis Detection (MASO-Epi and
MCASO-Epi)

Input
        Data: Simulated dataset, N: number of Atoms
        m: interaction order
        max_iter: Maximum iterations
        α: Depth weight, β: multiplier weight
Fitness Functions
        AIC score and K2 score
Output
        Significant disease-related SNP pairs
-------------------------------------------------------------------------
Screen Stage
Step 1: Initialize the positions and velocity of atoms, randomly.
Step 2: Each atom is assigned with a random position (MASO-Epi). In MCASO-Epi,
each
      atom is initialized with ChebyShev chaotic map.
Step 3: while t < max_iter do
          for i= 1 to N
             For every atom in the solution space considers a combination of
             SNPs. Select a SNP combination and generate a local solution
             based on the AIC score and  K2 score for the SNPs.
             The SNPs with non-dominated solutions are stored separately.
             Calculate interaction Force F, constraint Force G
             Determine the acceleration of the atoms using the equation (6)
             Update Position and Velocity of the Atom.
             The Atom evaluates new combinations of SNPs and Compared
             it with the previously stored solution space and update the current
             solution space.
          end for
      end while
   Step 4: The Non-dominated SNPs are return to the clean stage for choosing the best
candidate solution
Clean Stage
   for i = 1 to size (non-dominated SNPs)
   for j = i+1 to size (non-dominated SNPs)
   Epistasic_pair = G-test (xi , xj)
   end for
   end for
```

**Fig. 2** Pseudo code for ASO for epistasis detection

### *4.1 Simulated Datasets*

The ASMO-Epi algorithm is tested on simulated datasets of various disease models such as Disease Loci with Marginal effect (DME) models and Disease Loci without Marginal Effects (DNME) models. Here, the simulated datasets are generated using the software GAMETES_2.0 [14]. Each DNME and DME model contains 100 data sets with 1600 samples including 800 cases and 800 controls. Each sample holds

100 SNP loci, including two disease correlated SNP loci (MOP01 and MOP11) and 98 non-pathogenic SNPs.

#### 4.1.1 Disease Models with Marginal Effects (DMEs)

DME model depicts the interactive and marginal effects of the disease. The three DME models, namely additive, multiplicative and threshold models are chosen for analysis, which are presented in [15]. These three simulated models are generated with four MAF values (0.05, 0.1, 0.2, and 0.5) for this research. Therefore, totally $3 * 4 = 12$ sets of DME models are generated.

#### 4.1.2 Disease Loci Without Marginal Effects (DNME) Model

DNME model shows only interactive and no marginal effects of the disease [16]. A total of 40 DNME 2-locus models are generated using Gametes under varying $h^2$ and MAF values. The $h^2$ values are ranged from 0.025 to 0.2, and MAFs are set as 0.2 and 0.4.

### *4.2 Performance Metrics*

The efficacy of the proposed epistasis models is evaluated using statistical parameters power [8] for detecting true disease loci by rejecting the null hypothesis. The power is defined as:

$$Power = \frac{\#DS}{100}$$

where #DS represents the number of datasets containing successful detection of disease-related SNP in the 100 datasets with respect to the same criteria and the penetrance table.

### *4.3 Result and Discussion*

In this section, we evaluate the performance of MASO-Epi, MCASO-Epi and MACOED [9]. The parameter settings of MASO-Epi, MCASO-Epi and MACOED are expressed in Table 1.

**Table 1** Parameter settings

| Method name | Parameter |
|---|---|
| MASO-Epi and MCASO-Epi | Population size—100<br>Max. iterations—100<br>Alpha—0.5<br>Beta—0.4 |
| MACOED | As in [9] |

### 4.3.1 Experimental Results of DNME Models

Table 2 and line graph in Fig. 3 show the detection power of the MCASO-Epi, MASO-Epi, and MACOED for 40 DNME models. The MCASO-Epi approach produces 100% power for 26 models (Models 1–15 and Models 21–31), the MASO-Epi and MACOED did not yield 100% power for none of the 40 models. Apart from the 26 models, the power of the MCASO-Epi is also superior in Models 16–21 and Models 32–35 compared to MASO-Epi and MACOED. The lowest detection power is arrived by all the three approaches for the DNME models 36–40. Among these five models, the MCASO-Epi is more powerful than the other methods in models 36 and 38. Hence, the detection power of MCASO-Epi is superior for 37 models among the 40 models than others. The power of MCASO-Epi and MASO-Epi is the same in models 37, 39 and 40. In Model 37, MACOED is superior to MCASO-Epi and MASO-Epi. The line graph also shows that the efficacy of MACOED is superior to MASO-Epi. The maximum power difference between these two methods did not exceed 10. The highest power achieved by MACOED is 96 in Model 26, while MASO-Epi yielded 93 in Model 31. The lowest power achieved by MACOED is 8 in Model 26, while MASO-Epi and MCASO-Epi yielded 7 for the same model. The MACOED is superior to MASO-Epi in 28 DNME models. In seven models, MASO-Epi detection power is higher than MACOED. MASO-Epi and MACOED achieved similar power for five DNME models. The detection power of the DNME model revealed that MCASO-Epi has superior detection power than MASO-Epi and MACOED.

### 4.3.2 Experimental Results of DME Models

The power of MCASO-Epi, MASO-Epi and MACOED for 12 DME models is exhibited in Fig. 4 and the same is presented in Table 3. In the additive model, MCASO-Epi achieved 100% power for Models 3 and 4. The MACOED obtained the power of 84 for additive model 3, while MASO-Epi yielded a power of 90, which is 6% higher than MACOED. In additive model 1, all the three approaches did not detect any disease causative SNPs. These three approaches perform too poor in multiplicative models 1, 2 and 4. In Model 1 and Model 3, the three approaches did not find any disease correlated SNPs. In multiplicative model 2, MCASO-Epi obtained 100% of power, which is 10% superior to both MACOED and MASO-Epi. In multiplicative

**Table 2** Power comparison of DNME models

| Model | MCASO-Epi | MASO-Epi | MACOED |
|---|---|---|---|
| M1 | 100 | 90 | 92 |
| M2 | 100 | 87 | 88 |
| M3 | 98 | 86 | 88 |
| M4 | 97 | 84 | 86 |
| M5 | 100 | 87 | 93 |
| M6 | 100 | 88 | 93 |
| M7 | 100 | 89 | 86 |
| M8 | 100 | 89 | 93 |
| M9 | 100 | 85 | 92 |
| M10 | 100 | 88 | 93 |
| M11 | 100 | 90 | 89 |
| M12 | 100 | 86 | 92 |
| M13 | 100 | 89 | 90 |
| M14 | 100 | 91 | 88 |
| M15 | 100 | 89 | 93 |
| M16 | 54 | 56 | 46 |
| M17 | 44 | 40 | 38 |
| M18 | 41 | 35 | 36 |
| M19 | 70 | 58 | 64 |
| M20 | 82 | 65 | 72 |
| M21 | 100 | 86 | 89 |
| M22 | 100 | 89 | 87 |
| M23 | 100 | 89 | 89 |
| M24 | 100 | 88 | 96 |
| M25 | 100 | 89 | 86 |
| M26 | 100 | 88 | 90 |
| M27 | 100 | 84 | 93 |
| M28 | 100 | 86 | 86 |
| M29 | 100 | 91 | 91 |
| M30 | 100 | 86 | 94 |
| M31 | 100 | 93 | 91 |
| M32 | 99 | 85 | 89 |
| M33 | 96 | 91 | 85 |
| M34 | 96 | 85 | 90 |
| M35 | 88 | 79 | 89 |
| M36 | 15 | 14 | 14 |
| M37 | 7 | 7 | 8 |

(continued)

**Table 2** (continued)

| Model | MCASO-Epi | MASO-Epi | MACOED |
|---|---|---|---|
| M38 | 17 | 14 | 13 |
| M39 | 10 | 10 | 10 |
| M40 | 12 | 12 | 9 |

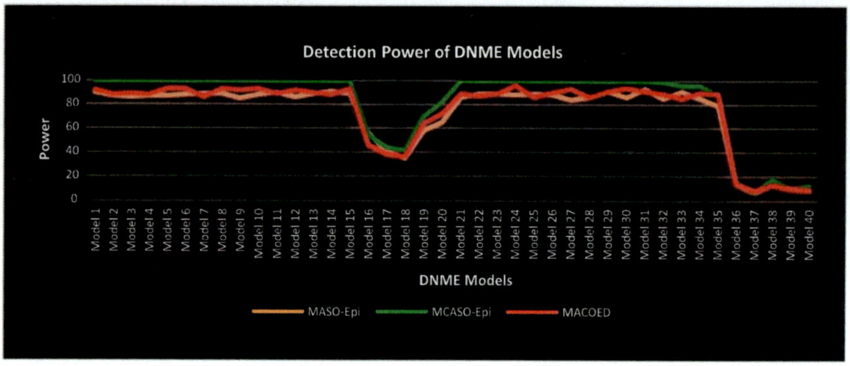


**Fig. 3** Performance evaluation of DNME models

**Fig. 4** Power comparison of DME models

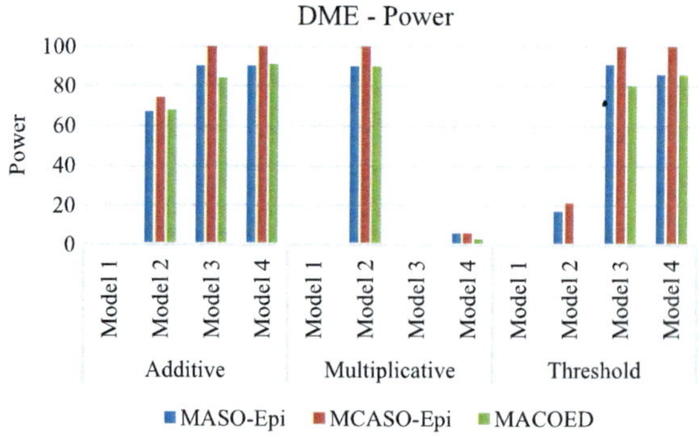


model 4, MCASO-Epi and MASO-Epi reached the same power value of 6, whereas MACOED yielded the power of 3. In threshold models, MCASO-Epi obtained 100% power for Models 3 and 4. The power of MCASO-Epi is superior to MACOED and MASO-Epi for all three models. The analysis proved that one of the proposed approaches MCASO-Epi has better detection power than MACOED in identifying 2-locus association.

**Table 3** DME power evaluation

| DME model | Model name | Power | | |
|---|---|---|---|---|
| | | MCASO-Epi | MASO-Epi | MACOED |
| Additive | M1 | 0 | 0 | 0 |
| | M2 | 74 | 67 | 68 |
| | M3 | 100 | 90 | 84 |
| | M4 | 100 | 90 | 91 |
| Multiplicative | M1 | 0 | 0 | 0 |
| | M2 | 100 | 90 | 90 |
| | M3 | 0 | 0 | 0 |
| | M4 | 6 | 6 | 3 |
| Threshold | M1 | 0 | 0 | 0 |
| | M2 | 21 | 17 | 0 |
| | M3 | 100 | 91 | 80 |
| | M4 | 100 | 86 | 86 |

## 5 Conclusion

The identification of possible genetic interactions is a critical and difficult problem in GWAS. In this article, we suggest a two-step approach called MCASO-Epi and MASO-Epi to detect epistasis effects. It employs Atom Search optimization in the screening stage to discover non-dominated SNPs and G-test in the clean stage to find the significant 2-locus SNP combination. Experimental findings proved that MCASO-Epi is superior to MASO-Epi and MACOED for DNME and DME models. The future outlook of the work consists of the following aspects: the detection time of the proposed methods can be reduced by hybridizing the methods with the fuzzy set, deep learning approaches, etc., and also discover high-order interactions. In addition, we anticipate validating our methodology over real genotype data.

## References


1. Wu T, Chen Y, Hastie T, Sobel E, Lange K (2009) genomewide association analysis by lasso penalized logistic regression. Bioinformatics 25:714–721. https://doi.org/10.1093/bioinformatics/btp041
2. Wan X, Yang C, Yang Q, Xue H, Tang NLS, Yu W (2009) Predictive rule inference for epistatic interaction detection in genome-wide association studies. Bioinformatics 26:30–37. https://doi.org/10.1093/bioinformatics/btp622
3. Zhang Y, Liu JS (2007) Bayesian inference of epistatic interactions in case-control studies. Nat Genet 39:1167–1173. https://doi.org/10.1038/ng2110
4. Visscher PM, Brown MA, McCarthy MI, Yang J (2012) Five years of GWAS discovery. Am J Hum Genet 90:7–24. https://doi.org/10.1016/j.ajhg.2011.11.029

5. Genetics home reference (2019) What are single nucleotide polymorphisms (SNPs)? http://ghr.nlm.nih.gov/primer/genomicresearch/snp
6. Mackay TFC, Moore JH (2014) Why epistasis is important for tackling complex human disease genetics. Genome Med 6:42. https://doi.org/10.1186/gm561
7. Kim H, Jeong H Bin, Jung HY, Park T, Park M (2019) Multivariate cluster-based multifactor dimensionality reduction to identify genetic interactions for multiple quantitative phenotypes. Biomed Res Int. https://doi.org/10.1155/2019/4578983
8. Tuo S, Zhang J, Yuan X, He Z, Liu Y, Liu Z (2017) Niche harmony search algorithm for detecting complex disease associated high-order SNP combinations. Sci Rep 7:1–18. https://doi.org/10.1038/s41598-017-11064-9
9. Jing P-J, Shen H-B (2014) MACOED: A multi-objective ant colony optimization algorithm for SNP epistasis detection in genome-wide association studies. Bioinformatics 31. https://doi.org/10.1093/bioinformatics/btu702
10. Sun Y, Shang J, Liu JX, Li S, Zheng CH (2017) EpiACO—a method for identifying epistasis based on ant Colony optimization algorithm. BioData Min 10:1–17. https://doi.org/10.1186/s13040-017-0143-7
11. Li X, Zhang S, Wong K-C (2018) Nature-inspired multiobjective epistasis elucidation from genome-wide association studies. IEEE/ACM Trans Comput Biol Bioinforma, pp 1. https://doi.org/10.1109/TCBB.2018.2849759
12. Sitarcik J, Lucka M (2019) epiBAT: Multi-objective bat algorithm for detection of epistatic interactions
13. Zhao W, Wang L, Zhang Z (2019) A novel atom search optimization for dispersion coefficient estimation in groundwater. Futur Gener Comput Syst 91:601–610. https://doi.org/10.1016/j.future.2018.05.037
14. Urbanowicz RJ, Kiralis J, Sinnott-Armstrong NA, Heberling T, Fisher JM, Moore JH (2012) GAMETES: a fast, direct algorithm for generating pure, strict, epistatic models with random architectures. BioData Min 5. https://doi.org/10.1186/1756-0381-5-16
15. Chen Q, Zhang X, Zhang R (2019) Privacy-preserving decision tree for epistasis detection. Cybersecurity 2. https://doi.org/10.1186/s42400-019-0025-z
16. Velez DR, White BC, Motsinger AA, Bush WS, Ritchie MD, Williams SM, Moore JH (2007) A balanced accuracy function for epistasis modeling in imbalanced datasets using multifactor dimensionality reduction. Genet Epidemiol 31:306–315. https://doi.org/10.1002/gepi.20211

# Chapter 3
# A New Algorithm for Color-Image Encryption Using 3D-Lorenz Chaotic Map and Random Modulus Decomposition in Transform Domain

**Anand B. Joshi, Dhanesh Kumar, Sonali Singh, and Keerti Srivastava**

**Abstract** This paper proposes a new algorithm for encryption and decryption of digital color-image using 3D-Lorenz chaotic map (LCM) and random modulus decomposition (RMD) in 2D-fractional discrete cosine transform domain (FrDCT) domain. The color-image to be encrypted is converted to its indexed formats by extracting its color map. A 3D-LCM is used for generating a keystream, and this keystream is used to generate public key as well as secret keys. The indexed image is combined with the public key as a complex matrix. Now, this complex matrix is encrypted using 2D-FrDCT with fractional order $a$ and $b$. Subsequently, the RMD is implemented to obtain the trapdoor one-way function. Finally, a scrambling scheme is used to randomize the encrypted image and private key. In the decryption, encrypted image and private are calculated by the inverse pixels scrambling process. The retrieval data is transformed by 2D-inverse fractional discrete cosine transform (IFrDCT). The experimental results and the security analysis are given to validate the feasibility and robustness of the proposed method.

## 1 Introduction

The security of the digital images is a major issue in the present era of digital communication. The problem of security can be solved using encryption methods. Some classic image encryption methods such as optical transforms and chaotic maps play

A. B. Joshi (✉) · D. Kumar · S. Singh
Department of Mathematics and Astronomy, University of Lucknow, Lucknow, MP, India
e-mail: anandiitd.joshi@gmail.com

D. Kumar
e-mail: dhaneshkumar.lu@gmail.com

S. Singh
e-mail: singhsonali09192@gmail.com

K. Srivastava
Department of Mathematics, CIPET, Lucknow, UP, India
e-mail: keerti.cipet@gmail.com



D. Singh et al. (eds.), *Proceedings of International Conference on Scientific and Natural Computing*, Algorithms for Intelligent Systems,
https://doi.org/10.1007/978-981-16-1528-3_3

a vital role in protecting images due to the higher level of security as compared to other methods. Most of the schemes [1–6] described above are symmetric encryption algorithm. Asymmetric encryption methods have attracted much attention in the last few decades. Much work has been done on asymmetric image encryption in [7–11].

In paper [7], Chen et al. proposed an asymmetric cryptosystem for color-image encryption using chaotic Ushiki map and equal modulus decomposition in fractional Fourier transform domains. In this method, for the design of asymmetric approach, the trapdoor one-way function is calculated by the equal modulus decomposition. Whereas Sui et al. [8] proposed an asymmetric multiple-image encryption scheme based on coupled logistic maps in fractional Fourier transform domain. In this method for the design of asymmetric approach, in the encryption process, three random phase functions are used as encryption keys to retrieve the phase-only functions of plain images. Simultaneously, three decryption keys are generated in the encryption process.

Yadav et al. [9] also proposed an asymmetric encryption algorithm for color images based on fractional Hartley transform. In this technique, for the design of asymmetric cryptosystem, amplitude, and phase-truncation approach are used. Also in [11], Huang et al. proposed asymmetric cryptosystem by using amplitude and phase-truncation technique.

In this paper, we have proposed a new asymmetric scheme for digital color-image encryption and decryption using 3D-LCM and RMD in 2D-FrDCT domain. The color-image to be encrypted is converted to its indexed formats by extracting its color map. A 3D-LCM is used for generating a keystream, and this keystream is used to generate public key as well as secret keys. The indexed image is combined with the public key as a complex matrix. Now, this complex matrix is encrypted using 2D-FrDCT with fractional order $a$ and $b$. Subsequently, the RMD is implemented to obtain the trapdoor one-way function. Finally, a scrambling scheme is used to randomize the encrypted image and private key.

The remaining sections of this paper are organized as follows. Section 2 presents the fundamental knowledge of 2D-FrDCT, 3D-LCM and RMD. Section 3 discusses the proposed digital color-image encryption and decryption method. Section 4 discusses the experimental results. Security analysis is given in Sect. 5. Finally, the conclusion of the proposed method is given in Sect. 6.

## 2 Fundamental Knowledge

### 2.1 2D-Fractional Discrete Cosine Transform

The fractional discrete cosine transform is a generalization of the discrete cosine transform (DCT). A DCT expresses a finite sequence of data points in terms of a sum of cosine functions oscillating at different frequencies. The DCT, first proposed by Ahmed [12] in 1972, is a widely used transformation method in signal processing and data compression. The 2D-DCT [13] of any 2D-signal $I_{m,n}$ of size $M \times N$ is given as follows:

$$I'_{p,q} = \alpha_p \alpha_q \sum_{m=0}^{M-1} \sum_{n=0}^{N-1} I_{m,n} \cos\frac{\pi(2m+1)p}{2M} \cos\frac{\pi(2n+1)q}{2N}, \quad (1)$$

where $0 \leq p \leq M-1, 0 \leq q \leq N-1, 0 \leq m \leq M-1$ and $0 \leq n \leq N-1$,

$$\alpha_p = \begin{cases} \frac{1}{\sqrt{M}} & \text{if } p = 0 \\ \frac{2}{\sqrt{M}} & \text{otherwise} \end{cases}, \alpha_q = \begin{cases} \frac{1}{\sqrt{N}} & \text{if } q = 0 \\ \frac{2}{\sqrt{N}} & \text{otherwise.} \end{cases}$$

The frequency domain array $I'_{p,q}$ in Eq. 1 is also of the same size as that of spatial domain array $I_{m,n}$. In matrix form, it can be written as follows:

$$I'_{p,q} = C I_{m,n}, \quad (2)$$

where $C$ is the DCT kernel-II matrix or transformation matrix of type-II. Comparing Eqs. 1 and 2, the DCT kernel matrix $C$ can be given by Eq. 3,

$$C = \left\| \frac{1}{\sqrt{M}} \beta_p \cos\left(2\pi \frac{(2m+1)p}{4M}\right) \right\|, \quad (3)$$

where $\|.\|$ represents $M \times M$ matrix, $0 \leq m, p \leq M-1$ and $\beta_0 = 1, \beta_p = \sqrt{2}$ for $p \geq 1$.

The FrDCT is derived based on the eigen-decomposition and eigenvalue substitution of Eq. 3, which is expressed as

$$C = UDU^* = \sum_m U_m e^{i\phi_m}, \quad (4)$$

where $U$ is a unitary matrix, composed of columns (eigenvectors) $u_m$, $u_m^* u_n = \delta_{mn}$, $U_m = u_m u_m^*$ and $D$ is the diagonal matrix with diagonal entries $\lambda_m$ (eigenvalues), $\lambda_m = e^{i\phi_m}$ with $0 < \phi_m < \pi$.

The FrDCT matrix $C_a$ can be written by substituting the eigenvalues $\lambda_m$ with their $a$th powers $\lambda_m^a$ as follows:

$$C_a = U D^a U^*, \quad (5)$$

where $a$ is an order of FrDCT.

For an image $I_{m,n}$, 2D-FrDCT of fractional orders $a$ and $b$ is expressed as

$$I'_{p,q} = C_a I_{m,n} C_b^T, \quad (6)$$

where $C_b^T$ is the transpose of $C_b$.

The 2D-IFrDCT can be calculated by Eq. 7,

$$I_{m,n} = C_{-a} I'_{p,q} C^T_{-b}. \tag{7}$$

## 2.2 3D-Lorenz Chaotic Map

The equation of 3D-Lorenz chaotic map [14, 15] is described as following:

$$\dot{x} = \lambda(y - x), \quad \dot{y} = rx - xz - y, \quad \dot{z} = xy - cz, \tag{8}$$

where $\lambda, r,$ and $c$ are control parameters. When $9 < \lambda < 10, 24.74 < r < 30,$ and $2 < c < 3$, the system is chaotic. The trajectory of Lorenz system can be obtained by the fourth order Runge-Kutta algorithm. The 3D-LCM creates three chaotic sequences which are entirely differ from one to other.

## 2.3 Random Modulus Decomposition Method

The RMD is a kind of unequal modulus decomposition, which is different from equal modulus decomposition. In RMD, the two-dimensional data is randomly divided into two complex value masks. Therefore, one two-dimensional image can be divided into two statistically independent masks randomly. To simplify the illustration, one complex number in two-dimensional Cartesian coordinate system is considered as a vector $Z(x, y)$ and it is divided into two vectors randomly as shown in Fig. 1.

Suppose that $Z(x, y)$ in Fig. 1 is an image, then $C_1(x, y)$ and $C_2(x, y)$ is expressed by Eqs. 9 and 10,

$$C_1(x, y) = \frac{A(x, y).sin[\beta(x, y)]}{sin[\alpha(x, y) + \beta(x, y)]}.\exp\{i[\phi(x, y) - \alpha(x, y)]\}, \tag{9}$$

$$C_2(x, y) = \frac{A(x, y).sin[\alpha(x, y)]}{sin[\alpha(x, y) + \beta(x, y)]}.\exp\{i[\phi(x, y) + \beta(x, y)]\}, \tag{10}$$

where the function $\alpha(x, y)$, $\beta(x, y)$ and $\phi(x, y)$ is given below,

$$\alpha(x, y) = 2\pi \times rand(x, y), \ \beta(x, y) = 2\pi \times rand(x, y), \ \phi(x, y) = angle\{Z(x, y)\},$$

where the function $rand(x, y)$ generates a random matrix whose element are distributed normally in the interval $[0, 1]$ and $angle$ finds the argument of the complex number. Besides, the amplitude of $Z(x, y)$ is given by $A(x, y) = |Z(x, y)|$.

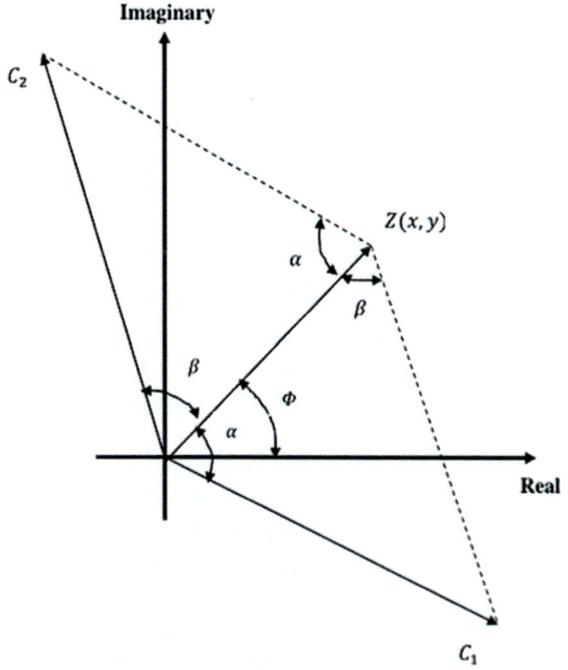


**Fig. 1** Vector representation of the decomposition process in RMD

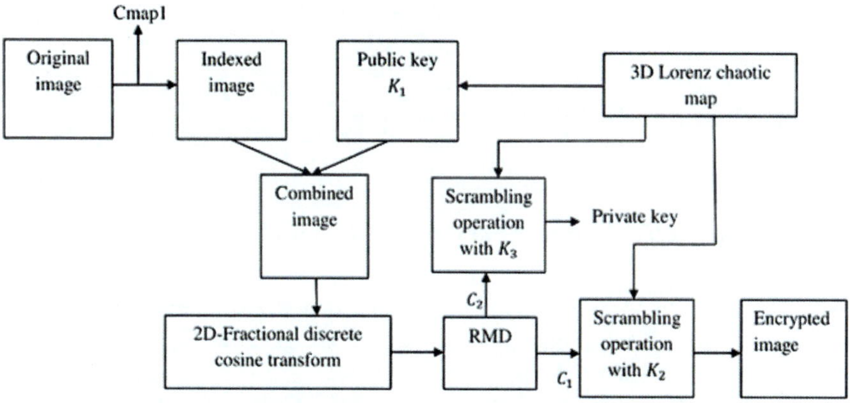


**Fig. 2** Block diagram of the proposed color-image encryption method

## 3 Color-Image Encryption and Decryption Process

Figure 2 show block diagram of our color-image encryption method. In the first step, color-image is converted to their indexed formats by extracting their color map and then use 2D-FrDCT for encryption and decryption.

### 3.1 Key Generation

Step 1: Iterate Eq. 8, $k$ times in advance to eliminate the transient response, to generate random sequences $x = \{x_1, x_2, x_3, ..., x_{MN}\}$, $y = \{y_1, y_2, y_3, ..., y_{MN}\}$ and $z = \{z_1, z_2, z_3, ..., z_{MN}\}$, respectively, each of size $\max\{1 \times MN\}$.

Step 2: Converting the sequences $x$, $y$ and $z$ into integers as

$$X = \text{floor}(x \times 10^{14}) \bmod MN, \tag{11}$$

$$Y = \text{floor}(y \times 10^{14}) \bmod MN, \tag{12}$$

$$Z = \text{floor}(z \times 10^{14}) \bmod MN, \tag{13}$$

where floor($s$) returns $s$ to the nearest integers less than or equal to $s$ and $u \bmod v$ returns the remainder after division.

Step 3: Sort the sequences 11–13 and get three sorted sequences $\overline{X}, \overline{Y}$ and $\overline{Z}$. Find the positions of the values of $\overline{X}, \overline{Y}$ and $\overline{Z}$ in $X$, $Y$ and $Z$ and mark down the transform positions i.e. $P = \{P(i) : i = 1, 2, 3, ..., MN\}$, $Q = \{Q(i) : i = 1, 2, 3, ..., MN\}$ and $R = \{R(i) : i = 1, 2, 3, ..., MN\}$, where $X(P(i)) = \overline{X}(i), Y(Q(i)) = \overline{Y}(i)$ and $Z(R(i)) = \overline{Z}(i)$.

Step 4: Now, transform the position sequences $P$, $Q$ and $R$ into matrices $M_1$, $M_2$ and $M_3$, each of size $M \times N$ and generate keys $K_1$, $K_2$ and $K_3$ as

$$K_1 = (M_1) \bmod 256, \tag{14}$$

$$K_2 = (M_2) \bmod 256, \tag{15}$$

$$K_3 = (M_3) \bmod 256, \tag{16}$$

where $K_1$ is public key and $K_2$, $K_3$ are secret keys.

### 3.2 Digital Color-Image Encryption Algorithm

The proposed method is an asymmetric color-image encryption method. The flowcharts of the proposed encryption algorithm are given in Fig. 2. The details of the process are given as follows:

Step 1: Let $CI_1$ color-image of size $M \times N$. First, Convert $CI_1$ to their indexed format $I_1$, and get corresponding colormaps cmap1.

Step 2: This indexed-color image $I_1$ is converted into complex matrix $C_m$ using public key $K_1$, given as

$$C_m = A_1 \times A_{phase}, \tag{17}$$

where

$$A_1 = I_1/255, \tag{18}$$

$$A_{phase} = \exp(i\pi(\frac{K_1}{255})), \tag{19}$$

where $\exp(x)$ is the exponential function of $x$ and $i$ is complex number satisfy $i^2 = -1$.

Step 3: Now, $C_m$ is encrypted by using 2D-FrDCT of fractional order $a$ and $b$ (given in Eq. 6), we get encrypted complex matrix $C'$, which is given by Eq. 20.

$$C' = C_a C_m C_b^T, \tag{20}$$

where $a$ and $b$ are secret keys.

Step 4: The result of the cosine transform is analytically decomposed into two masks $C_1$ and $C_2$ by using RMD.

Step 5: Finally, with the help of secret keys $K_2$ and $K_3$, the pixels position scrambling operation is performed according to $K_2$ and $k_3$ with $C_1$ and $C_2$ to obtain the cipher text $(C_t)$ and private key $(P_k)$. The 8-bit image equivalent of ciphertext is denoted by $E$ attained by using Eq. 21,

$$E = mod(abs(C_t), 256). \tag{21}$$

## 3.3 Decryption Algorithm

Accordingly, the private key and all the additional secret keys are required to retrieve the encrypted image. Since the proposed cryptosystem is asymmetric cryptosystem, the encrypted image cannot be retrieved without the private and public keys. The decryption process is given in the following steps.

Step 1: The receiver obtains an encrypted image $E$ of size $M \times N$. In this step, the encrypted image and private key are calculated by the inverse pixels scrambling operation using secret keys $K_2$ and $K_3$ to obtain $C_1$ and $C_2$.

Step 2: The complex matrix $C'$ is obtained by following Eq. 22,

$$C' = [C_1 + C_2]. \tag{22}$$

Step 3: In this step, $C'$ is transformed by using 2D-IFrDCT (given in Eq. 7), one calculates $C_m$ using Eq. 23,

$$C_m = C_{-a} C' C_{-b}^T. \tag{23}$$

Step 4: Now, using public key $K_1$ to obtain the decrypted data $(D_d)$ given by Eq. 24,

$$D_d = C_m \times \exp(-i\pi(\frac{K_1}{255})). \tag{24}$$

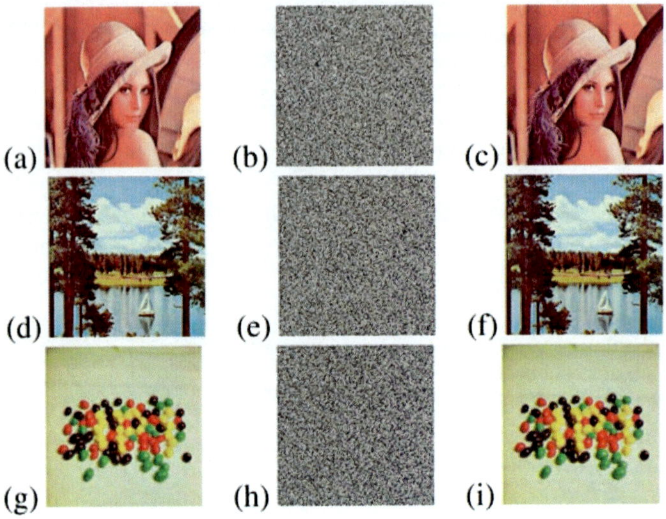


**Fig. 3** Experimental results of color-image encryption: **a** original Lena image, **b** encrypted image, **c** decrypted Lena image, **d** original Sailboat image, **e** encrypted image, **f** decrypted Sailboat image, **g** original Jellybeans image, **h** encrypted image, and **i** decrypted Jellybeans image

Step 5: Finally, After adding colormaps cmap1 in $D_d$, one gets decrypted color images.

## 4 Experimental Results

For experimental results of Fig. 3, the initial and control parameters of 3D-LCM are taken randomly as $x_0 = 1.6758$, $y_0 = 4.7854$, $z_0 = 2.3576$, $\lambda = 8.7523$, $r = 26.8756$, and $c = 2.5643$, respectively, and fractional orders of 2D-FrDCT are $a = 2.232$, and $b = 1.534$. Figure 3 illustrates method validation results for encryption and decryption. Clearly, the encrypted images are completely disordered and cannot be recognized.

## 5 The Security Analysis

A good encryption algorithm should resist all kinds of known attacks, such as an exhaustive attack, differential attack, and statistical attack In this section, we will discuss the security analysis of the proposed encryption algorithm using the images of Sect. 4.

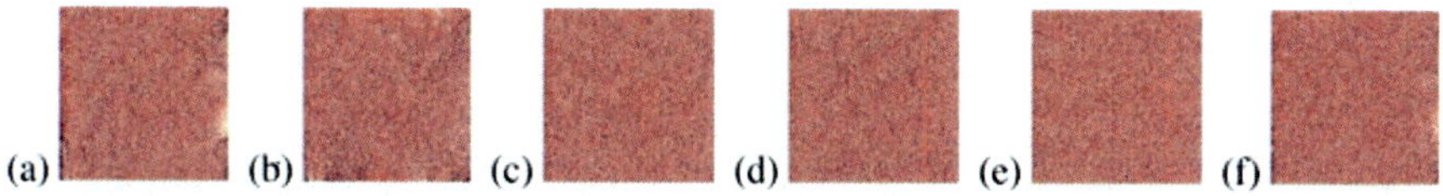


**Fig. 4** Decrypted image of Lena with incorrect keys: **a** $a = 2.233$, **b** $b = 1.535$, **c** $x_0 = 1.6759$, **d** $y_0 = 4.7855$, **e** $z_0 = 2.3577$, and **f** only one value change in private key

## *5.1 Resistance to Exhaustive Attack*

### 5.1.1 Keyspace Analysis

The security of an image cryptosystem depends heavily on its keyspace size. A large enough keyspace is essential for resisting the exhaustive attack efficiently. The proposed cryptosystem is very sensitive with respect to small changes in secret keys of 3D-LCM: the initial values $x_0$, $y_0$, $z_0$, and control parameters $\lambda$, $r$, $c$. The fractional-order parameters $a$ and $b$ of 2D-FrDCT are also very sensitive up to three decimal places. As the precision is about $10^{-15}$, the total keyspace of the proposed cryptosystem is about $10^{96} \approx 2^{319}$. Hence, the proposed cryptosystem has a good performance in resisting the exhaustive attack.

### 5.1.2 Key Sensitivity Analysis

The 3D-LCM is very sensitive with respect to the initial values and control parameters. If they have a slight difference, the decryption fails. The secret key sensitivity tests with respect to different parameters are shown in Fig. 4. The experimental shows that our proposed algorithm is sensitive to the secret key.

## *5.2 Differential Attack Analysis*

The discovery of differential cryptanalysis is usually attributed to Eli Biham and Adi Shamir [16, 17]. Differential attack play a crucial role in cryptology. Our proposed algorithm resist differential attack. The number of changing pixel rate (NPCR) and unified averaged changed intensity (UACI) [18] experimental results validate this fact. NPCR and UACI are defined in Eqs. 25 and 26, respectively,

$$\text{NPCR} = \frac{\sum_{i=1}^{M} \sum_{j=1}^{N} D(i, j)}{M \times N} \times 100\%, \tag{25}$$

$$\text{UACI} = \frac{1}{MN} \left[ \frac{\sum_{i=1}^{M} \sum_{j=1}^{N} |E(i, j) - E'(i, j)|}{255} \right] \times 100\%, \tag{26}$$

**Table 1** NPCR and UACI results performed on different standard images

| Standard image | Experimental values | |
|---|---|---|
| | NPCR (%) | UACI (%) |
| Lena | 99.6122 | 33.3843 |
| Sailboat | 99.6643 | 33.4432 |
| Jellybeans | 99.6492 | 33.4032 |

where $E$ and $E'$ are two encrypted images corresponding to original images with one pixel difference. $E(i, j)$ and $E'(i, j)$ denote the pixel values at the position $(i, j)$ in the both encrypted images, respectively. $M$ and $N$ are the size of the image, $D(i, j)$ is a bipolar array and given by Eq. 27,

$$D(i, j) = \begin{cases} 0 & \text{if} \quad E(i, j) = E'(i, j), \\ 1 & \text{if} \quad E(i, j) \neq E'(i, j). \end{cases} \tag{27}$$

From Table 1, the proposed method has high NPCR and suitable UACI values, which are close to standard values. The high NPCR values mean that position of each pixel is randomized and suitable UACI values represent almost all pixels in the encrypted image are changed. Based on a comparison of the experimental values and the theoretical values, we can say that our method passes both NPCR and UACI tests, so the proposed method is resistant to differential attack.

## 5.3 *Statistical Analysis*

### 5.3.1 Error Analysis, Entropy Analysis, Histogram Analysis, and Correlation Analysis

In this subsection, we have discussed mean square error (MSE), peak signal to noise ratio (PSNR), and structural similarity index metric (SSIM) analysis.

The MSE and PSNR between the original image $(I)$ and encrypted image $(E)$ are calculated by Eqs. 28 and 29,

$$\text{MSE}(I, E) = \frac{1}{MN} \sum_{p=1}^{M} \sum_{q=1}^{N} [I(p, q) - E(p, q)]^2, \tag{28}$$

$$\text{PSNR}(I, E) = 10 \log_{10} \frac{(255)^2}{\text{MSE}(I, E)}, \tag{29}$$

where $p$ and $q$ are the numbers of pixels of the frame.

**Table 2** MSE, PSNR, and SSIM values between original and corresponding encrypted image

| Image | MSE | PSNR | SSIM |
|---|---|---|---|
| Figure 3a, b | $8.9769e + 03$ | 8.9306 | 0.0011 |
| Figure 3d, e | $1.0046e + 04$ | 8.2163 | 0.0043 |
| Figure 3g, h | $1.1093e + 04$ | 7.8290 | 0.0037 |

**Table 3** Entropy value of the original and encrypted images

| Image | Entropy | |
|---|---|---|
| | Original image | Encrypted image |
| Lena | 7.2544 | 7.9787 |
| Sailboat | 7.3968 | 7.9757 |
| Jellybeans | 6.2700 | 7.9335 |

SSIM measures the similarity between two images. The Lesser the value of SSIM indicates more dissimilarity between both the images. The SSIM index between $I$ and $E$ is calculated by Eq. 30,

$$\mathrm{SSIM}(I, E) = \frac{(2\mu_I\mu_E + J_1)(2\sigma_{IE} + J_2)}{(\mu_I^2 + \mu_E^2 + J_1)(\sigma_I^2 + \sigma_E^2 + J_2)}, \tag{30}$$

where $\mu_I$ and $\mu_E$ are mean of the original image pixels and encrypted image pixels, respectively, $\sigma_I$ and $\sigma_E$ are the standard deviation of the original image pixels and the encrypted image pixels, respectively, $\sigma_{IE}$ is the covariance between the original image pixels and the encrypted image pixels, $J_1 = (k_1L)^2$, $J_2 = (k_2L)^2$ and $k_1 = 0.01$, $k_2 = 0.03$ and $L = 2^{\text{number of bits per pixel}} - 1$.

Table 2 shows the MSE, PSNR, and SSIM values between $I$ and $E$ (Fig. 3).

Entropy is a statistical measure of randomness of data. The entropy $H(z)$ of the data $z$ can be calculated by Eq. 31,

$$H(z) = -\sum_{i=1}^{M} P(z_i) log_2 P(z_i), \tag{31}$$

where $P(z_i)$ is the probability occurrence of the symbol $z_i$.

The value of the entropy of the original and encrypted images are shown in Table 3.

An image histograms can graphically exhibit the distribution of pixel values. The experimental results for histogram analysis are shown in Fig. 5.

The experimental results show that our algorithm is resistant to correlation attack. The experimental results are shown in Fig. 6.

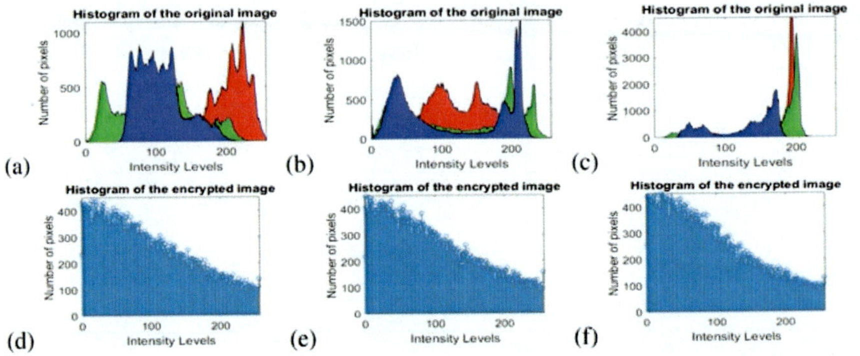


**Fig. 5** Histogram results: **a** histogram of Lena image, **b** histogram of Sailboat image, **c** histogram of Jellybeans image, **d** histogram of encrypted image of Lena, **e** histogram of encrypted image of Sailboat, and **f** histogram of encrypted image of Jellybeans

**Fig. 6** Correlations of two adjacent pixels of: **a** Original Lena image in H, V, and D direction, and **b** encrypted image of Lena in H, V, and D direction

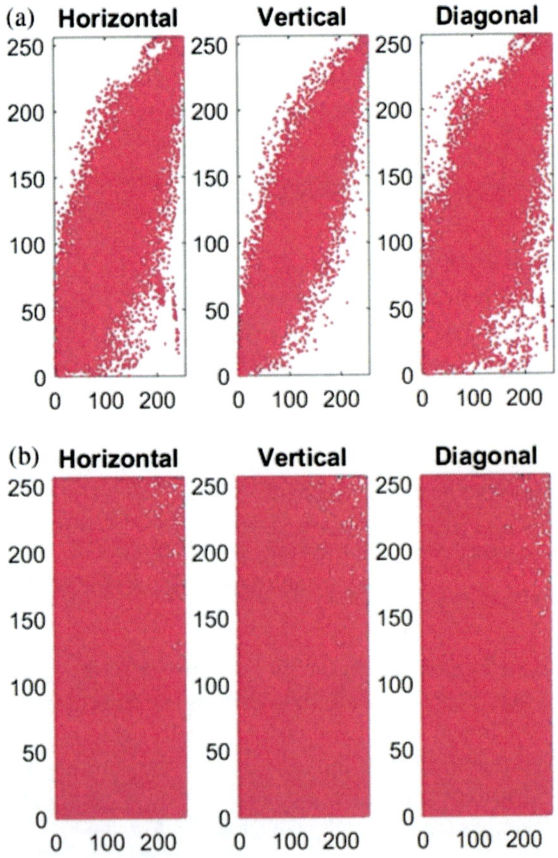

## 6 Conclusion

This manuscript proposes an asymmetric digital color-image encryption method using 3D-LCM and RMD in 2D-FrDCT. The proposed method is designed in such a way that it can handle images of any size. It is structured in such a way that the 3D-LCM parameters, initial conditions along with the independent fractional order of 2D-FrDCT, and private key serve as additional keys, therefore, it has an extensively huge keyspace and is highly sensitive to the keys. So, it is resistant against any brute force attack. The proposed method has ability to resist the histogram, entropy, and correlation-based attack. Also, the method resistant against differential attack. The experimental results provided confirm that the proposed method is highly secure and has good performance.

## References

1. Refregier P, Javidi B (1995) Optical image encryption based on input plane and Fourier plane random encoding. Opt Lett 20(7):767–769
2. Joshi AB, Kumar D, Gaffar A, Mishra DC (2020) Triple color image encryption based on 2D multiple parameter fractional discrete Fourier transform and 3D Arnold transform. Opt Lasers Eng 133:106139. https://doi.org/10.1016/j.optlaseng.2020.106139
3. Kumar D, Joshi AB, Mishra VN (2020) Optical and digital double color-image encryption algorithm using 3D chaotic map and 2D-multiple parameter fractional discrete cosine transform. Results Optics 1
4. Kumar D, Joshi AB, Gaffar A (2020) Double color-image encryption based on 2D fractional discrete Fourier transform and Arnold cat map. GANITA 70(2):129–141
5. Joshi AB, Kumar D (2019) A new method of multi color image encryption. In: IEEE conference on information and communication technology, Allahabad, India, pp 1–5. https://doi.org/10.1109/CICT48419.2019.9066198
6. Joshi AB, Kumar D, Mishra DC, Guleria V (2020) Colour-image encryption based on 2D discrete wavelet transform and 3D logistic chaotic map. J Mod Opt 67(10):933–949
7. Chen H, Liu Z, Zhu L, Tanougast C, Blondel W (2019) Asymmetric color cryptosystem using chaotic Ushiki map and equal modulus decomposition in fractional Fourier transform domains. Opt Lasers Eng 112:7–15
8. Sui L, Duan K, Liang J, Zhang Z, Meng H (2014) Asymmetric multiple-image encryption based on coupled logistic maps in fractional Fourier transform domain. Opt Lasers Eng 62:139–152
9. Yadav AK, Singh P, Saini I, Singh K (2019) Asymmetric encryption algorithm for colour images based on fractional Hartley transform. J Mod Opt 66(6):629–642
10. Joshi AB, Kumar D, Mishra DC (2020) Security of digital images based on 3D Arnold Cat map and elliptic curve. Int J Image Graphics 20(4):2150006
11. Huang ZJ, Cheng S, Gong LH, Zhou NR (2020) Nonlinear optical multi-image encryption scheme with two-dimensional linear canonical transform. Opt Lasers Eng 124:105821–105831
12. Ahmed N, Natarajan T, Rao KR (1974) Discrete cosine transform. IEEE Trans Comput 23(1):90–93. https://doi.org/10.1109/T-C.1974.223784
13. Kumar S, Panna B, Jha RK (2019) Medical image encryption using fractional discrete cosine transform with chaotic function. Med Biol Eng Comput 57(11):2517–2533
14. Steffi AA, Sharma D (2013) An image encryption algorithm based on 3D Lorenz map. Int J Adv Res Comput Sci 4(2):312–316
15. Zheng Y, Singh SH (1997) Adaptive control of chaos in Lorenz system. Dyn Control 7:143–154. https://doi.org/10.1023/A:1008275800168

16. Biham E, Shamir A (1991) Differential cryptanalysis of DES-like cryptosystems. J Cryptol 4(1):3–72
17. Biham E, Shamir A (1992) Differential cryptanalysis of the full 16-round DES. In: Annual international cryptology conference, pp 487–496. Springer, Heidelberg
18. Wu Y, Noonan JP, Agaian S (2011) NPCR and UACI randomness tests for image encryption. J Selected Areas Telecommun 1(2):31–38

# Chapter 4
# Optimal Capacity and Location of DGs in Radial Distribution Network Using Novel Harris Hawks Optimization Algorithm

**Moumita Ghosh, B. Tudu, and K. K. Mandal**

## 1 Introduction

The concept of DG plays a significant role in the modern electric utility grid. The accelerative demand for electricity has offered a challenge to maintain the power quality, efficiency, and economical aspects of the power utility system. Thus, the focus of research and industrial sectors has been shifted towards DG over centralized power system to remove the large-scale power outages and energy crisis. These small-scale power generating units are placed very close to the end-users or consumers. Therefore, the structural complexity, cost, and transmission loss of the cables in the transmission network system may be avoided. Global concerns about the limited resources and environmental degradation have further inspired to utilize renewable energy sources. Thus, the penetration of renewable DGs has gained a voluminous interest in modern times. However, as an adverse effect, the DG(s) installation can also cause a bidirectional power flow, poor voltage level, and power loss. To mitigate the aforesaid issues, there is a need for optimal allocation and sizing of DG units before installation.

A novel method for DG allocation was presented by Essallah et al. [1]. Here, a voltage stability margin index is introduced to determine the vulnerable buses. Voltage stability and load fluctuation were taken into account while selecting optimal DG location and size. The dynamic models for DG, loads, and network branches were presented by Nezhadpashaki et al. [2] to scan the effects of small-signal stability constraint on DG allocation in the radial distribution system (RDS). In [3], Multi-leader Particle Swarm Optimization (MLPSO) was presented for optimal placement of DG(s) and minimization of power loss to reduce the drawbacks such as premature convergence, the precise output, and the complicated optimization techniques. An

M. Ghosh (✉) · B. Tudu · K. K. Mandal
Power Engineering Department, Jadavpur University, Kolkata, India
e-mail: moumitaghosh1095@gmail.com



D. Singh et al. (eds.), *Proceedings of International Conference on Scientific and Natural Computing*, Algorithms for Intelligent Systems,
https://doi.org/10.1007/978-981-16-1528-3_4

analysis was carried out on the IEEE-33 bus and a real bus system in the Malaysian context. With the penetration of three DGs (at unity power factor), the loss reduction percentage becomes 67.40% and 80.32% in the two systems, respectively. The comparison with other optimization methods showed the significance of the MLPSO algorithm in optimal sizing and placement of DGs. A hybrid sequential quadratic programming based an active set method was harnessed by Angalaeswari et al. [4] to identify the optimal location and size of DG in RDSs. The objective was to minimize the real power loss. The hybrid PIPSO–SQP method was harnessed in standard IEEE buses under various loading conditions. In [5], optimal allocation of DG is performed by combining the clonal selection rule of the immune system with the particle swarm optimization. The effectiveness of the proposed method is tested on IEEE-33 RDS and IEEE 14 loop distribution system. An enhanced sunflower optimization is used in [6] where a mutation technique is added with the original sunflower optimization for updating the best plant in DG placement problem. A hybrid technique (joined execution of Cuckoo Search Algorithm and Grasshopper Optimization Algorithm) is proposed in [7] to optimize the size and position of DGs in the distribution system. Different load states of the system are executed to investigate system stability by loss minimization. Nonetheless, new and improved optimization techniques always play an important role in finding global solution with greater efficiency. The key contributions of this paper can be summarized as:

- Harris Hawks Optimization (HHO) has been applied to determine the optimal capacity and location of single and multiple DG units operating at different power factors.
- Backward/Forward Sweep Approach is utilized for the load flow analysis.
- The objective of this work is to minimize the net real power loss while improving the voltage profile of the system.
- The effectiveness of the HHO has been applied on standard IEEE 33 bus RDS.
- Finally, it has been shown that how the number of DG units and their operating power factors can affect the power loss, voltage profile, and power flow through the lines.

## 2 System Description and Problem Formulation

In this paper, the optimal size and location of DG units are formulated as an optimization problem. Prime objective of this work is to minimize net real power loss while improving the voltage profile of system. Total real power loss in the system is calculated by using 'Exact Loss Formula' [8] in Eq. (1)

$$P_{loss} = \sum\nolimits_{u=1,v=1}^{n} [A_{uv}(P_u P_v + Q_u Q_v) + B_{uv}(Q_u P_v - P_u Q_v)] \quad (1)$$

Here, $A_{uv}$ and $B_{uv}$ are expressed as

$$A_{uv} = \frac{r_{uv}}{V_u V_v} \cos(\delta_u - \delta_v) \tag{2}$$

$$B_{uv} = \frac{r_{uv}}{V_u V_v} \sin(\delta_u - \delta_v) \tag{3}$$

Suffix $u$ and $v$ indicate values at $u^{th}$ and $v^{th}$ nodes, respectively. $Vu$, $V_v$, $\delta_u$, and $\delta_v$ are bus voltage magnitude and voltage angel at $u^{th}$ and $v^{th}$ bus, respectively. Also, $P_u$, $P_v$, $Q_u$, and $Q_v$ are active and reactive power loss at $u^{th}$ and $v^{th}$ bus, respectively. $r_{uv}$ is the line resistance between $u^{th}$ and $v^{th}$ bus.

## 3 Constraints

### 3.1 Distributed Generator Size Constraint

In order to obtain a reasonable solution, size of distributed generators (DGs) must be within a range with respect to the load demand of the system [9].

$$10\%\, of \left(\sum\nolimits_{u=1}^{n} S_{du}\right) \le S_{dgu} \le 80\%\, of \left(\sum\nolimits_{u=1}^{n} S_{du}\right) \tag{4}$$

where $S_{dgu}$ and $S_{du}$ are net power generation and load demand at $u^{th}$ DG unit respectively. $n$ denotes total bus number in the system.

### 3.2 Power Balance

The power balance equation can be written as:

$$P_{dgu} = P_{du} + P_{loss} \tag{5}$$

where $P_{dgu}$, $P_{du}$, and $P_{loss}$ are active power generation, active power demand, and real power loss at $u^{th}$ DG unit respectively [10].

### 3.3 Voltage Constraint

The voltage constraint can be expressed as:

$$V_{u\_min} \le V_u \le V_{u\_max} \tag{6}$$

Here, $V_{u\_min}$ and $V_{u\_max}$ define the lower and upper limit of voltage at $u^{th}$ bus respectively. In this study, $V_{u\_max}$ and $V_{u\_min}$ are considered as 1.05 per unit (p.u) and 0.95 p.u respectively for direct comparison [9].

# 4 Harris Hawks Optimization (HHO)

Harris Hawks Optimization was developed by Heidari et al. [11]. It is a population based, gradient-free and nature-inspired intelligent optimization. A random population of hawks: $s_i$ $(i = 1, 2,\ldots, n)$ is initialized (from Eqs. (7) to (9)) by considering the constraints of Sect. 3. Then the fitness values (real power loss) of the hawks are obtained by executing the load flow analysis. Now, $s_{rabbit}$ (location of rabbit) is set as the best solution (corresponding to the lowest real power loss). $n$ is the total number of population of hawks. For each hawk ($s_i$), the Energy ($E$) is updated by using Eq. (10). Based on the value of $E$, exploitation phase is performed (Sect. 4.3). The location vector of hawks is updated until the stopping criterion is met. Then the best solution (location and size of DG(s)) is returned.

## 4.1 Exploration

Harris' hawks are assumed as candidate solutions and the best solution from each step is taken as the optimum one. Let us assume an equal chance '$c$' for each perching strategy of the hawks. They perch based on the family members' position (Eq. (7)) and the rabbit and on some random tall trees (Eq. (8)) [11].

$$s(t+1) = s_{rand}(t) - r_1|s_{rand}(t) - 2r_2 s(t)| c \geq 0.5 \tag{7}$$

$$s(t+1) = (s_{rabbit}(t) - s_m(t)) - r_3(lb + r_4(ub - lb)) c < 0.5 \tag{8}$$

where

$$s_m(t) = \frac{1}{n}\sum\nolimits_{i=1}^{n} s_i(t) \tag{9}$$

Here, $s(t)$ and $s(t + 1)$ are the position vector of hawks in $t^{th}$ and $(t + 1)^{th}$ iteration respectively. $s_{rabbit}(t)$ is the rabbit's position at $t^{th}$ iteration. $c$, $r_1$, $r_2$, $r_3$, and $r_4$ are random numbers inside (0,1), whereas $lb$ and $ub$ are the upper and lower limits of variables (here, upper and lower limit of DG's size and location in RDSs). $s_{rand}(t)$ is a randomly chosen hawk from current population inside the range ($lb$, $ub$). $s_m$ is the average position of the hawks.

## 4.2 Condition of Transition from Exploration to Exploitation

Based on the escaping energy ($E$) of prey (rabbit), HHO transits between exploration and exploitation phase [11]:

$$E = 2E_0 * \left(1 - \frac{t}{iter_{max}}\right) \quad (10)$$

$E_0$ is the initial energy of prey and it randomly changes inside interval $(-1, 1)$ during every iteration. $iter_{max}$ is the maximum number of iterations.

## 4.3 Exploitation

In this phase, hawks do the surprise pounce on the prey (rabbit) found in the exploration phase but rabbits try to escape from this situation. Here, '$f$' is a chance of a rabbit in escaping.

**Soft Besiege.** When $|E| \geq 0.5$ and $f \geq 0.5$, the rabbit (prey) has sufficient energy, so it tries to escape by random jumps but finally it fails [11]. It is shown below [11]:

$$s(t+1) = \Delta s(t) - E|Js_{rabbit}(t) - s(t)| \quad (11)$$

$$\Delta s(t) = s_{rabbit}(t) - s(t) \quad (12)$$

$$J = 2(1 - r_5) \quad (13)$$

Here, $\Delta s(t)$ is the difference between the location of rabbit and the hawks' location in the $t^{th}$ iteration, $r_5$ is a random number inside a range of (0,1), and $J$ is the jump strength of rabbit.

**Hard Besiege.** When $|E| < 0.5$ and $f \geq 0.5$, the rabbit gets tired and it has a low escaping energy. The positions are modified by the following equation [11]:

$$s(t+1) = s_{rabbit}(t) - E|\Delta s(t)| \quad (14)$$

**Soft Besiege with Progressive Rapid Dives.** When $|E| \geq 0.5$ but $f < 0.5$, the rabbit has sufficient energy for a successful getaway. To perform a soft besiege, it is assumed that the hawks can evaluate their next move by the following equation [11]:

$$u = s_{rabbit}(t) - E|Js_{rabbit}(t) - s(t)| \quad (15)$$

If it is seen that the rabbit is doing more deceptive dives, hawks start some rapid dives based on the Levy Flight ($LF$) based patterns, which is shown below:

$$v = u + x \times LF(dim) \tag{16}$$

Here, $x$ is the random vector of size $(1 \times dim)$ where '*dim*' is the dimension of the problem. *LF* is defined as follows:

$$LF(x) = 0.01 \times \frac{y \times \sigma}{|z|^{1/\beta}} \tag{17}$$

$$\sigma = \left( \frac{\Gamma(1+\beta) \times \sin\left(\frac{\pi\beta}{2}\right)}{\Gamma\left(\frac{1+\beta}{2}\right) \times \beta \times 2^{\left(\frac{\beta-1}{2}\right)}} \right)^{1/\beta} \tag{18}$$

$y$, $z$ are random variables with the range of $(0,1)$ and $\beta$ is a constant of value $1.5$.

Therefore, the final rule for updating the hawks' positions in the soft besiege is shown by the following equations:

$$s(t+1) = u \quad if\ F(u) < F(s(t)) \tag{19}$$

$$s(t+1) = v \quad if\ F(v) < F(s(t)) \tag{20}$$

$u$ and $v$ are obtained using Eqs. (15) and (16).

**Hard Besiege with Progressive Rapid Dives**. When $|E| < 0.5$ and $f < 0.5$, the prey has not sufficient energy to escape. The hard besiege is executed before the sudden attack to catch the rabbit [11].

$$s(t+1) = u \quad if\ F(u) < F(s(t)) \tag{21}$$

$$s(t+1) = v \quad if\ F(v) < F(s(t)) \tag{22}$$

$$u = s_{rabbit}(t) - E|Js_{rabbit}(t) - s_m(t)| \tag{23}$$

$$v = u + x \times LF(dim) \tag{24}$$

Here, $s_m(t)$ is obtained using Eq. (9).

## 5 Results and Discussion

The optimal allocation and sizing of the DG unit(s) are carried out on a standard IEEE-33 bus RDS. This network contains total 33 buses, 32 lines. Total load of

**Table 1** Loss and voltage profile without the Installation of DG Unit(s)

| Parameters | Value |
|---|---|
| Active (real) power loss (kW) | 210.0135 |
| Minimum voltage (p.u) | 0.9039 |

**Table 2** Various parameters of Harris Hawks Optimization

| Settings | Value |
|---|---|
| Number of hawks (n) | 30 |
| Maximum number of iterations ($iter_{max}$) | 500 |
| Dimension of optimization ($dim$) | 30 |

the system is 3.715 MW and 2.3 MVAR, and its voltage is 12.66 kV. The standard IEEE 33 bus RDS data are collected from a standard literature [12]. Table 1 shows the value of active power loss (in kW) and minimum bus voltage (in per unit (p.u)) before the installation of DG(s) for IEEE-33 bus radial distribution network. Prior to the DG installation, total real (or active) and reactive power loss are 210.0135 kW and 142.3644 kVAR, respectively.

Harris Hawks Optimization (HHO) technique is now applied to obtain optimal capacity and location of DG(s) to minimize the power loss. The parameters selected for the present optimization (HHO) are given in Table 2.

The DG units' locations, corresponding sizes, active power loss and voltage (in p.u ) after the installation of DG(s) using HHO are summarized in Table 3.

From Table 3, it can be observed that with the HHO, the optimal location of single DG with unity p.f is 28 with active (or real) power capacity equal 1.2196 MW. It leads to reduce the power loss from 210.0135 kW to 110.8482 kW where the loss reduction (LR) reaches 47.22%. Again, the optimal location of single DG with 0.866 p.f is 3 with active (or real) power capacity equal 0.7149 MW. It further reduces

**Table 3** Results obtained in HHO for Installation of DG Unit(s)

| Case | Optimal sizing and location of DG (s) | | Power loss (kW) | LR (%) | Min. voltage (p.u) |
|---|---|---|---|---|---|
| | Bus no | Size of DG unit (MW) | | | |
| 1 DG at unity p.f | 28 | 1.2196 | 110.8482 | 47.22 | 0.9226 |
| 1 DG at 0.866 p.f | 3 | 0.7149 | 68.5893 | 67.34 | 0.9077 |
| 3 DGs at unity p.f | 22 | 1.0337 | 76.1983 | 63.72 | 0.9474 |
| | 6 | 2.8472 | | | |
| | 21 | 1.4580 | | | |
| 3 DGs at 0.866 p.f | 6 | 1.3892 | 20.5505 | 90.21 | 0.9444 |
| | 24 | 0.6027 | | | |
| | 3 | 1.6915 | | | |

the power loss to 68.5893 kW where the LR reaches 67.34%. Under multiple DGs penetration, the optimal locations of three DGs with unity p.f are 22, 6, and 21 with active (or real) power capacities equal 1.0337 MW, 2.8472 MW, and 1.4580 MW respectively. It leads to reduce the real power loss from 210.0135 kW to 76.1983 kW where the LR reaches 63.72%.

Again, the optimal locations of three DGs with 0.866 p.f. are 6, 24, and 3 with active (or real) power capacities of 1.3892 MW, 0.6027 MW, and 1.6915 MW respectively. It further reduces the power loss from to 20.5505 kW where the LR reaches 90.21%. While the bus voltage reaches to its minimum value of 0.9039 per unit at bus number 18, but after the installation of one DG with unity and 0.866 power factor, bus voltage level improves to a value of 0.9226 per unit and 0.9077 per unit respectively. Similarly, with the implementation of three DGs with unity and 0.866 power factor, bus voltage level further improves to a value of 0.9474 per unit and 0.9444 per unit respectively.

Figure 1 depicts the active power loss is reduced drastically after the installation of DG(s), operated at different power factors.

Also, active power loss is the lowest with 3 DGs at 0.866 p.f. From Fig. 2, it is shown that the voltage profile of the system improves while installing DG(s). It can be seen that with the variation of operating power factor and number of DG(s), voltage profile improves.

It is seen from Fig. 3 that power flow (branch power flow) from branch number 1 to branch number 6 was quite high initially but decreases after the installation of DG(s). The efficacy of the HHO algorithm has been applied in IEEE 33 bus RDS and it is shown that the HHO has outperformed the methods reported in Table 4 in terms of real power loss reduction. Thus, the real-life preying strategy of HHO has improved the quality of solution (DG size and location). The HHO has performed

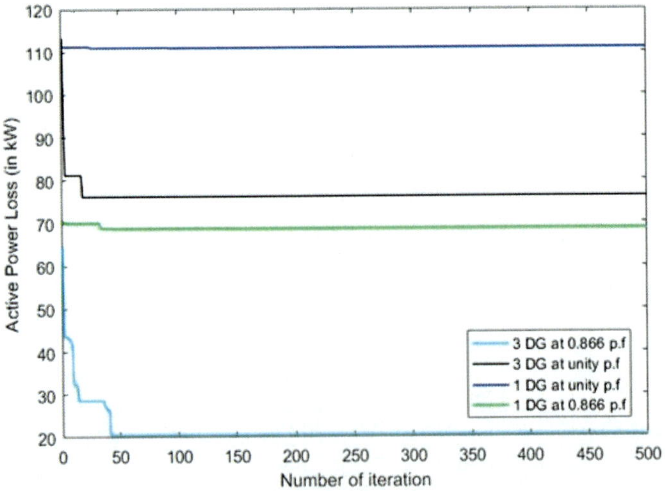


**Fig. 1** Active power loss versus number of iterations

 

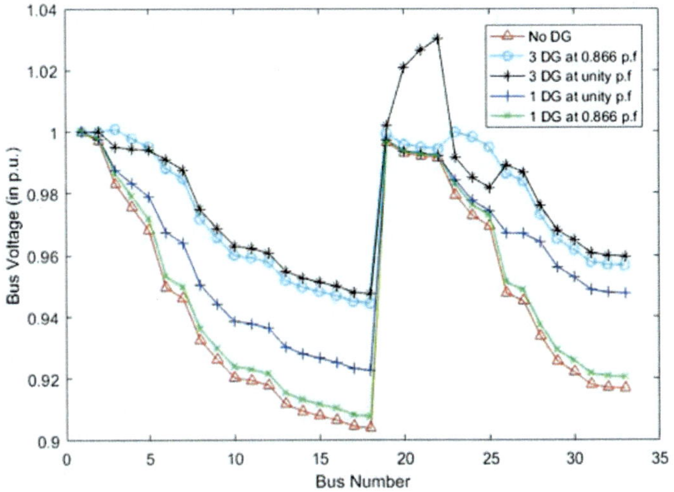


**Fig. 2** Bus voltage versus bus number for IEEE 33 bus system

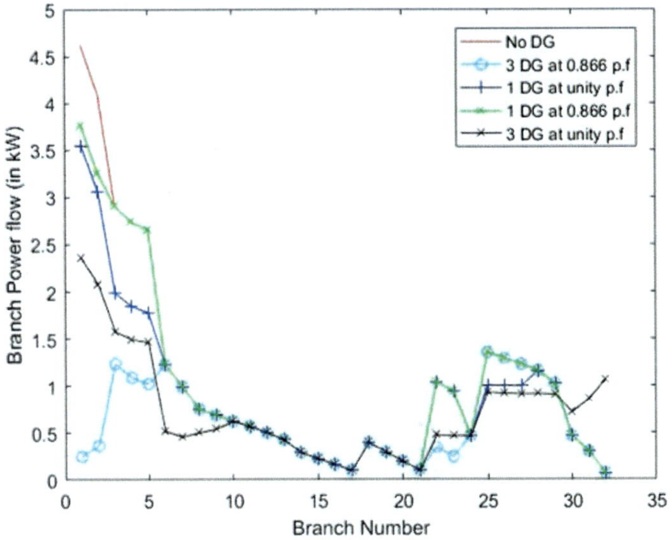


**Fig. 3** Branch power flow versus branch number for IEEE 33 bus system

quite well in terms of minimized real (or active) power loss, improved voltage profile, controlled power flow and overall system's reliability.

**Table 4** Optimal DG allocation for IEEE 33-Bus system using different optimization techniques

| Method | Reference | Year | Optimal sizing and location of DG (s) | | LR (%) |
|---|---|---|---|---|---|
| | | | Bus No | Size of DG unit (MW) | |
| **HHO** | **Proposed** | – | **22, 6, 21** | **1.0337, 2.8472, 1.4580** | **63.72** |
| SFLA | [13] | 2019 | 28, 30, 14 | 0.5639, 0.3182, 0.5144 | 58.86 |
| FWA | [13] | 2019 | 14, 18, 32 | 0.5897, 0.1895, 1.0146 | 56.24 |
| HPSO | [14] | 2017 | 8, 13, 31 | 0.56, 0.56, 0.79 | 60.26 |

## 6 Conclusion

In this paper, optimal sizing and placement of DG Unit(s) in IEEE-33 bus RDS have been carried out using novel Harris Hawks Optimization. From the results, it can be concluded that the proposed algorithm is efficient enough to reduce the active (real) power loss and improve the bus voltage profile. Also, the penetration of DG (single and multiple) operated at different power factors has a considerable effect on the active power loss, voltage profile and branch power flow of the system. Thus, the overall system performance is improved.

**Acknowledgements** Authors would like to acknowledge and thank Jadavpur University for extending the required support and facility to carry out this research work and this work is financially supported by TEQIP-III, JU.

## References

1. Essallah S, Khedher A, Bouallegue A (2019) Integration of distributed generation in electrical grid: optimal placement and sizing under different load conditions. Comput Electr Eng 79:1–14
2. Nezhadpashaki MA, Karbalaei F, Abbasi S (2020) Optimal placement and sizing of distributed generation with small signal stability constraint. Sustain Energy Grids Netw 23:1–10
3. Karunarathne E, Pasupuleti J, Ekanayake J, Almeida D (2020) Optimal placement and sizing of DGs in distribution networks using MLPSO algorithm. Energies 13(23):1–25
4. Angalaeswari S, Sanjeevi kumar P, Jamuna K, Leonowicz Z (2020) Hybrid PIPSO-SQP algorithm for real power loss minimization in radial distribution systems with optimal placement of distributed generation. Sustainability 12(14):1–21
5. Bhadoria VS, Pal NS, Shrivastava V (2019) Artificial immune system based approach for size and location optimization of distributed generation in distribution system. Int J Syst Assur Eng Manage 10(3):339–349
6. Nguyen TT (2021) Enhanced sunflower optimization for placement distributed generation in distribution system. Int J Electr Comput Eng 11(1):107–113
7. Suresh MC, Edward JB (2020) A hybrid algorithm based optimal placement of DG units for loss reduction in the distribution system. Appl Soft Comput J 91:1–15
8. Hung DQ, Mithulananthan N, Bansal RC (2013) Analytical strategies for renewable distributed generation integration considering energy loss minimization. Appl Energy 105:75–85
9. Rao BH, Sivanagaraju S (2012) Optimum allocation and sizing of distributed generations based on clonal selection algorithm for loss reduction and technical benefit of energy savings. In:

International conference on advances in power conversion and energy technologies (APCET), pp 1–5
10. Injeti SK, Kumar NP (2013) A novel approach to identify optimal access point and capacity of multiple DGs in a small, medium and large scale radial distribution systems. Int J Electr Power Energy Syst 45(1):142–151
11. Heidari AA, Mirjalili S, Faris H, Aljarah I, Mafarja M, Chen H (2019) Harris hawks optimization: algorithm and applications. Future Gener Comput Syst 97:849–872
12. Kumar KS, Jayabarathi T (2012) Power system reconfiguration and loss minimization for an distribution systems using bacterial foraging optimization algorithm. Int J Electr Power Energy Syst 36(1):13–17
13. Onlam A, Yodphet D, Chatthaworn R, Surawanitkun C, Siritaratiwat A, Khunkitti P (2019) Power loss minimization and voltage stability improvement in electrical distribution system via network reconfiguration and distributed generation placement using novel adaptive shuffled frogs leaping algorithm. Energies 12(3):1–12
14. Tolba MA, Tulsky VN, Diab AA (2017) Optimal allocation and sizing of multiple distributed generators in distribution networks using a novel hybrid particle swarm optimization algorithm. In: IEEE conference of russian young researchers in electrical and electronic engineering (EIConRus), pp 1606–1612

# Chapter 5
# Model Order Reduction Using Grey Wolf Optimization and Pade Approximation

**Pranay Bhadauria and Nidhi Singh**

## 1 Introduction

The approximation of complex system to simple model is very common nowadays. The mathematical modeling is the first step in analyzing any real problem. Approximation of large-scale systems or model reduction has been analyzed and different methods have been suggested for obtaining reduced order models in the time domain, frequency domain or combination of both. It must be noted that in spite of having different techniques, no single model is applicable for all objectives.

The recommended technique is a combination of Pade approximation and Grey wolf optimization technique for order reduction of large-scale SISO system. The Grey wolf optimization technique is used for finding the coefficients of denominator polynomial and for determining the coefficients of numerator polynomial, Pade approximation technique is used [1–14]. The Pade approximation technique has a disadvantage that if the original system is stable it is not necessary that the reduced model may be stable it can be unstable also.

## 2 Problem Description

Consider a linear dynamic SISO system, given by the transfer function

$$G(s) = \frac{N(s)}{D(s)} = \frac{x_{n-1}s^{n-1} + \ldots + x_1 s + x_0}{y_n s^n + \ldots + y_1 s + y_0} \tag{1}$$

P. Bhadauria (✉) · N. Singh
Department of Electrical Engineering, Gautam Buddha University, Greater Noida 201312, UP, India
e-mail: pranayesp4u@gmail.com



D. Singh et al. (eds.), *Proceedings of International Conference on Scientific and Natural Computing*, Algorithms for Intelligent Systems,
https://doi.org/10.1007/978-981-16-1528-3_5

where $x_i$; $0 \le i \le n-1$ and $y_i$; $0 \le i \le n$ are scalar constants.

The analogous $r^{th}$ (r < $n$) order reduced model is synthesized as

$$G_r(s) = \frac{N_r(s)}{D_r(s)} = \frac{m_{r-1}s^{r-1} + \ldots + m_1 s + m_0}{n_r s^r + \ldots + n_1 s + n_0} \tag{2}$$

where $m_i$; $0 \le i \le r-1$ and $n_i$; $0 \le i \le r$ are scalar constants.

The purpose of this paper is to determine the $r$th order reduced model in the form of (2) from (1) such that it preserves the necessary features of the original system.

## 3 Explanation of the Methods

The recommended technique follows two steps:

### *3.1 Step 1: Grey Wolf Optimization Technique is Used to Determine the Denominator of Rth Order Reduced Model*

Grey wolf are mostly founded in Canada, which are considered as apogee carnivore, which follow a very strict community commanding chain. The commanding position in the Grey wolves is given to alphas, which can be male or a female. The alphas are the decision-makers of the group. The next position in the chain is given to beta, which assists the alphas in taking decisions or other group action [12–14].

The lowest position in the chain of Grey wolves is given to omega. They are considered not so influential member of the group but the whole group sees internal struggle and complications in case of losing the omega. There is one other category of Grey wolf, which is delta. Delta wolves are inferior to alphas and beta but superior to omega. First, the Grey wolves track, chase and reach the prey. Second, they surround and tease the prey and lastly raid the prey. As told earlier, Grey wolves surround prey during the raid. The following equation suggests the mathematical model of surrounding action:

$$\vec{A} = \left| \vec{B}\,\vec{S}_p(i) - \vec{S}(i) \right| \tag{3}$$

$$\vec{S}(i+1) = \vec{S}_p(i) - \vec{C}.\vec{A} \tag{4}$$

where $i$ illustrates the ongoing iteration. $\vec{C}$ and $\vec{B}$ are the coefficient vectors, $\vec{S}_p$ is the prey position vector and $\vec{S}$ illustrates the Grey wolf position vector. The $\vec{C}$ and

$\overrightarrow{B}$ vectors are determined as follows:

$$\overrightarrow{C} = 2\overrightarrow{a}.r_1 - \overrightarrow{a} \tag{5}$$

$$\overrightarrow{B} = 2.r_2 \tag{6}$$

$$\overrightarrow{A_\alpha} = \left|\overrightarrow{B_1}\overrightarrow{S_\alpha} - \overrightarrow{S}\right| \tag{7}$$

$$\overrightarrow{A_\beta} = \left|\overrightarrow{B_2}\overrightarrow{S_\beta} - \overrightarrow{S}\right| \tag{8}$$

$$\overrightarrow{A_\delta} = \left|\overrightarrow{B_3}\overrightarrow{S_\delta} - \overrightarrow{S}\right| \tag{9}$$

Thus using Eq. (5)

$$\overrightarrow{S_1} = \overrightarrow{S_\alpha} - \overrightarrow{C_1}.\left(\overrightarrow{A_\alpha}\right) \tag{10}$$

$$\overrightarrow{S_2} = \overrightarrow{S_\beta} - \overrightarrow{C_2}.\left(\overrightarrow{A_\beta}\right) \tag{11}$$

$$\overrightarrow{S_3} = \overrightarrow{S_\delta} - \overrightarrow{C_3}.\overrightarrow{(A_\delta)} \tag{12}$$

and finally

$$\overrightarrow{S}(i+1) = \frac{\left(\overrightarrow{S_1} + \overrightarrow{S_2} + \overrightarrow{S_3}\right)}{3} \tag{13}$$

## 3.2 *Step 2: For Determining the Numerator of Rth Order Reduced Model, Pade Approximation is Used [1–5]*

Power Series expansion of the original $r$th order system about s = 0 is given as:

$$\mathrm{F(s)} = \frac{\sum_{i=0}^{r-1} u_i s^i}{\sum_{i-0}^{r} v_i s^i} = e_0 + e_1 s + e_2 s^2 + \ldots \tag{14}$$

The power series expansion coefficient is determined as follows:

$$e_0 = u_0 \tag{15}$$

$$e_i = \frac{1}{v_0}\left[u_i - \sum_{k=1}^{i} v_k e_{i-k}\right] \quad i > 0 \tag{16}$$

$$e_i = 0 \quad i > r - 1 \tag{17}$$

The reduced model of r⁺ order is given as:

$$F_r(s) = \frac{N_r(s)}{D_r(s)} = \frac{\sum_{i=0}^{r-1} m_i s^i}{\sum_{i=0}^{r} n_i s^i} \tag{18}$$

Here $D_r(s)$ is known through Grey wolf optimization technique.

For Eq. 18 representing $F_r(s)$ to be Pade approximants of Eq. (14) representing $F(s)$, we have

$$\begin{aligned} m_0 &= n_0 e_0 \\ m_1 &= n_0 e_1 + n_1 e_0 \\ m_{r-1} &= n_0 e_{r-1} + n_1 e_{r-2} + \ldots\ldots + n_{r-1} e_1 + n_r e_0 \end{aligned} \tag{19}$$

By solving the above r linear equations, the $m_k$; $K = 0, 1, 2, \ldots, r-1$ can be determined.

# 4 Illustrative Examples

## 4.1 *Example 1: Examine an Eighth Order System Taken from [7, 11, 16–19]*

$$G_8(s) = \frac{\begin{array}{c} 18s^7 + 514s^6 + 5982s^5 + 36380s^4 + 122664s^3 \\ +222088s^2 + 185760s + 40320 \end{array}}{\begin{array}{c} s^8 + 36s^7 + 546s^6 + 4536s^5 + 22449s^4 + 67284s^3 \\ +118124s^2 + 109584s + 40320 \end{array}}$$

GWO reduces the denominator to second-order model given as $D_2(s) = 1.4098s^2 + 9.6400s + 7.4108$ [12].

Now the reduced second-order model can be written in the form of

$$F_2(s) = \frac{m_0 + m_1 s}{7.4108 + 9.6400s + s^2}.$$

Now using **Step 2**.

$m_0 = n_0 e_0 = 7.4108$

$m_1 = n_0 e_1 + n_1 e_0 = 23.64937632$

The reduced second-order model is given as:

$$G_2(s) = \frac{7.4108 + 23.64937632s}{7.4108 + 9.6400s + 1.4098s^2}$$

Comparison of the eighth order model and second-order reduced model in terms of step response has been shown in Fig. 1. The ISE of the second-order reduced model has been compared with other existing techniques in Table 1.

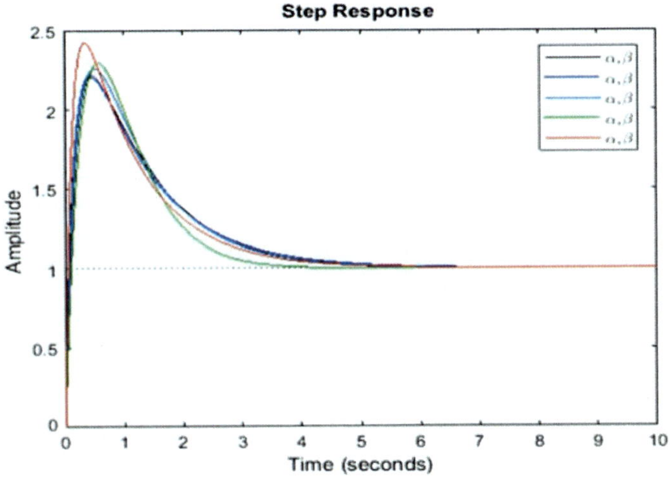


**Fig. 1** Step response of eighth order model and second-order reduced model

**Table 1** Comparison of the existing methods with example 1 in terms of ISE

| Techniques of order reduction | Second-order reduced models | ISE |
|---|---|---|
| Proposed method | $\frac{7.4108+23.64937632s}{7.4108+9.6400s+1.4098s^2}$ | $7.7816 \times 10^{-4}$ |
| Singh et al. [6] | $\frac{14.0097s+4.5}{s^2+5.5s+4.5}$ | $10.08 \times 10^{-3}$ |
| Prasad and Pal [15] | $\frac{17.98561s+500}{s^2+13.24571s+500}$ | 1.4584 |
| Pal [9] | $\frac{151776.576s+40320}{65520s^2+75600s+40320}$ | 1.6509 |
| Gutman et al. [10] | $\frac{4[133747200s+203212800]}{85049280s^2+552303360s+812851200}$ | 1.3760 |
| Mukherjee et al. [11] | $\frac{11.3909s+4.4357}{s^2+4.2122s+4.4357}$ | $5.6897 \times 10^{-2}$ |
| Mittal et al. [16] | $\frac{7.0908s+1.9906}{s^2+3s+2}$ | $2.689 \times 10^{-1}$ |
| Parmar et al. [17] | $\frac{24.1144s+8}{s^2+9s+8}$ | $4.8090 \times 10^{-2}$ |
| Shamash [7] | $\frac{6.7786s+2}{s^2+3s+2}$ | $27.92 \times 10^{-2}$ |
| Sikander and Prasad [18] | $\frac{16.504s+5.462}{s^2+6.197s+5.462}$ | $1.390 \times 10^{-2}$ |
| Khanam and Parmar [19] | $\frac{16.14315s+5.03619}{0.95698s^2+6.55351s+5.03619}$ | $7.22248 \times 10^{-4}$ |

**Table 2** Second-order reduced model compared with other existing techniques

| Techniques of order reduction | Second-order reduced models | ISE |
|---|---|---|
| Proposed | $\frac{0.17800768s+0.9029}{0.7246s^2+9.4786s+0.9029}$ | $4.257 \times 10^{-4}$ |
| Singh et al. [13] | $\frac{0.0496s+0.168}{s^2+1.78s+0.168}$ | $3.68 \times 10^{-3}$ |
| Vishwakarma [11] | $\frac{8s+1}{100.805s^2+16.2254s+1}$ | $128 \times 10^{-3}$ |

It can be identified from Table 1 that the recommended method gives low value of ISE as compared to the existing techniques.

### 4.2 *Example 2: Examine an Eighth Order System Taken from [6, 13]*

$$G_6(s) = \frac{\begin{array}{c}2s^5 + 3s^4 + 16s^3\\ +20s^2 + 8s + 1\end{array}}{\begin{array}{c}2s^6 + 33.6s^5 + 155.9s^4 + 209.5s^3\\ +102.4s^2 + 18.3s + 1\end{array}}$$

GWO reduces the denominator to second-order model given as $D_2(s) = 0.7246s^2 + 9.4786s + 0.9029$ [5].

Now the reduced second-order model can be written in the form of

$$F_2(s) = \frac{m_0 + m_1 s}{0.7246s^2 + 9.4786s + 0.9029}$$

Now using **Step 2**.

$m_0 = n_0 e_0 = 0.9029$.

$m_1 = n_0 e_1 + n_1 e_0 = 0.17800768$.

The reduced 2nd order model is given as:

$$G_2(s) = \frac{0.9029 + 0.17800768s}{0.9029 + 9.4786s + 0.7246s^2}$$

Comparison of the sixth order model and second-order reduced model in terms of step response has been shown in Fig. 2. The ISE of the second-order reduced model has been compared with other existing techniques in Table 2.

It can be identified from Table 2 that the recommended method gives low value of ISE as compared to the existing techniques.

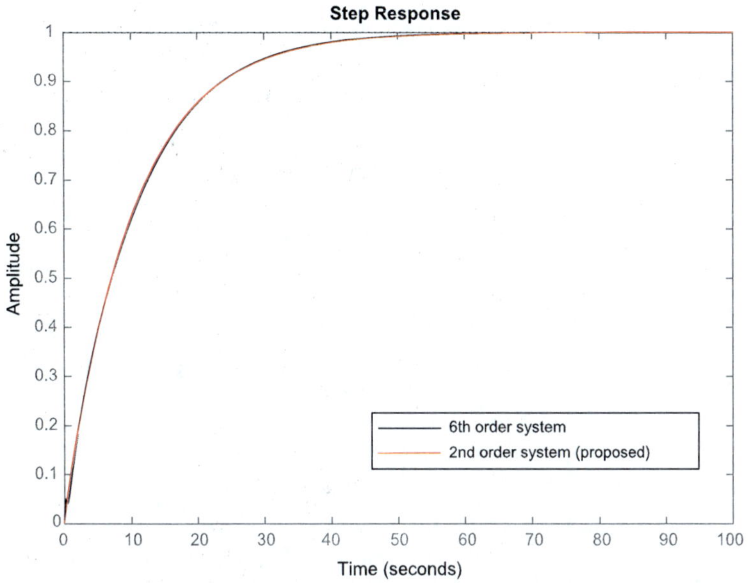


**Fig. 2** Sixth order model and second-order reduced model step response

# 5 Conclusions

A hybrid technique based on Grey wolf optimization and Pade approximation has been recommended for order reduction of SISO stable system. In this method, coefficient of the denominator polynomial for reduced model is determined by using Grey wolf optimization. The numerator of the reduced model is obtained by using Pade approximation. The recommended method is explained with the help of two examples and it has been noticed from Tables 1 and 2 that this technique is superior to many other extant order reduction methods available in survey. This hybrid method conserves model stability and evades steady-state error between the step responses of the original and reduced models.

# References

1. Shamash Y (1975) Linear system reduction using pade approximation to allow retention of dominant modes. Int J Control 21(2):257–272
2. Lucas TN (1988) Differentiation reduction method as a multipoint Pade approximant. Electron Lett 24(1):60–61
3. Ismail O, Bandyopadhyay B, Gorez R (1997) Discrete interval system reduction using Pade approximation to allow retention of dominant poles. IEEE Trans Circ Syst I: Fund Theory Appl 44(11):1075–1078

4. Wan B (1981) Linear model reduction using mihailov criterion and pade approximation technique. Int. J Control 33(6):1073–1089
5. Vishwakarma CB, Prasad R (2008) Clustering method for reducing order of linear system using pade approximation. IETE J Res 54(5):326–330
6. Singh J, Chatterjee K, Vishwakarma CB (2014) System reduction by eigen permutation algorithm and improved pade approximations. IJMCPECE World Acad Sci Eng Technol 8(1):180–184
7. Shamash Y (1975) Linear system reduction by pade approximation to allow retention of dominant modes. Int J Control 21:257–272
8. Parmar G, Mukherjee S, Prasad R (2007) System reduction using eigen spectrum analysis and pade approximation technique. Int J Comput Math 84(12):1871–1880
9. Pal J (1979) Stable reduced order pade approximants using the Routh Hurwitz array. Electron Lett 15(8):225–226
10. Gutman PO, Mannerfelt CF, Molander P (1982) Contributions to the model reduction problem. IEEE Trans Automat Control AC-27(2): 454–455
11. Mukherjee S, Satakshi M, Mittal RC (2005) Model order reduction using response matching techniques. J Franklin Inst 342:503–519
12. Bhatnagar U, Gupta A (2017) Application of grey wolf optimization in order reduction of large scale LTI systems. In: 2017 4th IEEE Uttar Pradesh section international conference on electrical, computer and electronics (UPCON), Mathura, pp 686–691
13. Prajapati AK, Prasad R (2018) Model order reduction by using the balanced truncation and factor division methods. IETE J Res. https://doi.org/10.1080/03772063.2018.1464971.
14. Mirjalili S, Mirjalili SM, Lewis A (2014) Grey wolf optimizer. Adv Eng Softw 69:46–61
15. Prasad R, Pal J (1991) Stable reduction of linear systems by continued fractions. J Inst Eng India IE(I) J-EL 72:113–116
16. Mittal AK, Prasad R, Sharma SP (2004) Reduction of linear dynamic system using an error minimization technique. J Inst Eng India IE(I) J. EL 84:201–206
17. Parmar G, Mukherjee S, Prasad R (2007) System reduction using eigen spectrum analysis and Padeapproximation technique. Int J Comput Math 84(12):1871–1880
18. Sikander A, Prasad A (2017) A new technique for reduced-order modelling of linear time-invariant system. IETE J Res 1–9
19. Khanam I, Parmar G (2017) Application of stochastic fractal search in order reduction of large scale LTI systems. In: IEEE international conference on computer, communications and electronics (comptelix 2017), Manipal University Jaipur and IRISWORLD, pp 190–194

# Chapter 6
# Modeling the Effect of Malicious Objects in Sensor Networks and Its Control by Anti-Malicious Software

**Shyam Sundar, Ram Naresh, Amit K. Awasthi, and Atul Chaturvedi**

## 1 Introduction

Now a days, wireless sensor network (WSN) is the fast-growing area of research. Development of wireless sensor network was motivated by military applications such as battlefield surveillance but later, such types of networks are widely used in home monitoring, agriculture, disaster relief operation, railways and many more. Sensor nodes are the small devices, which consist of four basic parts; sensing unit, processing unit, communication unit and power unit. Sensor nodes have limited communication and computation capabilities. Security is very challenging issue for resource-constrained sensor network. Sensor nodes send sensed data to the sink node either directly or via neighboring node. The user collects data from sink node while the receiver has to process the packet that arrives. Although, there are various authentication frameworks taking part in the process, attackers can create a fake identity and send malicious packets. A virus is a hidden malicious program that gets intentionally transmitted over the sensor network to make some changes in network/node to destroy the stability of the network/node. When the programs run, the viruses get activated along with the programs and start infecting associated programs that come into their contacts [1]. Thus, these viruses spread over the network via communication and make other nodes infected. To clean the sensor node, anti-virus software is

S. Sundar (✉) · A. Chaturvedi
Department of Mathematics, Pranveer Singh Institute of Technology, Kanpur 209305, India
e-mail: ssmishra15@gmail.com

R. Naresh
Department of Mathematics, School of Basic & Applied Sciences, H B Technical University, Kanpur 208002, India

A. K. Awasthi
Department of Mathematics, School of Vocational Studies and Applied Sciences, Gautam Budh University, Gautam Budh Nagar, Greater Noida 201308, India



D. Singh et al. (eds.), *Proceedings of International Conference on Scientific and Natural Computing*, Algorithms for Intelligent Systems,
https://doi.org/10.1007/978-981-16-1528-3_6

used. This software either cleans, quarantines, or deletes the local instance of virus at node where this is installed. Thus, it helps to make network virus free by isolating a particular node. It is noted here that the anti-malicious software must be updated regularly so that it can recognize the latest malicious objects.

Mathematical modeling is a very important tool to study the dynamics of the propagation of malicious objects in WSN. There is a close resemblance between malicious objects and biological viruses as it propagates from one node to other in the same way as a biological virus from one person to other. Therefore, these malicious objects are termed as computer viruses. In recent years, some mathematical models have been proposed and analyzed to study the dynamics of the virus infection in various networks, in computer networks [1–12] as well as in WSN [13–26]. In particular, Yuan and Chen [10] presented a network epidemic e-SEIR model and discussed the spread of virus in a computer network by using the stability theory of differential equations. Mishra and Jha [4] proposed a model for the transmission of malicious objects in a computer network. Mishra and Saini [13] have also studied SEIRS type model for the spread of computer viruses by considering the effect of delay in the transmission process. Newman et al. [8] have studied the spread of computer viruses in e-mail network. Khayam et al. proposed a worm propagation model, the topologically aware worm propagation model (TWPM) for WSN and they consider susceptible or infected two possible states [14]. Akdere et al. [15] show the applicability of epidemic theory on WSN and a comparative analysis of epidemic algorithms for data dissemination has been done. Tang [16] proposed a modified SI model to improve the anti-virus capability of networks by leveraging sleep mode. Zhu et al. [17] developed a delay reaction–diffusion model and described the process of malware propagation for mobile WSN. Singh et al [18] proposed a worm propagation analysis for WSN.

In most of the studies, the effect of anti-virus software on infected network has been modeled and analyzed without taking a separate equation for anti-malicious software. Therefore, in this paper, our focus is to model the sensor network infected by malicious objects and we show that how anti-virus program could clean the system considering a separate equation for anti-malicious software. In the rest of the paper, we have developed a mathematical model under some crucial assumptions in Sect. 2, equilibrium and stability analysis is shown in Sects. 3 and 4, respectively. Numerical simulation is carried out in Sect. 5. In Sect. 6, the results and findings of the proposed work are given as conclusion.

## 2 Mathematical Model

Consider a computer network immune system affected by malicious objects. To develop a mathematical model, we consider four dependent variables namely; susceptible computers, infected computers, malicious objects and anti-malicious software. Let $N$, $I$ and $V$ be the number density of susceptible computers, infected computers and malicious objects, respectively, and $A$ be the potency of anti-malicious software.

To model the phenomenon, we have made the following assumptions,

(i) Computers are connected to a network with a constant rate.
(ii) Growth rate of infected computers is in direct proportion to the densities of susceptible computers as well as malicious objects.
(iii) Growth rate of malicious objects is in direct proportion to the density of infected computers.
(iv) The rate of increase in the potency of anti-malicious software is in direct proportion to the density of infected computers.

Let $Q$ be the installation rate of computers in a network system. Susceptible computers become infected when they come in contact with malicious objects and it is assumed to be in direct proportion to the number density of susceptibles as well as the malicious objects (i.e. $\lambda_1 NV$), $\lambda_1$ being the coefficient by which susceptibles become infected due to interaction with malicious objects and $\pi_1$ $(0 \leq \pi_1 \leq 1)$ denotes the rate at which susceptibles become infected when interact with malicious objects. Further, it is assumed that the malicious objects are cleaned by anti-malicious software and infected computers are recovered. Thus, the recovered computers leave the infected class with a rate $\lambda_2 IA$ and come back into susceptible class. The constant $\lambda_2$ is the diminishing rate coefficient of infected computers due to their recovery by anti-malicious software. It is further assumed that infected computers are recovered at a rate proportional to the densities of infected computers and the anti-malicious software's (i.e. $\pi_2\lambda_2 IA$) enhancing the growth of susceptible computers, $0 \leq \pi_2 \leq 1$ being the recovery rate coefficient. The constant $d$ denotes the rate by which susceptible or infected computers get damaged and are removed from the network system. The growth of malicious objects is in the direct proportion to the number density of infected computers (i.e. $\beta_1 I$), $\beta_1$ being the growth rate of malicious objects in a computer network system, $\beta_2$ is the rate by which malicious objects are not able to infect the network. It is assumed that malicious objects get deactivated or cleaned at a rate $\beta_3$ due to interaction with anti-malicious software. Potency of anti-malicious software, to be introduced in a computer network system, is assumed to be increased at a rate proportional to the infected computers (i.e. $\gamma_1 I$), $\gamma_2$ is the rate by which anti-malicious software fails to work efficiently and $\gamma_3$ is the rate by which anti-malicious software is not able to deactivate or clean malicious objects.

In view of the above, the dynamics of the spread of malicious objects in a computer network system is assumed to be governed by the following system of differential equations,

$$\frac{dN}{dt} = Q - \lambda_1 NV + \pi_2\lambda_2 IA - dN \tag{1}$$

$$\frac{dI}{dt} = \pi_1\lambda_1 NV - \lambda_2 IA - dI \tag{2}$$

$$\frac{dV}{dt} = \beta_1 I - \beta_2 V - \beta_3 VA \tag{3}$$

$$\frac{dA}{dt} = \gamma_1 I - \gamma_2 A - \gamma_3 I A \tag{4}$$

$$N(0) > 0, \;\; I(0) \geq 0, \;\; V(0) \geq 0, \;\; A(0) > 0$$

In the model, all the constants are assumed to be positive.

**Remark 1** It is remarked here that if rate of interaction of susceptible computers with malicious objects (i.e. $\lambda_1$) is very large so that $\frac{dN}{dt}$ may become negative and then all the susceptible computers would be infected. Further, if $\lambda_2$ is very large, $\frac{dI}{dt}$ will be negative and the malicious objects can be cleaned completely from the computer network system.

Here, the symbols are also described in the following Table 1.

## 3 Equilibrium Analysis

The model (1)–(4) consists of two equilibria namely,

(i) Axial equilibrium $E_0\left(\frac{Q}{d}, 0, 0, 0\right)$
(ii) Interior equilibrium $E^*(N^*, \;\; I^*, \;\; V^*, \;\; A^*)$

The existence and uniqueness of $E_0$ are obvious.

The positive solution of $E^*(N^*, \;\; I^*, \;\; V^*, \;\; A^*)$ is given by the following algebraic equations,

$$Q - \lambda_1 N V + \pi_2 \lambda_2 I A - dN = 0 \tag{5}$$

$$\pi_1 \lambda_1 N V - \lambda_2 I A - dI = 0 \tag{6}$$

$$\beta_1 I - \beta_2 V - \beta_3 V A = 0 \tag{7}$$

$$\gamma_1 I - \gamma_2 A - \gamma_3 I A = 0 \tag{8}$$

Using (8) in (7), we get

$$\beta_1 \gamma_3 I^2 + \beta_1 \gamma_2 I - \beta_2 \gamma_2 V - (\beta_2 \gamma_3 + \beta_3 \gamma_1) V I = 0 \tag{9}$$

From Eqs. (5) and (6), we get

$$\begin{aligned} &\pi_1 \lambda_1 \gamma_2 Q V + (\pi_1 \lambda_1 \gamma_3 Q - d\lambda_1 \gamma_2) V I - (\lambda_1 \lambda_2 \gamma_1 (1 - \pi_1 \pi_2) + d\lambda_1 \gamma_3) V I^2 \\ &-(d\lambda_2 \gamma_1 + \gamma_3 d^2) I^2 - \gamma_2 d^2 I = 0 \end{aligned} \tag{10}$$

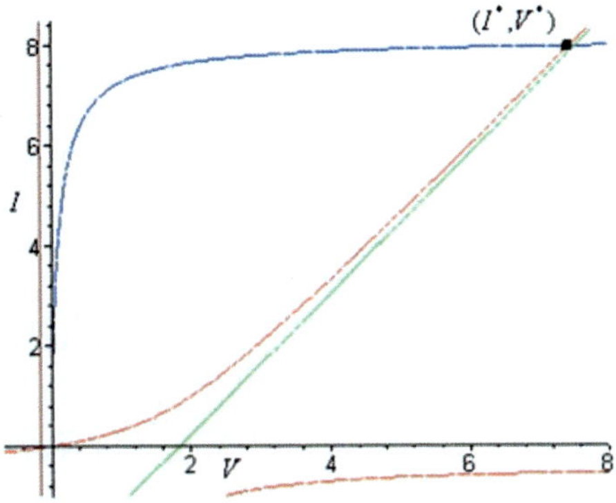


**Fig. 1** Existence of $(I^*,\ V^*)$

From (9), we note that.

(i) $V = 0 \Rightarrow I = 0,\ I = -\frac{\gamma_2}{\gamma_3}$

(ii) $\frac{dI}{dV} = \frac{\beta_2\gamma_2 + (\beta_2\gamma_3 + \beta_3\gamma_1)I}{\beta_1\gamma_3 I^2 + \beta_2\gamma_2 V} I > 0$ in first quadrant.

(iii) Asymptote are $I = -\frac{\beta_2\gamma_2}{(\beta_2\gamma_3 + \beta_3\gamma_1)},\ I = \frac{(\beta_2\gamma_3 + \beta_3\gamma_1)}{\beta_1\gamma_3} V - \frac{\beta_3\gamma_1\gamma_2}{\gamma_3(\beta_2\gamma_3 + \beta_3\gamma_1)}$

From (10), we note that.

(i) $V = 0 \Rightarrow I = 0,\ I = -\frac{\gamma_2 d}{\gamma_1\lambda_2 + \gamma_3 d}$

(ii) $\frac{dI}{dV} = \frac{(d\lambda_2\gamma_1+\gamma_3 d^2)I^2+\gamma_2 d^2 I}{\pi_1\lambda_1\gamma_2 QV+(\lambda_1\lambda_2\gamma_1(1-\pi_1\pi_2)+d\lambda_1\gamma_3)VI^2+(d\lambda_2\gamma_1+\gamma_3 d^2)I^2}\frac{I}{V} > 0$ in first quadrant.

(iii) Asymptote is $V = -\frac{(d\lambda_2\gamma_1+\gamma_3 d^2)}{(\lambda_1\lambda_2\gamma_1(1-\pi_1\pi_2)+d\lambda_1\gamma_3)}$

In view of the above analysis, the existence of $E^*$ is shown in Fig. 1 in the interior of the first quadrant in $I - V$ plane where two isoclines, given by (9) and (10), intersect at a unique point. Knowing the values of $I^*$ and $V^*$, the values of $N^*$ and $A^*$ can easily be obtained from the Eqs. (5) and (8) simultaneously.

## 4 Stability Analysis

In this section, first, we have established the local stability behavior of equilibria $E_0$ and $E^*$ to check the feasibility of the model system and then nonlinear stability behavior of equilibrium $E^*$.

The local stability character of $E_0$ is checked by determining the sign of eigenvalues of Jacobian matrix of the model system (1)–(4) corresponding to $E_0$ whereas the local stability character of $E^*$ is investigated by considering a suitable Lyapunov's function.

To establish the local stability behavior of equilibria, we compute the following Jacobian matrix $M$ for model system (1)–(4),

$$M = \begin{bmatrix} -(\lambda_1 V + d) & \pi_2\lambda_2 A & -\lambda_1 N & \pi_2\lambda_2 I \\ \pi_1\lambda_1 V & -(\lambda_2 A + d) & \pi_1\lambda_1 N & -\lambda_2 I \\ 0 & \beta_1 & -(\beta_2 + \beta_3 A) & -\beta_3 V \\ 0 & \gamma_1 - \gamma_3 A & 0 & -(\gamma_2 + \gamma_3 I) \end{bmatrix}$$

Let $M_0$ and $M^*$ be the Jacobian matrices corresponding to equilibria $E_0$ and $E^*$.

Corresponding to $E_0\left(\frac{Q}{d}, 0, 0, 0\right)$, two eigenvalues of $M_0$ are $-d$ and $-\gamma_2$, remaining two are given by the following algebraic equation,

$$x^2 + (d + \beta_3)x + \left(d\beta_3 - \beta_1\pi_1\lambda_1\frac{Q}{d}\right) = 0 \tag{11}$$

From (11), we note the following,

(i) Both of the roots of (11) will be negative, if

$$d\beta_3 - \beta_1\pi_1\lambda_1\frac{Q}{d} > 0 \tag{12}$$

(ii) Both of the roots of (11) will be of opposite sign, if

$$d\beta_3 - \beta_1\pi_1\lambda_1\frac{Q}{d} < 0 \tag{13}$$

It is easy to see that, all the eigenvalues of Jacobian matrix $M_0$ will be negative provided the condition (12) is satisfied and in this case equilibrium $E_0$ is locally stable. Thus, the system will settle down to the equilibrium. The equilibrium $E_0$ is unstable if condition (13) is satisfied and system will never settle down to the equilibrium.

**Theorem 4.1** The equilibrium $E_0$ is locally stable provide the condition (12) is satisfied otherwise it is unstable.

Local and nonlinear stability behavior of the equilibrium $E^*$ is given in the form of following theorems,

**Theorem 4.2** The equilibrium $E^*$, if it exists, is locally stable provided the following inequalities are satisfied,

$$\pi_1\pi_2\lambda_1\lambda_2 A^* V^* < \frac{1}{12}(\lambda_1 V^* + d)(\lambda_2 A^* + d) \tag{14}$$

$$\pi_1\pi_2\lambda_1(\gamma_1 - \gamma_3 A^*)I^* V^* < \frac{4}{9}A^*(\lambda_1 V^* + d)(\gamma_2 + \gamma_3 I^*) \tag{15}$$

$$C_0 \max\left\{\frac{3\lambda_1}{\lambda_1 V^* + d}, \frac{4\pi_1\pi_2\lambda_2 A^*}{(\lambda_2 A^* + d)V^*}\right\} < C_1 \min\left\{\frac{(\lambda_2 A^* + d)}{4\beta_1^2}, \frac{(\gamma_2 + \gamma_3 I^*)\lambda_2 I^*}{3(\beta_3 V^*)^2(\gamma_1 - \gamma_3 A^*)}\right\} \tag{16}$$

where $C_0 = \frac{\lambda_1 {N^*}^2}{\beta_2 + \beta_3 A^*}$ and $C_1 = \frac{(\beta_2 + \beta_3 A^*)\pi_2\lambda_2 A^*}{\pi_1\lambda_1 V^*}$.

For proof of the theorem, see Appendix A.

To establish nonlinear stability behavior of the equilibrium $E^*$ for model system (1)–(4), we need the bounds of dependent variables stated in the form of the following lemma,

**Lemma 4.1** The set.

$$\Omega = \left\{(N, I, V, A) \in R_+^4 : 0 < N + I \le \frac{Q}{d},\ 0 \le V \le \frac{\beta_1}{\beta_2}\frac{Q}{d},\ 0 < A \le \frac{\gamma_1}{\gamma_2}\frac{Q}{d}\right\}$$

is the region of attraction for all solutions of the model system (1)–(4) initiating in the interior of positive octant.

***Proof*** From Eqs. (1) and (2), we have.

$$\frac{d(N+I)}{dt} = Q - (1-\pi_1)\lambda_1 NV - (1-\pi_2)\lambda_2 I A - d(N+I) \le Q - d(N+I)$$

which implies that $0 < N + I \le \frac{Q}{d}$.
From Eq. (3), we have, $\frac{dV}{dt} \le \beta_1 I - \beta_2 V$.
which implies that $0 \le V \le \frac{\beta_1}{\beta_2}\frac{Q}{d}$.
From Eq. (4), we have, $\frac{dA}{dt} \le \gamma_1 I - \gamma_2 A$
which implies that $0 < A \le \frac{\gamma_1}{\gamma_2}\frac{Q}{d}$.
Hence the lemma.

**Theorem 4.3** The equilibrium $E^*$, if it exists, is nonlinearly stable inside the region of attraction $\Omega$ provided the following inequalities are satisfied,

$$\pi_1\pi_2\lambda_1\lambda_2 A^* V^* < \frac{d^2}{12} \tag{17}$$

$$\pi_1\pi_2\lambda_1\left(\frac{Q}{d}\right)^2(\gamma_1 - \gamma_3 A^*)V^* < \frac{4}{9}d\gamma_2 A^* I^* \tag{18}$$

$$C_2 \max\left\{3\lambda_1 {N^*}^2, 4\pi_1\pi_2\lambda_2\left(\frac{Q}{d}\right)^2\left(\frac{A^*}{V^*}\right)\right\} < C_3 \min\left\{\frac{d}{4\beta_1^2}, \frac{\gamma_2\lambda_2 I^*}{3(\beta_3 V^*)^2(\gamma_1 - \gamma_3 A^*)}\right\} \tag{19}$$

where $C_2 = \frac{\lambda_1}{\beta_2 d}$ and $C_3 = \frac{\beta_2\pi_2\lambda_2 A^*}{\pi_1\lambda_1 V^*}$.

For proof of the theorem, see Appendix B.

These theorems imply that under certain conditions, the system would remain in equilibrium. The number density of infected computers increases as the interaction rate with malicious objects increases. Further, if we introduce suitable anti-malicious software, the number density of infected computers decreases.

**Remark 2** We note that, for small values of $\pi_1$ or $\lambda_1$ the possibility of satisfying local stability conditions (14)–(16) and nonlinear stability conditions (17)–(19) is more plausible. Further, if $\pi_1 = 0$ or $\lambda_1 = 0$ then all the local and nonlinear stability conditions are satisfied automatically. This implies that $\pi_1$ and $\lambda_1$ have a destabilizing effect on the model system.

## 5 Numerical Simulation

In this section, we have studied the feasibility of equilibrium analysis and the behavior of various parameters on the model system considering the following set of parameter values. The model system is integrated by the fourth-order Runge–Kutta method with the help of Maple software.

$$Q = 10,\ \lambda_1 = 0.1,\ \lambda_2 = 0.3, \pi_1 = 0.7, \pi_2 = 0.5,\ d = 0.01$$

$$\beta_1 = 0.5,\ \beta_2 = 0.1,\ \beta_3 = 0.1,\ \gamma_1 = 0.3,\ \gamma_2 = 0.15,\ \gamma_1 = 0.05$$

The equilibrium $E^*$ for the above set of parameters is

$$N^* = 20.1820,\ \ I^* = 7.9738,\ \ V^* = 7.4386,\ \ A^* = 4.3597$$

The eigenvalues corresponding to the equilibrium $E^*$ are found to be $-2.1432$,$-0.2920 + 0.4203\,i$, $-0.2920 - 0.4203\,i$, $-0.4290$. It is evident that all the eigenvalues are either negative or have negative real part indicating local stability behavior of $E^*$. The nonlinear stability behavior of $E^*$ in $N - I$ and $I - V$ plane is shown in Figs. 2 and 3, respectively. It can easily be seen that all the trajectories having different initial starts approach to equilibrium $E^*$ showing that $E^*$ is nonlinearly asymptotically stable.

The variation of susceptibles $N$ and infecteds $I$ at different values of $\pi_1$ (rate by which susceptible computers become infected) with time $'t'$ is shown in Figs. 4

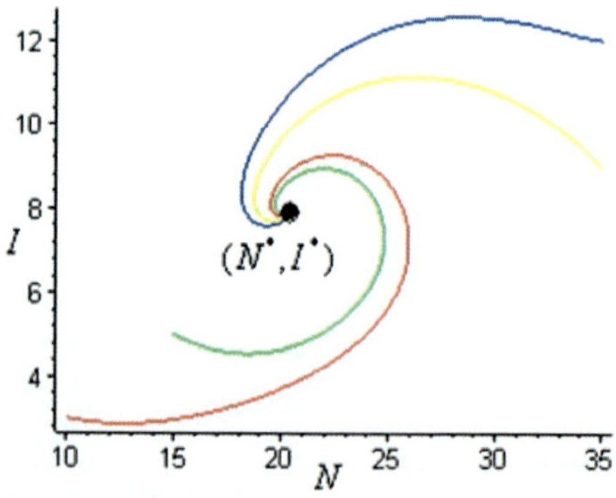


**Fig. 2** Nonlinear stability in $N - I$ plane

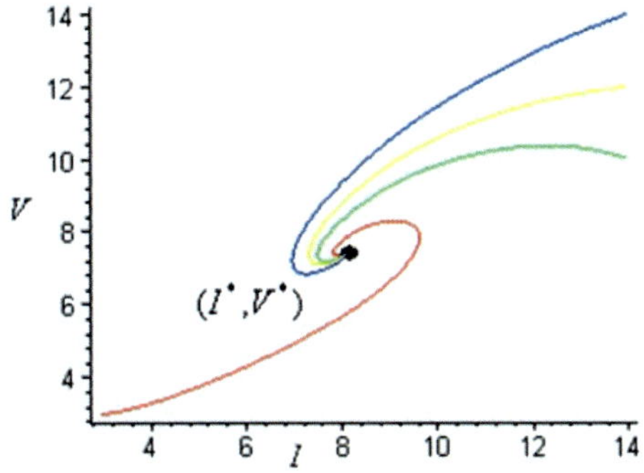


**Fig. 3** Nonlinear stability in $I - V$ plane

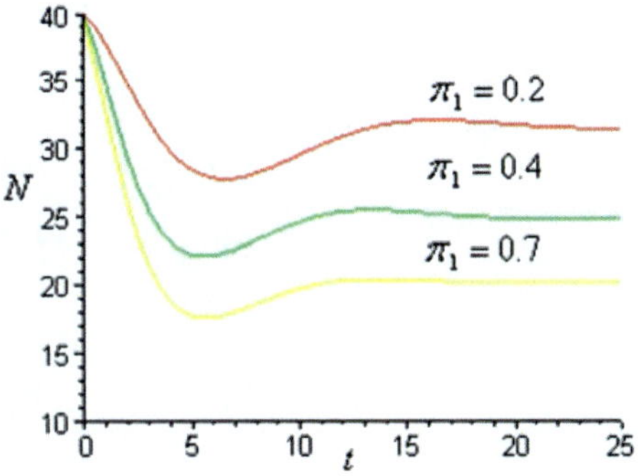


**Fig. 4** Variation of $N$ with time '$t$' at different values of $\pi_1$

and 5 respectively. It is apparent from these figures that as the rate of infection of susceptibles to infected increases the number density of susceptibles decreases but that of infected increases. To study the effect of $\gamma_1$ (updating rate of anti-malicious software) on different variables viz. susceptibles, infected and malicious objects, we have plotted the variation of these variables for different values of $\gamma_1$ with time $'t'$ in Figs. 6, 7 and 8 respectively. From these figures, it is seen that the number density of

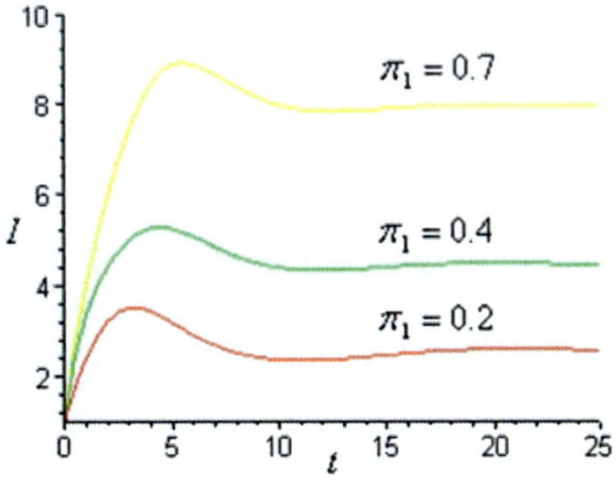


**Fig. 5** Variation of $I$ with time '$t$' at different values of $\pi_1$

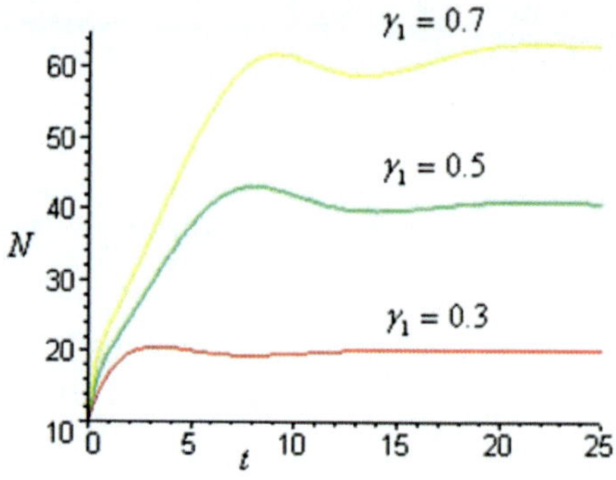


**Fig. 6** Variation of $N$ with time '$t$' at different values of $\gamma_1$

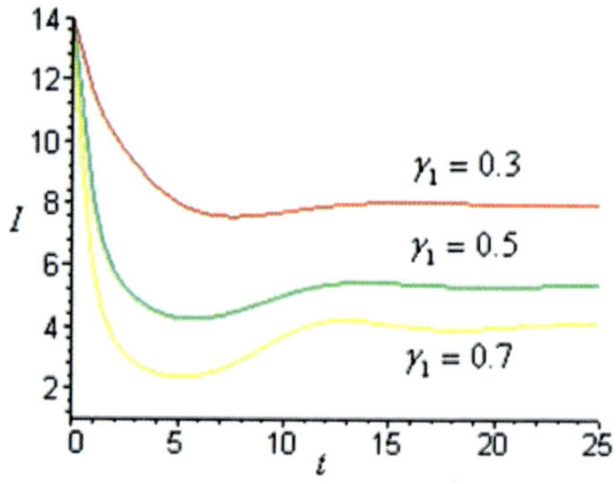


**Fig. 7** Variation of $I$ with time '$t$' at different values of $\gamma_1$

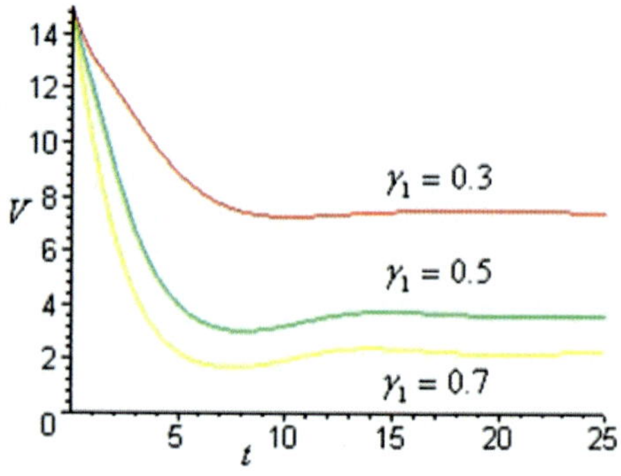


**Fig. 8** Variation of $V$ with time '$t$' at different values of $\gamma_1$

susceptibles increases while the number density of infected computers and malicious objects decreases as the updating rate of anti-malicious software increases.

# 6 Conclusion

The security and protection of WSN are crucial. The spread of malicious objects in WSN is becoming perilous day by day. After the revolution in sensor networks and introduction of malicious objects, it becomes very difficult to secure and protect WSN against attack of malicious objects. The effect of malicious objects in WSN can only be diminished by continuous updating of anti-malicious software. Therefore, in this paper, we have introduced and investigated a mathematical model to study the effect of anti-malicious software on the spread of malicious objects in WSN. The model is analyzed using stability theory of ordinary differential equations. Local and global stability criteria of equilibrium $E^*$ have been established. It is shown that under suitable conditions, the equilibrium $E^*$ is locally and none linearly asymptotically stable. It is, further, shown that the number density of susceptible nodes increases whereas the number density of infected systems and malicious objects decreases as the updating rate of anti-malicious software increases.

**Table 1** List of symbols

| Symbol | Detail |
|---|---|
| $Q$ | Rate at which external computers are connected to a network |
| $N$ | Number density of susceptible computers |
| $I$ | Number density of infected computers |
| $V$ | Number density of malicious objects |
| $A$ | Potency of anti-malicious software |
| $\lambda_1$ | Attack rate of malicious objects on susceptible computers |
| $\lambda_2$ | Rate of interaction of infected computers with anti-malicious software |
| $0 \leq \pi_1 \leq 1$ | Rate by which susceptible computers become infected |
| $0 \leq \pi_2 \leq 1$ | Rate by which infected computers are recovered |
| $d$ | Rate at which a computer is removed from the network |
| $\beta_1$ | Rate of introduction of malicious objects |
| $\beta_2$ | Rate by which malicious objects are not able to infect the network |
| $\beta_3$ | Rate by which malicious objects get deactivated or cleaned when come in contact with anti-malicious software |
| $\gamma_1$ | Updating rate of anti-malicious software |
| $\gamma_2$ | Rate by which anti-malicious software fails to work efficiently |
| $\gamma_3$ | Rate by which anti-malicious software is not able to deactivate or clean malicious objects |

## Appendix A

To establish the local stability behavior, we have consider the following positive definite function,

$$V = \frac{1}{2}k_1N_1^2 + \frac{1}{2}k_2I_1^2 + \frac{1}{2}k_3V_1^2 + \frac{1}{2}k_4A_1^2 \tag{A1}$$

where $N_1$, $I_1$, $V_1$ and $A_1$ are small perturbations about the equilibrium $E^*$ and the constants $k_i$ $(i = 1, 2, 3, 4)$ are to be chosen appropriately.

Differentiating (A1) with respect to $'t'$ and using the linearized system of the model (1)–(4) corresponding to $E^*$

$$\begin{aligned}\frac{dV}{dt} =& -k_1(\lambda_1V^* + d)N_1^2 - k_2(\lambda_2A^* + d)I_1^2 - k_3(\beta_2 + \beta_3A^*)V_1^2 - k_4(\gamma_2 + \gamma_3I^*)A_1^2 \\ &+ (k_1\pi_2\lambda_2A^* + k_2\pi_1\lambda_1V^*)N_1I_1 - (k_1\lambda_1N^*)N_1V_1 + (k_1\pi_2\lambda_2I^*)N_1A_1 \\ &+ (k_2\pi_1\lambda_1N^*)I_1V_1 + (k_3\beta_1)I_1V_1 + (-k_2\lambda_2I^* + k_4(\gamma_1 - \gamma_3A^*))I_1A_1 - (k_3\beta_3V^*)V_1A_1\end{aligned}$$

Now $\frac{dV}{dt}$ will be negative definite provided the following conditions are satisfied,

$$(k_1\pi_2\lambda_2A^* + k_2\pi_1\lambda_1V^*)^2 < \frac{k_1k_2}{3}(\lambda_1V^* + d)(\lambda_2A^* + d) \tag{A2}$$

$$k_1(\lambda_1N^*)^2 < \frac{k_3}{3}(\lambda_1V^* + d)(\beta_2 + \beta_3A^*) \tag{A3}$$

$$k_1(\pi_2\lambda_2I^*)^2 < \frac{4k_4}{9}(\lambda_1V^* + d)(\gamma_2 + \gamma_3I^*) \tag{A4}$$

$$k_2(\pi_1\lambda_1N^*)^2 < \frac{k_3}{4}(\lambda_2A^* + d)(\beta_2 + \beta_3A^*) \tag{A5}$$

$$k_3\beta_1^2 < \frac{k_2}{4}(\lambda_2A^* + d)(\beta_2 + \beta_3A^*) \tag{A6}$$

$$(-k_2\lambda_2I^* + k_4(\gamma_1 - \gamma_3A^*))^2 < \frac{k_2k_4}{3}(\lambda_2A^* + d)(\gamma_2 + \gamma_3I^*) \tag{A7}$$

$$k_3(\beta_3V^*)^2 < \frac{k_4}{3}(\beta_2 + \beta_3A^*)(\gamma_2 + \gamma_3I^*) \tag{A8}$$

Equation (A2) can be written as,

$$(k_1\pi_2\lambda_2A^* - k_2\pi_1\lambda_1V^*)^2 + 4k_1k_2\pi_1\pi_2\lambda_1\lambda_2A^*V^* < \frac{k_1k_2}{3}(\lambda_1V^* + d)(\lambda_2A^* + d) \tag{A9}$$

Choosing $k_1 = 1$ and $k_2 = \frac{\pi_2\lambda_2 A^*}{\pi_1\lambda_1 V^*}$, Eq. (A9) will reduce to,

$$\pi_1\pi_2\lambda_1\lambda_2 A^* V^* < \frac{1}{12}(\lambda_1 V^* + d)(\lambda_2 A^* + d)$$

Choosing $k_4 = \frac{\lambda_2 I^*}{\gamma_1 - \gamma_3 A^*} k_2$ from (A7), it satisfies automatically.
Choosing $k_3$ from (A3), (A5), (A6) and (A8) such that,

$$C_0 \max\left\{\frac{3\lambda_1}{\lambda_1 V^* + d}, \frac{4\pi_1\pi_2\lambda_2 A^*}{(\lambda_2 A^* + d)V^*}\right\} < k_3 < C_1 \min\left\{\frac{(\lambda_2 A^* + d)}{4\beta_1^2}, \frac{(\gamma_2 + \gamma_3 I^*)\lambda_2 I^*}{3(\beta_3 V^*)^2(\gamma_1 - \gamma_3 A^*)}\right\}$$

where $C_0 = \frac{\lambda_1 {N^*}^2}{\beta_2 + \beta_3 A^*}$ and $C_1 = \frac{(\beta_2+\beta_3 A^*)\pi_2\lambda_2 A^*}{\pi_1\lambda_1 V^*}$.
Using $k_1$ and $k_4$ in (A4), we get

$$\pi_1\pi_2\lambda_1(\gamma_1 - \gamma_3 A^*)I^* V^* < \frac{4}{9}A^*(\lambda_1 V^* + d)(\gamma_2 + \gamma_3 I^*)$$

Thus, $\frac{dV}{dt}$ will be negative definite provide the conditions (14), (15) and (16) are satisfied and hence the theorem.

## Appendix B

To establish the nonlinear stability behavior, we have consider the following positive definite function,

$$U = \frac{1}{2}m_1(N - N^*)^2 + \frac{1}{2}m_2(I - I^*)^2 + \frac{1}{2}m_3(V - V^*)^2 + \frac{1}{2}m_4(A - A^*)^2 \tag{B1}$$

where, the constants $m_i$ $(i = 1, 2, 3, 4)$ are to be chosen appropriately.

Differentiating (B1) with respect to $'t'$ and using the linearized system of the model (1)–(4) corresponding to $E^*$

$$\begin{aligned}\frac{dU}{dt} =& - m_1(\lambda_1 V + d)(N - N^*)^2 - m_2(\lambda_2 A + d)(I - I^*)^2\\ &- m_3(\beta_2 + \beta_3 A)(V - V^*)^2\\ &- m_4(\gamma_2 + \gamma_3 I)(A - A^*)^2 + (m_1\pi_2\lambda_2 A^* + m_2\pi_1\lambda_1 V^*)(N - N^*)(I - I^*)\\ &- (m_1\lambda_1 N^*)(N - N^*)(V - V^*) + (m_1\pi_2\lambda_2 I)(N - N^*)(A - A^*)\\ &+ (m_2\pi_1\lambda_1 N)(I - I^*)(V - V^*) + (m_3\beta_1)(I - I^*)(V - V^*)\\ &+ (-m_2\lambda_2 I^* + m_4(\gamma_1 - \gamma_3 A^*))(I - I^*)(A - A^*)\\ &- (m_3\beta_3 V^*)(V - V^*)(A - A^*)\end{aligned}$$

Now $\frac{dU}{dt}$ will be negative definite provided the following conditions are satisfied,

$$(m_1\pi_2\lambda_2A^* + m_2\pi_1\lambda_1V^*)^2 < \frac{m_1m_2}{3}(\lambda_1V + d)(\lambda_2A + d) \tag{B2}$$

$$m_1(\lambda_1N^*)^2 < \frac{m_3}{3}(\lambda_1V + d)(\beta_2 + \beta_3A) \tag{B3}$$

$$m_1(\pi_2\lambda_2I)^2 < \frac{4m_4}{9}(\lambda_1V + d)(\gamma_2 + \gamma_3I) \tag{B4}$$

$$m_2(\pi_1\lambda_1N)^2 < \frac{m_3}{4}(\lambda_2A + d)(\beta_2 + \beta_3A) \tag{B5}$$

$$m_3\beta_1^2 < \frac{m_2}{4}(\lambda_2A + d)(\beta_2 + \beta_3A) \tag{B6}$$

$$(-m_2\lambda_2I^* + m_4(\gamma_1 - \gamma_3A^*))^2 < \frac{m_2m_4}{3}(\lambda_2A + d)(\gamma_2 + \gamma_3I) \tag{B7}$$

$$m_3(\beta_3V^*)^2 < \frac{m_4}{3}(\beta_2 + \beta_3A)(\gamma_2 + \gamma_3I) \tag{B8}$$

Equation (B2) can be written as,

$$(m_1\pi_2\lambda_2A^* - m_2\pi_1\lambda_1V^*)^2 + 4m_1m_2\pi_1\pi_2\lambda_1\lambda_2A^*V^* < \frac{m_1m_2}{3}(\lambda_1V + d)(\lambda_2A + d) \tag{B9}$$

Maximizing LHS and minimizing RHS, choosing $m_1 = 1$ and $m_2 = \frac{\pi_2\lambda_2A^*}{\pi_1\lambda_1V^*}$, Eq.\,B9) will reduces to,

$$\pi_1\pi_2\lambda_1\lambda_2A^*V^* < \frac{d^2}{12}$$

Choosing $m_4 = \frac{\lambda_2I^*}{\gamma_1-\gamma_3A^*}m_2$ from (B7), it satisfies automatically.
Using $m_1$ and $m_4$ in (B4), we get

$$\pi_1\pi_2\lambda_1\left(\frac{Q}{d}\right)^2(\gamma_1 - \gamma_3A^*)V^* < \frac{4}{9}d\gamma_2A^*I^*$$

Choosing $m_3$ from (B3), (B5), (B6) and (B8) such that,

$$C_2\max\left\{3\lambda_1N^{*^2}, 4\pi_1\pi_2\lambda_2\left(\frac{Q}{d}\right)^2\left(\frac{A^*}{V^*}\right)\right\} < C_3\min\left\{\frac{d}{4\beta_1^2}, \frac{\gamma_2\lambda_2I^*}{3(\beta_3V^*)^2(\gamma_1 - \gamma_3A^*)}\right\}$$

where $C_2 = \frac{\lambda_1}{\beta_2 d}$ and $C_3 = \frac{\beta_2 \pi_2 \lambda_2 A^*}{\pi_1 \lambda_1 V^*}$.

Thus, $\frac{dU}{dt}$ will be negative definite provide the conditions (17), (18) and (19) are satisfied within the region of attraction $\Omega$ and hence the theorem.

## References


1. Cohen F (1987) Computer viruses: theory and experiments. Comput Secur 6:22–35
2. Misra AK, Verma M, Sharma A (2014) Capturing the interplay between malware and anti-malware in a computer network. Appl Math Comp 229:340–349
3. Shukla JB, Singh G, Shukla P, Tripathi A (2014) Modeling and analysis of the effects of antivirus software on an infected computer network. Appl Math Comp 227:11–18
4. Mishra BK, Jha N (2010) SEIQRS model for the transmission of malicious objects in computer network. Appl Math Model 34:710–715
5. Mishra BK, Saini DK (2007) SEIRS epidemic model with delay for transmission of malicious objects in computer networks. Appl Math Comp 188:1476–1482
6. Saini DK (2011) A mathematical model for the effect of malicious object on computer network immune system. Appl Math Model 35:3777–3787
7. Ren J, Yang X (2011) Dynamics and optimal shelter for computer virus propagation in network. J Inf Comput Sci 8(9):1735–1745
8. Newman MEJ, Forrest S, Balthrop J (2002) Email networks and the spread of computer viruses. Phys Rev E 66:035101(R)
9. Ren J, Yang X, Zhu Q, Yang L-X, Zhang C (2012) A novel computer virus model and its dynamics. Nonlinear Anal Real World Appl 13:376–384
10. Yuan H, Chen G (2008) Network virus-epidemic model with the point-to-group information propagation. App. Math Comp 206(1):357–367
11. Yang L-X, Yang X, Zhu Q, Wen L (2013) A computer virus model with graded cure rates. Nonlinear Anal Real World Appl 14:414–422
12. Mishra BK, Jha N (2007) Fixed period of temporary immunity after run of anti-malicious software on computer nodes. Appl Math Comp 190:1207–1212
13. Mishra BK, Saini DK (2007) SEIRS epidemic model with delay for transmission of malicious objects in computer network. Appl Math Comp 188:1476–1482
14. Khayam SA, Radha H (2005) A topologically-aware worm propagation model for wireless sensor networks. In: 25th IEEE international conference on distributed computing systems workshops. IEEE, pp 210–216
15. Akdere M, Bilgin CC, Gerdaneri O, Korpeoglu I, Ulusoy O, Cetintemel U (2006) A comparison of epidemic algorithms in wireless sensor networks. Comput Commun 29(13):24502457
16. Tang S (2011) A modified si epidemic model for combating virus spread in wire- less sensor networks. Int J Wirel Inf Netw 18(4):319326
17. Zhu L, Zhao H, Wang X (2015) Bifurcation analysis of a delay reaction diffusion malware propagation model with feedback control. Commun Nonlinear Sci Numer Simul 22(1):747768
18. Singh A, Awasthi AK, Singh K, Srivastava PK (2017) Modeling and analysis of worm propagation in wireless sensor network, wireless personal communications. Springer
19. Zhou Z, Wang W, Li Y (2018) Virus propagation model for wireless sensor networks based on IPv6. Int J Online Biomed Eng 14(10):117–128
20. Zhang C (2018) Global behavior of a computer virus propagation model on multilayer networks. Secur Commun Netw . https://doi.org/10.1155/2018/2153195
21. Yang LX, Yang X, Wu Y (2017) The impact of patch forwarding on the prevalence of computer virus: A theoretical assessment approach. Appl Math Model 43:110–125
22. He Y, Wang T, Xie J, Zhang M (2020) Analysis of computer network attack based on the virus propagation model. EURASIP J Wirel Commun Netw 63. https://doi.org/10.1186/s13638-020-1660-5

23. Huang Z, Xin Xu, Ni J, Zhu H, Wang C (2019) Multimodal representation learning for recommendation in Internet of Things. IEEE Internet Things J 6(6):10675–10685
24. Chen Z, Cai H, Zhang Y, Wu C, Mu M, Li Z, Sotelo MA (2019) A novel sparse representation model for pedestrian abnormal trajectory understanding. Expert Syst Appl 138:112753
25. Schierhorn KL, Jolmes F, Bespalowa J, et al (2017) Influenza a virus virulence depends on two amino acids in the N-terminal domain of its NS1 protein to facilitate inhibition of the RNA-dependent protein kinase PKR. J Virol 91(10):JVI.00198–JVI.00117
26. Simonjan J, Taurer S, Dieber B (2020) A generalized threat model for visual sensor networks. Sensors 20:3629. doi:https://doi.org/10.3390/s20133629

# Chapter 7
# Application of Artificial Neural Network in Maximum Power Point Tracking for Different Radiation and Temperature

**Dilip Yadav and Nidhi Singh**

## 1 Introduction

A solar photovoltaic (PV) system has become an important source of energy as it is available in abundance, approximately 3.8 *1024 J of energy falls on the earth's surface, which can be used for the generation of electricity [1]. The real advantage of using the solar photovoltaic system is that it is static and noiseless. The worldwide concern on reducing the carbon content ($CO_2$) and keeping the safe way of generation has uplifted renewable energy [2]. The solar photovoltaic (PV) device is considered more efficient than the other renewable energy sources because of the environmental and economic benefits [3]. Semiconductor manufacturing technology has helped a lot in decreasing the overall cost of the PV system [4]. Solar cells produce a non-linear I–V curve because of their dependency on irradiation and temperature. The MPPT system samples the output of the cells and applies the appropriate load to obtain MPP for any given environmental conditions. Total power extracted from the solar irradiation is represented by Power (P) = F.F*$V_{OC}$*$I_{SC}$; where $V_{OC}$ is Open Circuit Voltage, $I_{SC}$ is Short Circuit Current and F.F is fill factor. These are useful parameters, as they help in taking out the conclusions on the electrical behavior of a cell that is operating at different conditions [5]. According to the Maximum Power Transfer Theorem, the circuit's maximum power can be obtained if the circuit's Thevenin Impedance is equal to system impedance [6, 7]. The P–V characteristics start showing multiple peaks in the PV curve when a non-uniform condition occurs [8]. Maximum Power Point Tracking (MPPT) is expected to optimally harness the available power under all operating conditions [9]. Factors that affect the efficiency

D. Yadav (✉) · N. Singh
Department of Electrical Engineering, Gautam Buddha University, Greater Noida, India
e-mail: dilipyadav.eng@gmail.com

N. Singh
e-mail: nidhi@gbu.ac.in



D. Singh et al. (eds.), *Proceedings of International Conference on Scientific and Natural Computing*, Algorithms for Intelligent Systems,
https://doi.org/10.1007/978-981-16-1528-3_7

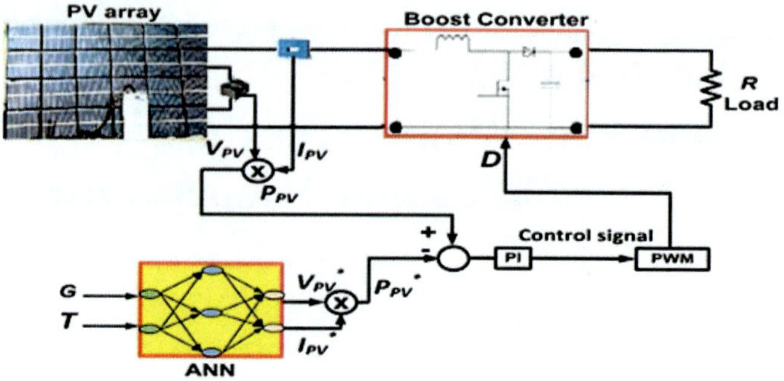


**Fig. 1** Stand-alone PV System with ANN-based MPPT technique and Boost converter

are cable thickness, module temperature, shading, dust particles, coal, etc. [10]. These factors have a high impact on designing a PV system to achieve optimum efficiency [11].

Various MPPT algorithms are available that can be classified into two main categories i.e., conventional-based MPPT [12, 13] and soft computing-based MPPT techniques [14–18]. Soft computing techniques have high tracking speed, a self-learning approach, and independent of a PV system. They are used to improve the performance of the PV module in terms of few ripples, high settling time, low cost, and should be able to provide the output with high efficiency [19, 20]. This paper presents a method to track the maximum power point voltage using an artificial neural network. MPPT system was added with the solar module, Boost Converter, and PI controller. In this paper, a set of real-time experimental data are considered, in terms of temperature, irradiation, open-circuit voltage, short circuit current, module temperature, maximum power, maximum voltage, and current at a different time frame. PV module contains 24 cells. For tracking the maximum power from the PV module, the ANN technique is used, which has two input parameters i.e., irradiation (G) and temperature (T) as shown in Fig. 1. The output of the ANN is taken as a reference power for calculating the error between the PV module power and ANN trained output power, error signal is given to the PI controller, which acts as a tuning parameter for the PV system. PI controller parameters are Kp (proportional gain factor) and Ki (integral gain), which decreases, the steady-state error of the system and increasing the settling time.

## 2 Overview of a PV System

### *2.1 Solar Module*

A solar cell is the smallest unit and it is interconnected in series and parallel to increase the voltage and current of the PV module and these modules form a solar

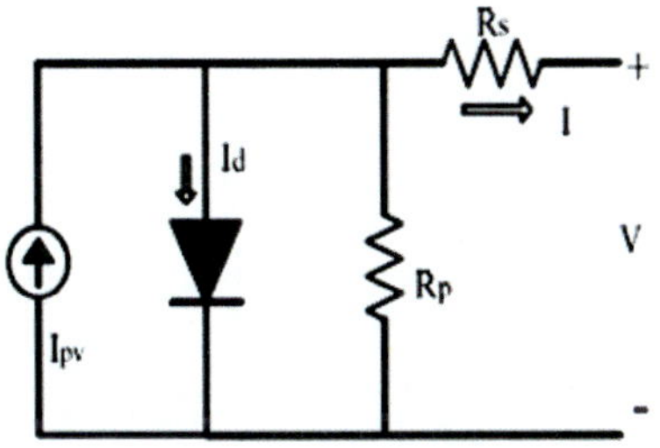


**Fig. 2** Single diode model

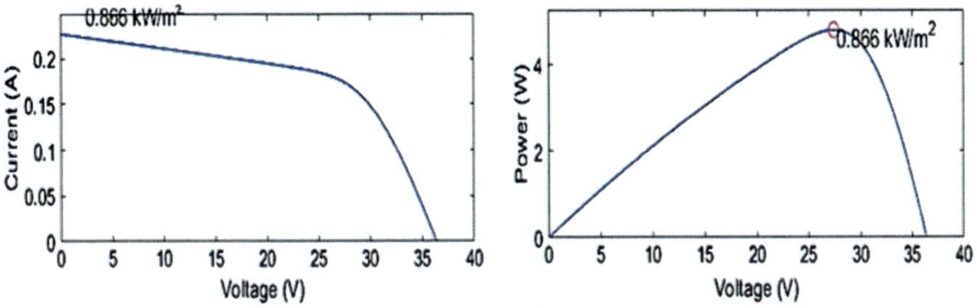


**Fig. 3** I–V and P–V characteristics of Solar Cell

module. Series connected resistance ($R_{se}$) opposes the flow of electrons [21]. The solar cell utilizes the irradiance of the solar energy (nonlinear) and converts it to DC [22]. Figure 2 represents the basic circuit of the single diode model and Fig. 3 shows the I–V and P–V characteristics of the PV module that is considered in this paper (6*6 V *4*0.1) amp).

In given Fig. 4, the temperature is kept constant (25 °C) and the irradiation has been varied at the range of 866, 739, 510, and 379 in W/m$^2$, similarly in Fig. 5, the irradiation is kept constant (866 W/m$^2$) and the temperature has been varied in the range of 20, 30, 40, and 50 °C. Open circuit voltage ($V_{oc}$) and short circuit current ($I_{sc}$) [23] are important parameters of the module, so these parameters were traced for 15 days for 24 PV cells. A total-cross-tied (TCT) connection is used in this paper, because of its advantage over other connections. The rating of a single cell is 0.6 W, 6 volts, and 0.1 amp, the tilt angle is 22° from the earth's surface, as shown in Fig. 6. Important parameters of a PV system are defined in Table 1.

### 2.2 *ANN-Based MPPT Utilizing G (Irradiation) and T (Temperature)*

An artificial neural network (ANN) is inspired by the concept of biological neural networks, it provides an alternative way of solving a complex problem that is

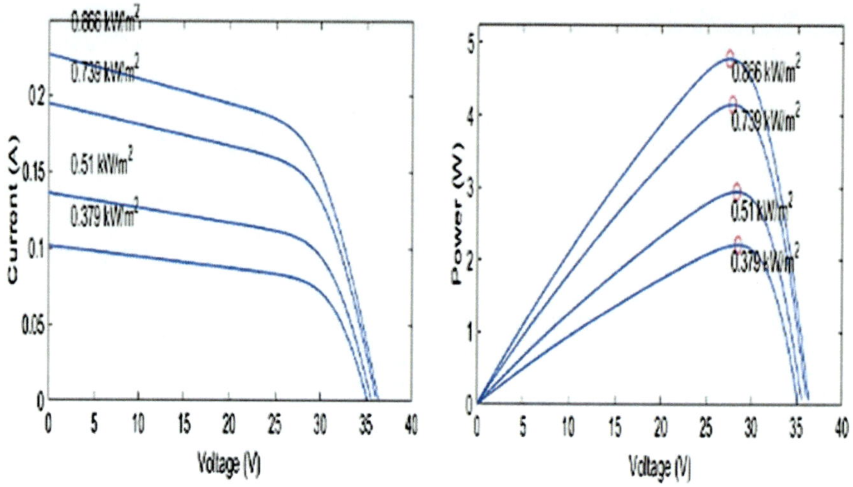


Fig. 4 Effect of irradiation

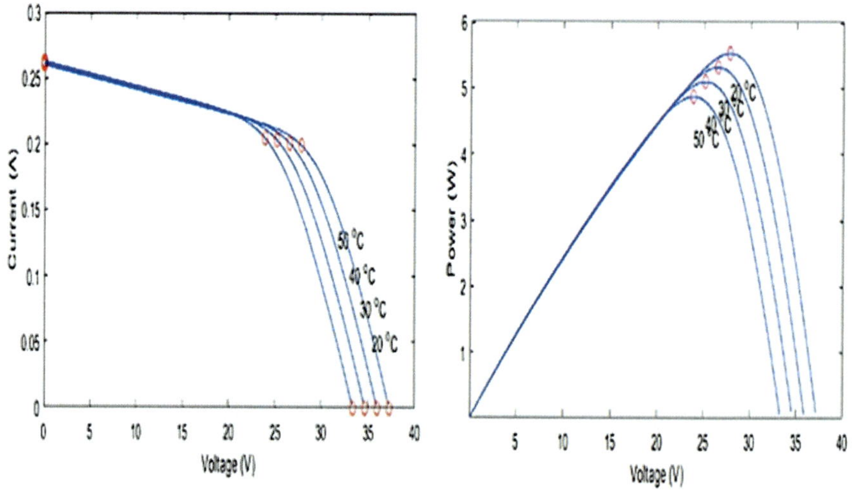


Fig. 5 Effect of temperature

Fig. 6 Solar Module considered for the experiment

**Table 1** Design parameter of the PV system

| Parameters | Values |
|---|---|
| Module open circuit voltage ($V_{oc}$) | 6.163*6 |
| Module short circuit current ($I_{sc}$) | 0.0625*4 |
| Calculated inductor (L) and capacitor (C) | 0.002 H, 5.35182 e-06 F |
| Proportional gain ($K_P$), integral gain ($K_I$), switching frequency ($F_S$) | 0.1, 0.005<br>20 k Hz |
| ANN input data set (G and T) | 195*2 |
| ANN output data (Power) | 195*1 |

nonlinear [24]. Artificial intelligence techniques have more advantages and applications than conventional-based MPPT techniques [25]. Designing of the ANN controller follows an essential three steps:

- **Input–output variable selection:** In this paper, two input parameters are irradiance (G) and temperature (T) with and one output parameter i.e., Power.
- **Network architecture selection:** The hidden layer used in this paper is 10 feed-forward layer neural networks. One hidden layer contains n nodes.
- **Network training and testing**: The data considered should have some percentage of training data set and testing data set for the ANN model [26].

$$F = \frac{1}{N}\sum_{i=1}^{N}\sqrt{(t_i - a_i)} \tag{1}$$

The performance of ANN is calculated in terms of Mean Squared Error (MSE). MSE is given by Eq. (1), MSE value should be as small as possible. Where $a$ is the present output, $t$ is the target to be attained.

### 2.3 *Conventional Boost Converter OR DC–DC Converter*

A conventional boost converter (Fig. 7) is mainly considered for stepping up the DC voltages [26]. The boost converter operates in two ways i.e., conduction and blockage mode. When the switch Q operates in forward biasing in the conduction mode, $dT_S$ and D (diode) operate in the reverse biased condition.

When the diode D is forward biased, then the switch in blocking mode is represented as (1–D) * $T_S$. From Fig. 7, the voltage balance equation can be given as:

$$V_0 = V_T/(1 - d)\text{and}i_0 = i_T(1 - d). \tag{2}$$

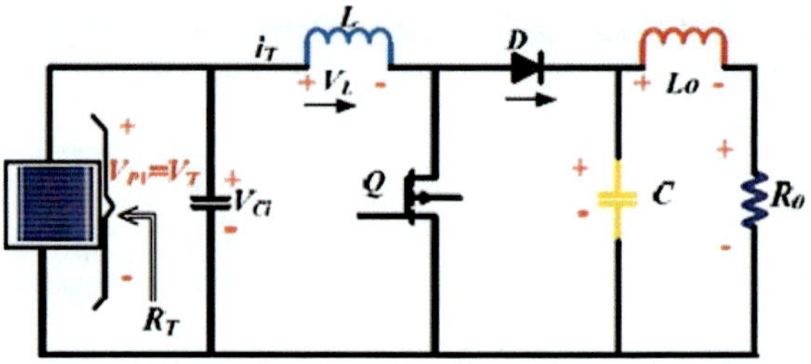


**Fig. 7** Conventional boost converter

Equation (2) gives the relationship between the output, input voltage with duty-cycle *d*.

## 3 Designing Using Matlab/Simulink

The complete model of the Photo Voltaic system is designed in Matlab and Simulation software. As per the experimental data set, the values of irradiation range from 61 to 866 W/m$^2$ and temperature ranges from 5 to 36 °C. For implementing it in simulation, few irradiation and temperature values are considered i.e., 866, 778, 700, 661, 572, 456, 380, 275, 165, 147 W/m$^2$ (irradiation) and 24, 27, 35, 36, 26, 21, 32, 32, 18, 18 °C (temperature), $V_{oc}$ is 36.978 volts $I_{sc}$ is 0.25 amp (Table 1) for the given module. In Fig. 8, the structure of neural networks is given. Input given to the neural network has two inputs i.e., module temperature and irradiance value. The number of hidden layers is calculated by using the trial-and-error method. In this paper, ANN is trained by using experimental data in terms of different irradiation and temperature.

The neural network is trained by using the tool "**nntool**" that is available in MATLAB as shown in Fig. 9. Another option is creating an **m-file** by using the command window. Bayesian Regulation (trainbr) is used for training the network. To train the neural network as per the experimental data, a total of 195 data were used. First input to ANN is the temperature that varies from 5 to 36 °C and second is irradiance that varies from 61 to 866 W/m$^2$. Data have been obtained in Greater Noida (28.4744 degrees north latitude, 77.5040 degrees east longitude. From 195

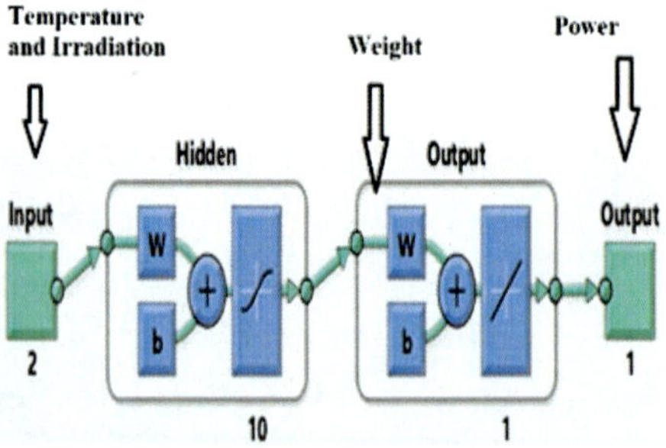


**Fig. 8** Neural network model

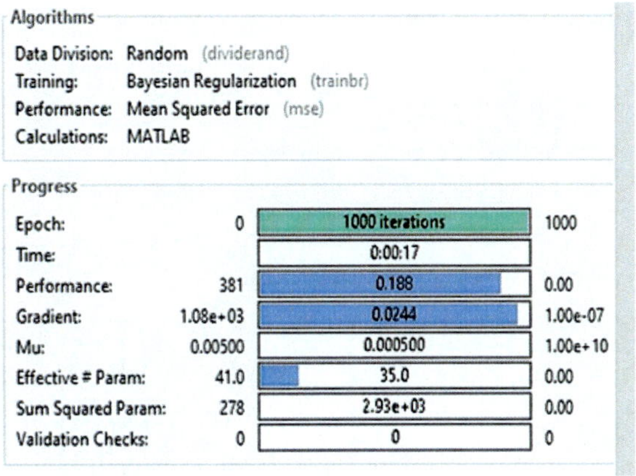


**Fig. 9** ANN training by MATLAB nntool

samples, 55% (107 samples) are taken for training, 20% (39) are taken for Validation and 25% (49) are taken for testing, this is known as Random division. The number of epochs for training is 10000 and the feedforward method is used for training the experimental data.

A complete model of the PV module with ANN, DC–DC converter section with a PI controller is shown in Fig. 10. Irradiation and temperature are input to the PV module and ANN. Input to the converter is voltage and current obtained by the module, irradiation and temperature are input to the ANN. After training of the data, the error signal is generated from the actual power from the panel to the ANN power after the training, and that error is given to the PI controller.

The generated output from the neural network is compared with the PV module output power and error is observed before the Gain block and passed through the PI controller. The output of the PI controller is used as a reference signal for generating necessary gate pulses for IGBT, which vary according to the input, ANN block with PI controller is shown in Fig. 11.

A boost converter work similarly to a step-up chopper i.e. used to step up the input voltage level. The basic circuit diagram for a boost converter using IGBT is given

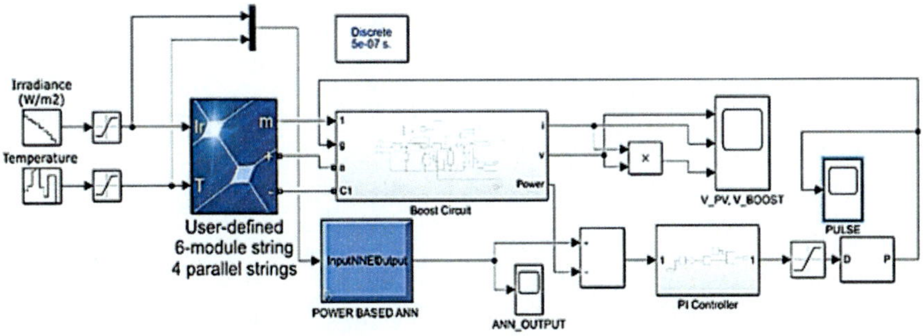


**Fig. 10** PV system Model with ANN-based MPPT and Boost Converter

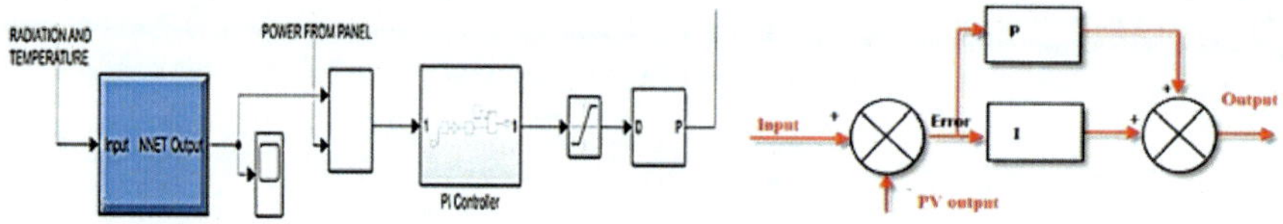


**Fig. 11** ANN Block with PI Controller

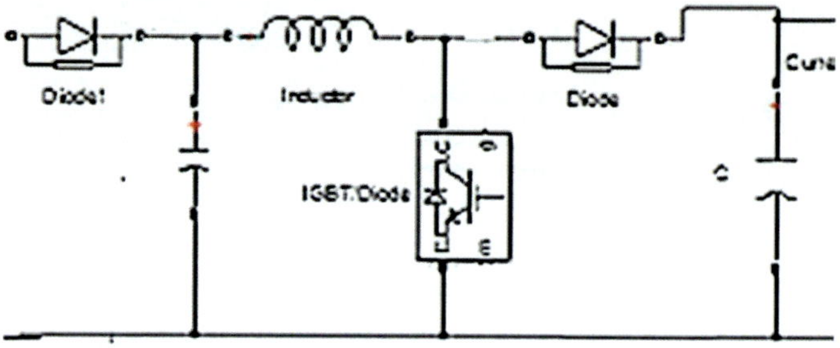


**Fig. 12** Boost Converter

below in Fig. 12. When the IGBT (Insulated gate bipolar transistor) is "on" current flows through the inductor and IGBT and energy get stored in the inductor. Now when IGBT is turned off then energy previously stored in the inductor is released through capacitor and load. The amount by which the output voltage gets boosted up depends upon the duty ratio.

A boost converter is generally used to transfer the power available from solar PV. The input current of the converter is shown below:

$$I_L = P/V_S \text{ and } \Delta \mathrm{I_L} = 20\text{–}40\,\% \text{ of the average value.} \tag{3}$$

$$\text{In boost mode } \mathrm{V_O} = \mathrm{V_S}/(1-\mathrm{d}). \tag{4}$$

$$\text{Duty ratio is } = V_o/V_{input} = 1/(l-d). \tag{5}$$

$$\text{Inductor value is given by} L_{min} = Vs * d / f * \Delta I_L. \tag{6}$$

Output capacitance for boost converter is calculated by using

$$C = d * I_0 / f * \Delta Vo \tag{7}$$

$$Efficiency(\eta) = P_{output} - P_{output} + Losses \tag{8}$$

where $V_O$ is the load voltage; $f$ is the switching frequency, $\Delta V_o$ is taken as 1%, d is duty cycle, $I\Delta_L$ is 0. 4. By using Eqs. (6) and (7), capacitor and inductor value has been calculated for the boost converter. Value of inductor and capacitor is given in Table 1.

## 4 Simulation Result

The simulation result is for 1 s, and the time period of the PV system is set to [0:0.1:1] sec, which represents the simulation time for the model. Different irradiation for the module is [866, 778, 700, 661, 572, 456, 380, 275, 165, 147] in W/m$^2$, and temperature is [24 27 35 36 26 21 32 32 18 18] in °C. The P–V and I–V curves for the given PV module are shown in Fig. 3. For 866 W/m$^2$ irradiation and 25 °C temperature, the value of $V_{oc}$ and $I_{sc}$ is taken by experiment. The output for the given module for the 10 different irradiations and temperature is shown in Fig. 13.

The upper graph represents the V_PV i.e., module voltage, I_PV is module current, and P_PV is the power of the module. The maximum value of voltage is drawn from the module during the time period of 0.2–0.3 s when the irradiation is 778 W/m$^2$ and temperature is 27 °C. $V_{max}$ during that period is 31.75 V, $I_{max}$ is 0.126 amp and $P_{max}$ is 4.52 W. At maximum irradiation of 866 W/m$^2$ and 24 °C, its value is 22.97 V, 0.19 amp, and 4.345 W (which is approximately equal to experimental data). The minimum value of voltage and current obtained in simulation is 10.30 V, 0.04019 amp, and 0.04 W during the time period of 0.9–1 s at 147 W/m$^2$ and 18 °C. Performance of the ANN is given by using mean squared error (MSE). MSE should have very less value at the end of the training. The performance plot of the ANN model in terms of the Mean Squared Error (MSE) is shown in Fig. 14. The regression plot for the module data is given in Fig. 15a, b, and c, which have a high correlation coefficient that has been obtained during training, testing, and validation

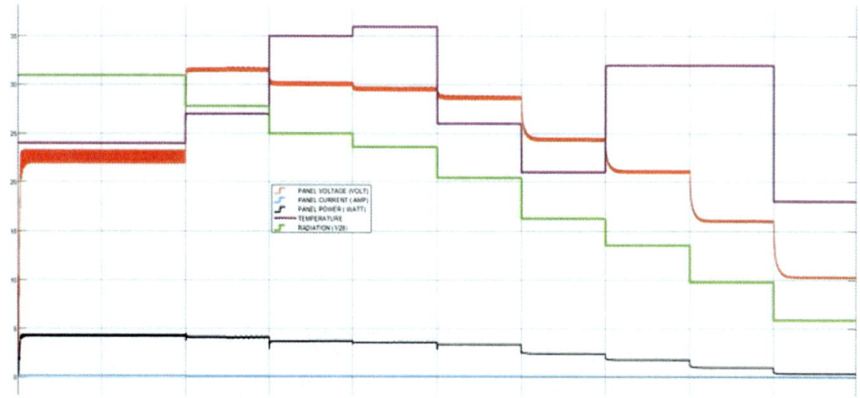


**Fig. 13** V_PV, I_PV, and P_PV from the module

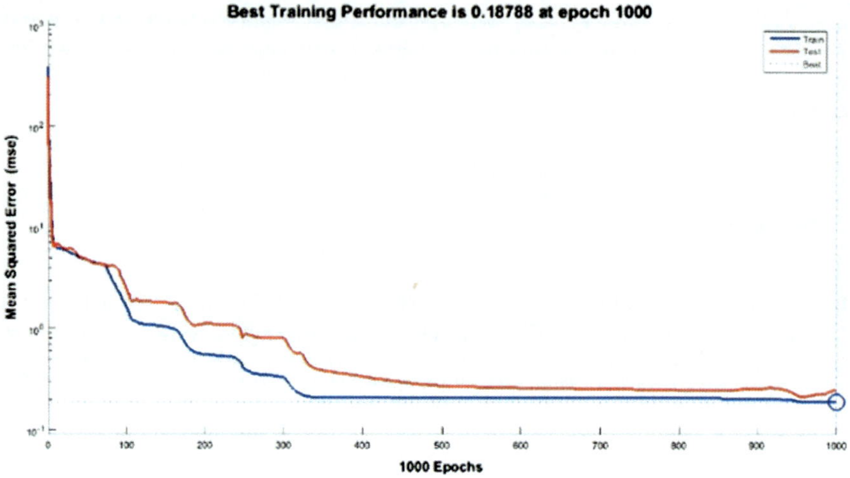


**Fig. 14** ANN best training performance at 1000 epoch

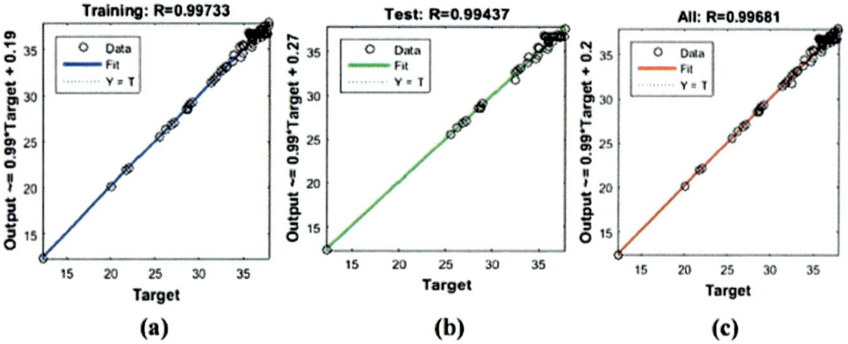


**Fig. 15** **a** Training: 0.99733 **b** Testing data have 0.99437 **c** All R that is about to be 0.99681

of the experimental data. The regression (R) values in this paper come out to be near 99.681%, which indicates a good relationship between outputs of the network and the desired targets.

The variation of trained output ANN is in terms of irradiation and temperature is shown in Fig. 16. The output generated through the neural network is compared with the PV module output power, their error is transferred to the PI controller. The PI controller output is used as a reference signal to produce the gate pulse for the converter power electronic device IGBT. The PV system designed in this paper is for the output voltage of 60 volts for the input voltage value of 36 volts (PV module).

Figure 17 shows the output of the boost converter for different irradiation and temperature for a maximum value of irradiation (866 W/m$^2$) and temperature (24 °C), the maximum power attained by the system was 4.17 W for the load of 1496 ohms.

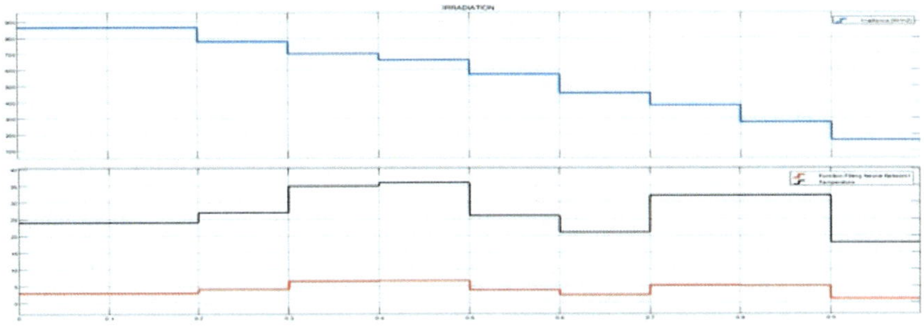

Fig. 16 The output from the ANN as per the trained irradiation and temperature

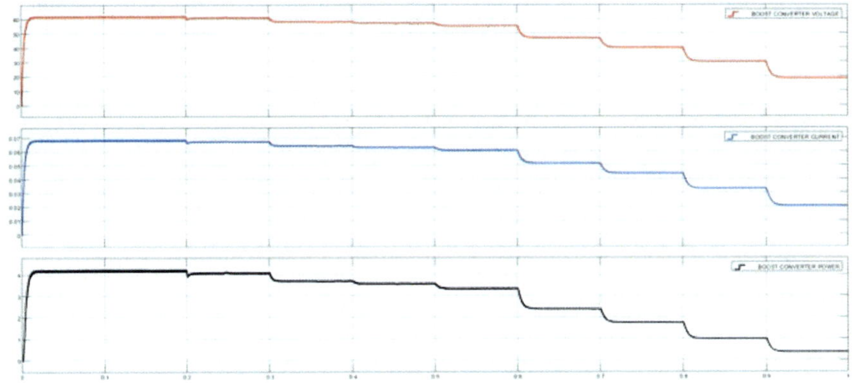

Fig. 17 The output from the boost converter voltage, current and power

Voltage and current during 0–0.2 s are 61.65 v, 0.0679 amp, and 4.18 W. For the lowest value of irradiation (147 W/m$^2$) and temperature (18 °C), the power attained by the system was 0.371 W, voltage and current 18.36 volts, 0.02 amp (0.9–1 s). At 0.2–0.3 s, the PV module's highest voltage was 31.75 (current 0.126, power 4.006) volt for irradiation (778w/m$^2$) and temperature (27 °C), which was boosted to 61 V (current 0.065, power 4.004 W), with an efficiency of 99.75%. Figure 18 shows the response after stimulation that is in terms of PV voltage, current, power, and boost converter voltage, current, and power for the different environmental conditions, irradiation and temperature were varied for a time period of 0.1 s. It can be seen that after using the ANN-based MPPT with PI controller, the output waveform has improved, and the ripples are very less in comparison to input values of the module, the curve looks more stable and with fewer transients. For the designed PV system, the efficiency is in the range of 98–99%, the output of the PV module and the converter with ANN- and PI controller are compared in Table 2. Table 3 is a brief comparison showing the effect of irradiation and temperature on module and boost converter voltage, current, and ripples.

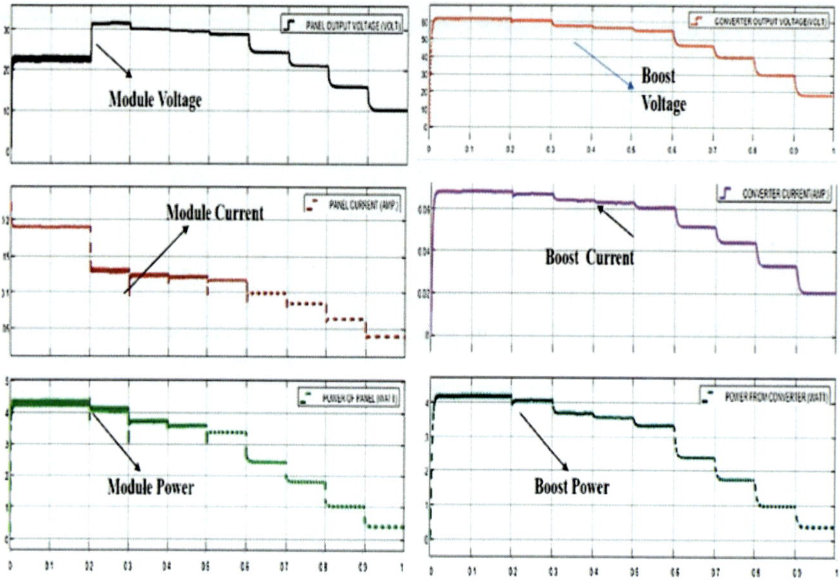


**Fig. 18** The output from Module and after using ANN-based MPPT and boost converter

**Table 2** The output of the PV module and converter with ANN + PI controller at STC

| Maximum | Voltage (V) | Current (amp) | Power ( W) | Efficiency (η) | |
|---|---|---|---|---|---|
| Module output | 23.51 | 0.258 | 5.099 | At STC 98.9% | At 778 w/m$^{2}$, 27 °C 99.75% |
| Ripples (output) | 0.72 | 0.0019 | 0.188 | | |
| Boost converter output | 67.42 | 0.075 | 5.045 | | |
| Ripples (output) | 0.19 | 0.0002 | 0.0286 | | |

## 5 Conclusion

In this paper, an ANN-based MPPT technique was implemented with a PI controller and boost converter. A total of 195 experimental data were used to train the ANN in terms of temperature and irradiation with power for the given PV module. The regression curve (99.6%) and performance (0.18788 at 1000 epos) prove that the set of data have very little error (RMS). The output obtained from the ANN was used to calculate the error and passed to the PI controller to reduce the steady-state error between the signals. PWM generator is used to generate the gate pulse, which is feed to the converter section. After using the ANN with PI, the output obtained at the boost converter terminal contains very few ripples and the waveform remains smooth in every environmental condition. Tables 2 and 3 give a brief analysis of the output obtained from the PV module and boost converter, it was concluded that with

**Table 3** Effect of irradiation and temperature at maximum/minimum value of irradiation, temperature

| Irradiation varies, Temperature fixed 25 °C | | | | | Temperature varies, irradiation fixed (866 W/m$^{2}$) | | | |
|---|---|---|---|---|---|---|---|---|
| | Maximum irradiation (866) | | Minimum irradiation (147) | | Maximum temp (36° C) | | Minimum temp (6 °C) | |
| Module Voltage (volt) | 24.8 | η=99% | 9.35 | η = 91% | 31.21 | η = 98% | 8.629 | η=85% |
| Module current (Amp) | 0.18 | | 0.035 | | 0.126 | | 0.210 | |
| Boost Voltage (volt) | 63.6 | | 16.48 | | 60.16 | | 37.24 | |
| Boost Current (Amp) | 0.07 | | 0.01818 | | 0.064 | | 0.0041 | |
| Ripples module (volt) | 0.45 | | 0.079 | | 0.32 | | 3.786 | |
| Ripples module (Amp) | $7.5*10^{-04}$ | | $1.8*10^{-05}$ | | $7.0*10^{-03}$ | | $4.81*10^{-3}$ | |
| Ripples Boost (volt) | 0.09 | | 0.010 | | 0.031 | | 0.4833 | |
| Ripples Boost (Amp) | $9.92*10^{-05}$ | | $1.8*10^{-04}$ | | $3.53*10^{-04}$ | | $5.3*10^{-4}$ | |

an increase in solar irradiance the output voltage of solar PV modules increases, and with the increase in temperature, current decreases. The ANN was varying the duty cycle of the system to keep the efficiency as maximum as possible. The PV system was designed to boost the input voltage (36 V) to 60 V, values of an L, C, $R_L$, and $f_s$ was calculated for the 60-volt system. For 31.75 V (module output voltage) was boosted to 61 volts, with a less transient value of 0.691 V. The effect of irradiation and temperature on the converter section is shown in Tables 3. An efficiency of 98.9% was traced at STC. Desired output was attained at 778 w/m$^2$ and 27 °C, with the power of 4.07 watts for the module and 4.06 w by boost, with an efficiency of 99.75. At fixed irradiation for maximum and minimum value of temperature efficiency is 98 and 85% and at a fixed temperature for maximum and minimum value of irradiation efficiency is 99 and 91%. So, this proves that ANN-based MPPT has better response and performance at a low and high value of irradiation and temperature. At a low value of temperature and a high value of irradiation, ripples increase.

## References

1. Hodge BM, Martinez-Anido CB, Wang Q, Chartan E, Florita E, Kiviluoma J (2018) The combined value of wind and solar power forecasting improvements and electricity storage. Appl Energy 2018(214):1–15
2. Curto D, Trapanese M (2018) A renewable energy mix to supply the Balearic Islands: sea wave, wind, and solar. In Proceedings of the 2018 IEEE international conference on environment and electrical engineering and 2018 IEEE industrial and commercial power systems Europe (EEEIC/I&CPS Europe), Palermo, Italy
3. Mahalakshmi R, Aswin KA, Kumar A (2014) Design of fuzzy logic based maximum power point tracking controller for a solar array for cloudy weather conditions. In: Proceedings of power and energy system conference
4. Kabir E, Kumar P, Kumar S, Adelodun AA, Kim KH (2018) Solar energy: potential and prospects. Renew Sustain Energy Rev 82:894–900
5. Divyansh M (2014) Maximum power point tracking with artificial neural network, Int J Emerg Sci Eng (IJESE). V-3 ISSN:2319–6378
6. Vitorino MA, Hartmann LV, Lima AMN (2007) Using the model of the solar cell for determining the maximum power point of photovoltaic systems. In: Proceedings of the European conference on power electronics and applications, pp 1–10
7. Sukhatme SP, Nayak JK (eds) (2008) Solar energy, 3rd edn. Tata McGraw-Hill Publication Co. Ltd. New Delhi, Chapter 9, Section 1, pp 313–331
8. Esram T, Chapman PL (2007) Comparison of photovoltaic array maximum power point tracking techniques. IEEE Trans Energy Convers 22:439–449
9. Bouselham L, Hajji M, Hajji B, Bouali H (2016) A new MPPT-based ANN for PV system under partial shading conditions. In: 8th International conference on sustainability in energy and buildings. Turin, ITALY Energy Procedia 11, pp 924–933
10. Yadav D, Singh Pal N, Ansari MA (2019) Analyzing the effects of coal & dust on solar panel & improving the F.F by using spray cooling. In: International conference on emerging trends in electronics. IT and Communication (Electrocon-2019) IILM Gr. Noida
11. Dewil T, Risma P, Oktarina Y (2019) A review of factors affecting the efficiency and output of a PV system applied in tropical climate. In: IOP conference series: earth and environmental science, vol 258, p 012039 https://doi.org/10.1088/1755-1315/258/1/012039
12. Femia N, Petrone G, Spagnuolo G, Vitelli M (2005) Optimization of perturb and observe maximum power point tracking method. IEEE Trans Power Electron 20:963–973

13. Lin CH, Huang CH, Du YC, Chen JL (2011) MPPT for the PV array using the fractional-order incremental conductance method. Appl Energy 88:4840–4847
14. Agha HS, Koreshi Z, Khan (2017) ANN-based MPPT for solar photovoltaics. In: International conference on information & communication technologies (ICICT), pp 150–155
15. El Khateb A, Rahim NA, Selvaraj J, Uddin MN (2014) Fuzzy-logic-controller-based SEPIC converter for maximum power point tracking. IEEE Trans:2349–2358
16. Yadav D, Singh N, Ansari MA (2019) Fuzzy logic based MPPT controller with bilevel measurements & peak finder panel. IEEE-COER-ICAIA
17. Tey KS, Mekhilef S, Seyedmahmoudian M, Horan AT, Stojcevski A (2018) Improved differential evolution-based MPPT algorithm using SEPIC for PV systems under partial shading conditions and load variation. IEEE Trans Ind Inform 14(10):4322–4333
18. Sundareswaran K, Vigneshkumar V, Palani S (2015) Development of a hybrid genetic algorithm/perturb and observe algorithm for maximum power point tracking in photovoltaic systems under non-uniform insolation. IET Renew Power Gener:9
19. Mathur D (2014) Maximum power point tracking with artificial neural network. Int J Emerg Sci Eng (IJESE) ISSN: 2319–6378
20. Ramaprabha R, Mathur BL, Sharanya M (2009). Solar array modeling and simulation of MPPT using neural network. In: International conference on control, automation, communication and energy conservation. INCACEC 2009, pp 1–5
21. Ayang A, Wamkeue R, Ouhrouche MG (2019) Maximum likelihood parameters estimation of a single-diode model of photovoltaic generator. Renew Energy:111–121
22. Nzimako O, Wierckx R (2017) Modeling and simulation of a grid integrated PV system using a real digital simulator. IEEE Trans Ind Appl 53:1326–1336
23. Vieira JAB, Mota AM (2008) Maximum power point tracker applied in batteries charging with PV panels. In: Proceedings of IEEE international symposium on industrial electronics, pp 202–207
24. Elobaid LM, Abdelsalam AK, Zakzouk EE (2015) Artificial neural network-based photovoltaic maximum power point tracking techniques: a survey. IET Renew Power Gener 9(8):1043–1063
25. Bendib B, Krim F, Balmily H, Almi MF, Bolouma S (2014) An intelligent MPPT approach based on neural-network voltage estimator and fuzzy controller, applied to a stand-alone PV system. In: Proceedings of IEEE industrial electronics (ISIE), vol. 10. pp 404–409
26. Basha CH, Rani C (2020) Different conventional and soft computing MPPT techniques for solar PV systems with high step-up boost converters. Energies (Published)

# Chapter 8
# Real Coded Genetic Algorithm for Selecting Optimal Machining Conditions

**Pinkey Chauhan**

## 1 Introduction

In the manufacturing industry, economic machining is a primary need due to high capital involved in these large-scale industries. These days the notion of computer numerically controlled (CNC) machining is introduced to deal with the problem for economic machining. During machining process, metals or wood parts are shaped to obtain desired product by removing unwanted material. The machining process of any desired product must satisfy the specified quality restrictions such as surface finish, accuracy and surface integrity. The objective is to minimize the unit cost or machining time according to available conditions. To meet these objectives, we have to determine the optimal machining conditions by optimizing machining parameters such as cutting speed, feed rate depth of cut and number of passes under several constraints imposed on machining processes. Although handbooks are available to provide recommended machining parameters but these handbooks do not satisfy the need of economic machining. To overcome this drawback, sufficient research work has been done and still going on, in this field to provide a reliable method that produces efficient solution. The analysis of single and multi-pass turning operations under practical constraints starts nearly three decades ago by [5] who considered production cost or machining time as minimization criteria. Some conventional optimization techniques such as dynamic programming [1] geometric programming [5] and Hook-Jeeves method [8] were also employed to optimize machining models. Some studies also concentrated on refining multi-pass machining models by adding some more real constraints and parameters that make simulated models more realistic [1, 9].

The complexity of machining models increases with an increase in the number of practical constraints and parameters. Conventional optimization techniques produce

P. Chauhan (✉)
Department of Mathematics, Jaypee Institute of Information Technology, Noida, India
e-mail: pinkeychauhan030@gmail.com

 
D. Singh et al. (eds.), *Proceedings of International Conference on Scientific and Natural Computing*, Algorithms for Intelligent Systems,
https://doi.org/10.1007/978-981-16-1528-3_8

local optimal solution and are not appropriate for these highly constrained problems. Therefore, heuristic methods are preferred to deal with such complex problems as they provide near global optimum solution. Recently, many non-traditional optimization techniques like Simulated annealing (SA) [3], Genetic algorithm [7], Ant colony optimization [11], and scatter searches [2], PSO [13] have been employed to determine optimal conditions for considered machining models. More recently, particle swarm optimization PSO has been used by [10] to optimize multi-pass turning operations.

The performance of LXPM-GA over benchmark functions [4] and machining models [14] encouraged us to implement it for optimizing machining parameters of considered machining model. In this study, we have considered LXPM for finding optimal machining condition for the model proposed by [3]. This model includes several constraints on tool life, surface finish, cutting forces machining power, chip tool interface along with parameter bounds. The objective function is to minimize the total unit production cost by determining the optimal number of passes, cutting speed, feed rate and depth of cut during roughing and finishing passes.

The remaining paper is organized as follows. Section 2 describes the considered machining model along with notations used in the model. Section 3 gives a brief introduction of employed methodology (LXPM-GA) and Sect. 4 comprises computational results along with the discussion of experimental results and Sect. 5 concludes the paper.

## 2 Mathematical Formulation of Multi-pass Turning Operation

Here in this paper, we have considered the machining model proposed by [3] for multi-pass turning operations, which involve multiple rough cuts and a single finish cut. The optimization criterion considered here is of minimization of total unit production cost. The machining parameters to be determined are cutting speed, feed rate depth of cut and number of rough cuts under constrained machining environment. The model is described as below.

### *2.1 The Cost Function*

The turning process is divided into multi-pass roughing and single-pass finishing. The components may be either a straight-cut type or of continuous forms. Here we have considered bar components. The unit production cost, *UC*, for multi-pass turning operations is divided into four basic cost elements and is given by:

$$UC = C_M + C_I + C_R + C_T.$$

All of these components are described as below.

1. Cutting cost by actual time in cut, $C_M$, is given by:

$$C_M = k_0 t_m$$

where, $t_m$ is the actual cutting time expressed as the sum of the times for roughing and finishing cutting phases:

$$t_m = t_{mr} + t_{ms}$$

where $t_{ms} = \frac{\pi DL}{1000 V_s f_s}$ and $t_{mr} = \frac{\pi DL}{1000 V_r f_r} \times n$
here the number of passes $(n)$ is given by:

$$n = \left(\frac{d_t - d_s}{d_r}\right)$$

2. Machine idle cost due to loading and unloading operations and idling tool motion,

$C_I$, is given by:

$$C_I = k_0 t_i$$

where, $t_i$ is machine idle time, and can be expressed as the sum of idle tool motion ($t_v$) and time due to loading and unloading operations, which is a constant term ($t_c$).

$$t_i = t_c + (h_1 L + h_2) \times (\frac{d_t - d_s}{d_r} + 1)$$

Tool replacement cost, $C_R$, is given by:

$$C_R = k_0 t_e (t_m / T_p)$$

where, $t_e$ is the time required to exchange a tool and $T_p$ is the tool life. Usually tool life (T) is determined using Taylor's tool life equation;

$$T = \frac{C^{1/\alpha}}{V^{1/\alpha} f^{1/\beta} d^{1/\gamma}} = \frac{C_0}{V^p f^q d^r}$$

Due to different machining conditions imposed on roughing and finishing processes, the wear rate changes for roughing and finish phases. In such cases, the tool life can be expressed as

$$T_p = T_r + T_s$$

where, $T_r = \frac{C_0}{V_r^p f_r^q d_r^r}$ and $T_s = \frac{C_0}{V_s^p f_s^q d_s^r}$
The tool cost, $C_T$ is given by:

$$C_T = k_t(t_m/T_p)$$

where, $k_t$ is cutting edge cost.

## 2.2 Constraints on Cutting Conditions

In the considered model, several constraints are imposed during roughing and finishing operations in the machining environment. The detailed description of these constraints for both roughing and finishing operations is given below.

*Rough Machining*:

1. Parameter bounds: The available range of cutting speed, feed rate and depth of cut are expressed in terms of lower and upper bounds, respectively, as below:

$$V_{rL} \leq \mathrm{V} \leq V_{rU}, f_{rL} \leq f \leq f_{rU}, d_{rL} \leq \mathrm{d} \leq d_{rU}.$$

2. Tool-life constraint: The constraint on tool life is expressed as

$$T_L \leq \mathrm{T_r} \leq T_U$$

3. Operating constraints: The constraints on machine operating conditions are as:

(i) Cutting force constraint: Cutting force constraint is taken as

$$F_r = k_1(f_r)^\mu (d_r)^\nu \leq F_u.$$

(ii) Power constraint: The power required during the cutting operation should not exceed the available power of the machine tool. The power is given as.

$$P_r = \frac{F_r V_r}{6120\eta} \leq P_U,$$

where, efficiency $\eta = 0.085$.

(iii) Stable cutting region constraint: Stable cutting region is given by

$$f_r(V_r)^\lambda (d_r)^\nu \geq SC$$

(iv) Chip-tool interface temperature constraint: This constraint is expressed as $Q_r = k_2(V_r)^\tau (f_r)^\varphi (d_r)^\delta \le Q_u$.

*Finish Machining:*

For finish machining operation, all the constraints other than the surface finish constraint are similar as for rough machining. Therefore, all finishing constraints are as:

Bounds on finishing parameters are

1. $V_{sL} \le \mathrm{V} \le V_{sU}, f_{sL} \le f \le f_{sU}, d_{sL} \le \mathrm{d} \le d_{sU}$
2. $T_L \le \mathrm{T_s} \le T_U$
3. $F_s = k_1(f_s)^\mu (d_s)^\nu \le F_u$
4. $P_s = \frac{F_s V_s}{6120\eta} \le P_U$
5. $f_s(V_s)^\lambda (d_s)^\nu \ge SC$
6. $Q_s = k_2(V_s)^\tau (f_s)^\varphi (d_s)^\delta \le Q_u$
7. In addition with these constraints imposed on finish machining, the constraint related to surface finish is given by

$$\frac{f_s^2}{8R} \le (SR)_U$$

The functional relationships between roughing and finishing parameters are as. $V_s \ge k_3 \mathrm{V_r}, f_r \ge k_4 f_s, \mathrm{d_r} \ge k_5 d_s$.

Finally, the bound on number of finish cuts $n = \left(\frac{d_t - d_s}{d_r}\right)$, which are considered similarly as rough cuts and is given by

$$\left(\frac{d_t - d_{sU}}{d_{rU}}\right) \le \left(\frac{d_t - d_s}{d_r}\right) \le \left(\frac{d_t - d_{sL}}{d_{rL}}\right)$$

All the notations used in modelling objective function and constraints are given in Table 5.

## 3 Laplace Crossover and Power Mutation Genetic Algorithm (LXPM-GA)

Genetic algorithms (GA) [6] are one of the oldest and perhaps the most frequently used population-based search algorithm for solving global optimization problems. In the present study, we have proposed the use of LXPM- GA [4], which is a real coded genetic algorithm. The crossover operator used in LXPM-GA is based on Laplace probability distribution and is named as LX while the mutation operator used in it is termed as power mutation. Its selection process is based on tournament selection. The computational steps of LXPM-GA are as follows:

1. Generate a suitably large initial set of random individuals within the domain prescribed only by the bounds on variable i.e. points satisfying.
2. Evaluate the fitness function of each individual in the population.
3. Check the stopping criterion, which is to terminate the program when reached up to the specified maximum number of generations. If stopping criteria are satisfied stop; else go to 4.
4. Apply tournament selection procedure on initial (old) population to construct mating pool.
5. Apply Laplace crossover and power mutation to all individuals in mating pool, with the probability of crossover and probability of mutation respectively, to make new population.
6. Increase generation; replace old population by new population; go to 2.

*Constraint handling:* For constraint handling parameter free, penalty function approach proposed by [12] is used in this algorithm. Fitness value, fitness (Xi) of an individual is evaluated using the following relation:

$$fitness(X_i) = \begin{cases} f(x_i), & if\ X_i\ is\ feasible \\ f_{worst} + \sum_{j=1}^{m} \left|\Phi_j(x_i)\right|, & otherwise \end{cases}$$

where, fworst is the objective function value of the worst feasible solution currently available in the population. Thus, the fitness of an infeasible solution not only depends on the amount of the constraint violation but also on the population of solutions at hand. However, the fitness of a feasible solution is always fixed and is equal to its objective function value. If there are no feasible solutions in the population, then fworst is set zero. Such a constraint handling scheme without the need for a penalty parameter is possible because GAs use a population of solutions in every iteration and comparison of solutions is possible using the tournament selection operator. In real-life optimization problems, the constraints are often non-commensurable. Therefore, constraints are normalized to avoid any sort of bias.

## 4 Experimental Results and Discussion

Here we have considered the experimental data for bar components provided by [3] and considered in [10, 11] which is shown in Table 4. The parameters used in LXPM-GA while optimizing the considered model are as: the initial population size is fixed to 20; number of simulations is taken as 50 with maximum number of generations set to 500 during each simulation. Table 1 shows the optimal unit production cost with non-dimensional constraint violation using LXPM-GA and other algorithms as quoted in the previous study [10]. The parameters to be determined to optimize machining conditions are feed rate, depth of cut, cutting speed and number of passes for roughing and finishing. Table 2 depicts the optimal value of these parameters using LXPM-GA and other algorithms as recorded in the literature. Table 3 presents the optimal value of unit production cost using LXPM-GA and along with cutting parameters for different values of the total depth of cut. The average computational time taken by LXPM-GA over 50 runs is 0.2665 s. Table 1 shows that optimal unit production using LXPM-GA is much better than PSO and simulated annealing (SA/PS) and also encouraging over other algorithms on the feasibility basis. From Table 3, we observe that the corresponding values of unit production cost using LXPM-GA significantly improve over PSO for all considered values of depth of cut. The convergence of total unit cost over maximum number of generations using LXPM-GA is shown in Fig. 1 that illustrates the algorithm converges up to 200 generations. Hence the proposed method produces good results with fewer computations (Table 5).

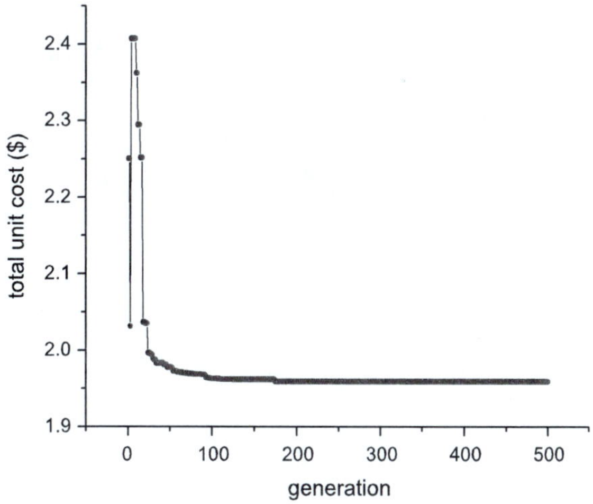


**Fig.1** Generation wise convergence graph of optimal unit cost using LXPM-GA for $d_t = 6.0$ mm

**Table 1** Optimal total unit cost and constraints at the optimum point

| Method | Optimum unit cost($) | Non-dimensional constraint violation |
|---|---|---|
| LXPM-GA | 1.9591 | 0 |
| PSO | 2.2721 | 0 |
| ACO | 1.8450 | 0.5396 |
| GA | 1.7842 | 0.5148 |
| SA/PS | 2.3135 | 0.0667 |

**Table 2** Optimal cutting parameters using different methods

| Method | $V_r$ | $f_r$ | $d_r$ | $V_s$ | $f_s$ | $d_s$ |
|---|---|---|---|---|---|---|
| LXPM-GA | 123.3430 | 0.5655 | 2.9999 | 169.9784 | 0.2262 | 2.9999 |
| PSO | 106.69 | 0.897 | 2 | 155.89 | 0.28 | 2 |
| ACO | 103.05 | 0.9 | 3 | 162.02 | 0.24 | 3 |
| GA | 114.22 | 0.7 | 2.9745 | 164.369 | 0.2978 | 2.963 |

**Table 3** Number of rough passes and optimal unit cost for different values of total depth of cut ($d_t$) for $d_{sL} = d_{rL} = 1$ mm

| Total depth of cut($d_t$) in mm | Maximum depth of cut (in mm) for $d_{sU} = d_{rU}$ | Bounds on n | $d_{opt}$ | | Unit cost in $ | |
|---|---|---|---|---|---|---|
| | | | PSO | LXPM-GA | PSO | LXPM-GA |
| 6 | 3 | $1 \le n \le 5$ | 2 | **2.99** | 2.272 | **1.9591** |
| 8 | 4 | $2 \le n \le 7$ | 3 | **2.66** | 3.306 | **2.438** |
| 10 | 5 | $2 \le n \le 9$ | 3 | **3.32** | 3.853 | **2.754** |
| 12 | 3 | $3 \le n \le 11$ | 5 | **2.99** | 3.846 | **3.237** |

**Table 4** Numerical data for cutting model

| Parameter | Value | Parameter | Value | Parameter | Value | Parameter | Value |
|---|---|---|---|---|---|---|---|
| $V_{rU}$ | 500 m/min | $d_t$ | 6 mm | $h_1$ | $7 \times 10^{-4}$ | $k_0$ | 0.5 $/min |
| $V_{rL}$ | 50 m/min | p | 5 | $h_2$ | 0.3 | $k_1$ | 108 |
| $V_{sU}$ | 500 m/min | q | 1.75 | $T_L$ | 25 mm | $k_2$ | 132 |
| $V_{sL}$ | 50 m/min | r | 0.75 | $T_U$ | 45 mm | $k_3$ | 1 |
| $f_{rL}$ | 0.1 mm/rev | u | 0.75 | $t_c$ | 0.75 min/piece | $k_4$ | 2.5 |
| $f_{rU}$ | 0.9 mm/rev | $\nu$ | 0.95 | $t_e$ | 1.5 min/edge | $k_5$ | 1 |
| $f_{rL}$ | 0.1 mm/rev | $\eta$ | 0.85 | $P_U$ | 5 kW | $k_t$ | 2.5 $/edge |
| $f_{rU}$ | 0.9 mm/rev | $\lambda$ | 2 | $F_U$ | 200 kgf | D | 50 mm |
| $d_{rL}$ | 1 mm | $\upsilon$ | −1 | $SR_U$ | 10 μm | R | 1.2 mm |
| $d_{rU}$ | 3 mm | $\tau$ | 0.4 | $Q_U$ | 1000 °C | L | 300 mm |
| $d_{sL}$ | 1 mm | $\varphi$ | 0.2 | SC | 140 | | |
| $d_{sU}$ | 3 mm | $\delta$ | 0.105 | $C_0$ | $6 \times 10^{11}$ | | |

**Table 5** Nomenclature of variables used in model

| |
|---|
| $C_I$—Machine idle cost due to loading and unloading operations and tool idle motion ($/piece) |
| $C_M$—Cutting cost by actual time in cut($/piece) |
| $C_R$—Tool replacement cost ($/piece) |
| $C_T$—Tool cost ($/piece) |
| $V_r$,$V_s$—Cutting speeds in rough and finish machining respectively (m/min) |
| $f_r$, $f_s$—Feed rates in rough and finish machining respectively (mm/rev) |
| $d_r$,$d_s$ —Depth of cut in rough and finish machining respectively (mm) |
| $V_{rL}$, $V_{rU}$—Lower and upper bound of cutting speed in rough machining (m/min) |
| $f_{rL}$, $f_{rU}$—Lower and upper bound of feed rate in rough machining (m/min) |
| $d_{rL}$, $d_{rU}$ —Lower and upper bound of depth of cut in rough machining (m/min) |
| $V_{sL}$, $V_{sU}$—Lower and upper bound of cutting speed in finish machining (m/min) |
| $f_{sL}$, $f_{sU}$—Lower and upper bound of feed rate in finish machining (m/min) |
| $d_{sL}$, $d_{sU}$ —Lower and upper bound of depth of cut in finish machining (m/min) |
| $d_t$—Depth of material to be removed (mm) |
| D,L —Diameter and length of work piece (mm) |
| $k_0$—Direct labor cost, including overheads ($/piece) |
| $k_t$ —Cutting edge cost ($/piece) |
| $t_{mr}$ , $t_{ms}$ , $t_m$—Rough machining time, finish machining time and actual machining time, respectively (min) |
| $t_c$, $t_v$, $t_i$ —Preparation time (such as loading and unloading time), idle tool motion time (such as tool Travel and tool approach/depart time) and total machine idle time (min) |
| $t_e$, $t_r$— Time required to exchange a tool and tool replacement time, respectively (min) |
| $h_1$, $h_2$ —Constants pertaining to tool travel and approach/depart time, respectively (min) |
| $T_r$, $T_s$ —Expected tool life for rough and finish machining, respectively (min) |
| $T_p$—Tool life of weighted combination of $T_r$ and $T_s$ (min) |
| $T_U$, $T_L$ —Upper and lower bounds for tool life, respectively (min) |
| p,q,r ,$C_0$— $p=1/\alpha$, $q= 1/\beta$, $r = 1/\gamma$, $C_0 = C^{1/\alpha}$ , respectively are constants of tool life equation |
| SR—Maximum allowable surface roughness (mm) |
| SC—Limit of stable cutting |
| R—Nose radius of cutting tool (mm) |
| $F_r$, $F_s$, $F_U$ —Cutting forces in rough and finish machining and maximum allowable cutting force (kgf) |
| $P_r$, $P_s$, $P_U$—Cutting powers for rough and finish machining and maximum allowable cutting power (kW) |
| $Q_r$, $Q_s$—Chip tool interface temperature for rough and finish machining ($^0$C) |
| $Q_U$ —Maximum allowable Chip tool interface temperature ($^0$C) |
| $k_1$, $\mu$, $\nu$—Constants of cutting force equation |
| $k_2$, $\tau$, $\varphi$, $\delta$—Constants related to chip–tool interface temperature equation |

(continued)

**Table 5** (continued)

| |
|---|
| $k_3$, $k_4$, $k_5$—Constants of roughing and finishing parameter relations $k_3, k_4, k_5 \geq 1$ |
| $\lambda$, $\nu$—Constants related to expression of stable cutting |
| n—Number of rough cuts (an integer) |
| UC —Unit production cost except material cost |
| $n_i$, $d_r$—Possible value of n and $d_r$ under a given X |
| $X = \{x_1, x_2, x_3, \ldots, x_m\}$ is machining parameter set |

## 5 Conclusion

This paper proposes the parameter optimization of a complex and highly constrained cutting model with the aim of minimizing total production cost using a real-coded genetic algorithm namely LXPM-GA. The results of the employed algorithm are compared with PSO, Binary GA, ACO and SA. The results obtained using LXPM-GA are promising on the quality and feasibility basis over other algorithms recorded in the literature. The proposed algorithm finds significantly better optimal solution in very less computational time and iterations. Therefore, LXPM-GA is a reliable and efficient algorithm for solving such complex machining problems and could be recommended to solve other machining models of the higher degree of complexity.

## References

1. Agapiou JS (1992) The optimization of machining operations based on a combined criterion: part I the use of combined objectives in single-pass operations. ASME J Eng Ind 114:500–507
2. Chen MC (2004) Optimizing machining economics models of turning operations using the scatter search approach. Int J Prod Res 42:2611–2625
3. Chen MC, Tsai DM (1996) A simulated annealing approach for optimization of multi-pass turning operations. Int J Prod Res 34:2803–2825
4. Deep K, Thakur M (2007) A new mutation operator for real coded genetic algorithms. Appl Math Comput 193:21–23
5. Ermer DS (1971) Optimization of constrained machining economics problem by geometric programming. ASME J Eng Ind 93:1067–1072
6. Holland HJ (1975) Adaption in natural and artificial systems. University Michigan Press, Ann Arbor, USA.
7. Onwubolu GC, Kumalo T (2001) Multi-pass turning operations optimization based on genetic algorithms. Proc Inst Mech Eng 215:117–124
8. Ruy M, Krasteva E, Doytchinov S (1995) Computer-aided selection of optimum machining parameters in multi pass turning. Int J Adv Manuf Technol 10:19–26
9. Shin YC, Joo YS (1992) Optimization of machining conditions with practical constraints. Int J Prod Res 30:2907–2919
10. Srinivas J,Giri R, Yang SH (2009) Optimization of multi pass turning using particle swarm intelligence. Int J Adv Manuf Technol 40:56–66
11. Vijayakumar K, Prabhaharan G, Asokan P, Saravanan R (2003) Optimization of multi-pass turning operations using ant colony system. Int J Mach Tools Manuf 43:1633–1639

12. Deb K (2000) An efficient constraint handling method for genetic algorithms. Comp Meth Appl Mech Eng 186:311–338
13. Chauhan P, Pant M, Deep K (2015) Parameter optimization of multi-pass turning using chaotic PSO. Int J Mach Learn Cyber 6:319. 337https://doi.org/10.1007/s13042-013-0221-1
14. Chauhan P, Pant M, Deep K (2011) Optimizing Cnc turning process using real coded genetic algorithm and differential evolution. Trans Evol Algorm Cont Optim 2:157–165

# Chapter 9
# Biogeography-Based Optimization Algorithm for Solving Emergency Vehicle Routing Problem in Sudden Disaster

Vanita Garg, Anjali Singh, and Divesh Garg

## 1 Introduction

Optimization is the method of discovering the best solution under the constraints of limited resources. Mathematically, an optimization problem is defined as:

$$\text{Max or (Min) } z = c\,(x_1, x_2, \ldots, x_n)$$

$$\text{Subject to the constraints } a_1\,(x_1, x_2, \ldots, x_n)\,\{\leq, =, \geq\}b_1$$
$$a_2\,(x_1, x_2, \ldots, x_n)\,\{\leq, =, \geq\}b_2$$
$$a_m\,(x_1, x_2, \ldots, x_n)\,\{\leq, =, \geq\}b_m$$
$$x_j \geq 0, \quad j = 1, 2, 3, \ldots,$$

Optimization methods can be categorized in two ways: traditional and non-traditional methods. Traditional methods are based on calculus techniques, which require a function that is continuous or differentiable; however, non-traditional methods are famously known as nature-inspired optimization techniques. These techniques are derivative-free methods that solve an optimization problem without having any prior information of continuity of the function. Thus, non-traditional methods play a vital role in solving various real-life problems. Genetic Algorithm

V. Garg (✉) · A. Singh · D. Garg
Department of Mathematics, Galgotias University, Noida, India
e-mail: vanitagarg16@gmail.com

A. Singh
e-mail: as1754002@gmail.com

D. Garg
e-mail: divesh.garg@galgotiasuniversity.edu.in



D. Singh et al. (eds.), *Proceedings of International Conference on Scientific and Natural Computing*, Algorithms for Intelligent Systems,
https://doi.org/10.1007/978-981-16-1528-3_9

(GA), Particle Swarm Optimization (PSO), Glow swarm Optimization (GWO), Dragonfly Algorithm (DA), Differential Evolution (DE), etc. are some of the names of nature-inspired optimization techniques. These techniques are derived from a natural phenomenon and a solution is updated in each iteration based on the idea of this phenomenon.

In this paper, a nature-inspired optimization technique, Biogeography-Based Optimization, is implemented to solve the routing problem of vehicles. Vehicle routing is considered as a crucial problem when relief materials for sudden disasters need to be deported and employed in the least time possible.

## 2 Biogeography-Based Optimization

In Simon, 2009 gave the idea of natural biogeography and generalized that to get an algorithm of optimization. Simon compared and distinguished BBO with algorithms in the literature. At last, he studied and applied BBO to solve the real-world problem such as how species move from one island to another island based on their habitat suitability. Basic terminology of the BBO algorithm is given as follows:

Some islands are more suitable for habitation than others. HSI high suitability index is the measurement of a habitat's suitability for living organisms. Some features that determine HIS are called Suitability variable index and are given as follows: vegetation, weather conditions, natural disaster prediction, geographical position of the land, etc.

A law that governs the migration of species leads to formulation of BBO algorithm. Living organisms try to move to a better habitat with high HSI, which confirms the accumulation of the population to the best habitat or solution.

The immigration rate λ and the emigration rate μ are defined as follows:

$$\left.\begin{aligned} \lambda_k &= I(1-\tfrac{k}{n}) \\ \mu_k &= E\tfrac{k}{n} \end{aligned}\right\} \tag{1}$$

Let us consider the probability that the habitat contains exactly s species and changes from time to time as follows:

$$P_s(t+\Delta t) = P_s(t)(1-\lambda_s\Delta t - \mu_s\Delta t) + P_{s-1}\lambda_{s-1}\Delta t + P_{s+1}\mu_{s+1}\Delta t \tag{2}$$

where $P_s(t) = there are s species at time t$

$$\dot{P}_s = \begin{cases} -(\lambda_s + \mu_s)P_s + \mu_{s+1}P_{s+1}, \; s = 0 \\ -(\lambda_s + \mu_s)P_s + \lambda_{s-1}P_{s-1} + \mu_{s+1}P_{s+1}, \; 1 \le s \le s_{\max} \\ -(\lambda_s + \mu_s)P_s + \lambda_{s-1}P_{s-1}, \; s = s_{\max} \end{cases} \tag{3}$$

As habitat suitability improves (see Fig. 1),

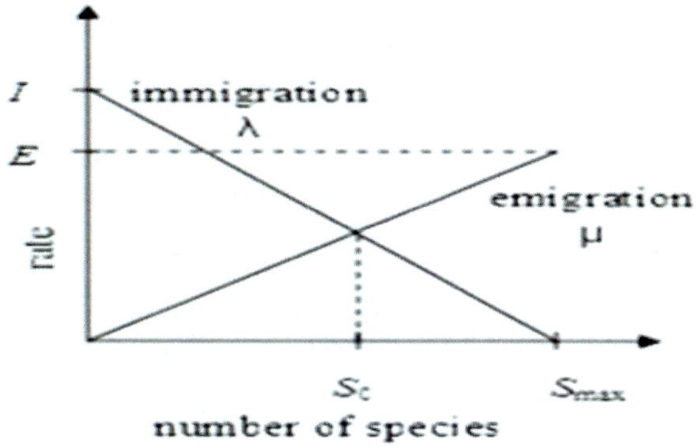


**Fig. 1** Illustration of immigration and emigration rate

The species count increases
Emigration increases (more species exit the habitat)
Immigration decreases (fewer species come into the habitat)

Figure 1 illustrates a model of species abundance in a single habitat.
The main operators of BBO are given as follows:

## *2.1 Migration*

In BBO algorithm, the migration operator helps in exploitation of the search space of the solution. This is required for changing the solution towards its best version. Mathematically, a solution or a habitat is represented as n- tuple of real numbers. The real numbers are called SIV. The function value at the vector is considered as its HSI.

## *2.2 Mutation*

A habitat or island can immediately lose its species count due to a sudden natural calamity. Mutation operator in BBO is responsible for exploration of the search space. Mutation probability based on random distribution is used in BBO.

## *2.3 Elitism*

Elitism is included in BBO algorithm to save the two best solutions from each iteration and then these chosen solutions will directly be added to the population pool.

## 3 Literature Survey

Garg and Deep, (2016) applied Laplace crossover operator on IEEE CEC 2014 benchmark functions and applied the algorithm on camera calibration problem. A detailed state of the art of BBO algorithm is also given in Garg and Deep, (2015).

Zhang et al. (2020) presented improved Laplacian Biogeography-Based Optimization. First, an active two differential perturbing operator is projected. Second, a new operator using two best solutions is incorporated. An improved Laplace operator is framed for the last habitats' enhancing to improve the operability and secure the convergence speed. The proposed algorithm is applied on IEEE CEC-2013 and applied to QAP. The outcomes on Quadratic Assignment Problems prove that ILxBBO is more reasonable compared with Improved Firefly Algorithm (IFA), LxBBO and Improved Particle Swarm Optimization (IPSO).

Wang et al. in 2019, presented monarch butterfly optimization. The goal of this paper was twofold: First, by studying the displacing behavior of the monarch butterflies, a new algorithm, called MBO, was proposed. In MBO, there are two ways to update the positions of the monarch butterflies. The migration operator and butterfly-adjusting operator help to obtain the balance between exploration and exploitation of the search space. Second, they compared and contrasted MBO with other population-based optimization methods.

Wen et al. (2017), presented Enhanced BBO using Quantitative Orthogonal learning and multi-topology. Two faults of biogeography-based optimization (BBO) are found by examining the governing migration operator's characteristics. To cure this, a new migration operator is proposed by improving topology and copy mode is presented to BBO. Moreover, diversity mechanism is also proposed. The new results show that the given approach has benefits regarding solution quality, convergence performance, as compared with standard BBO algorithms.

Zheng, et al. (2014) proposed a new algorithm, named Eco geography-based optimization (EBO), EBO hires a local topology to distinguish non-neighbors and neighbors of each habitat and describes two migration operators to enhance distribution between the habitats without harming the exploitation skill of BBO. A real-world presentation has authorized the EBO on an operational problem in an emergency.

In Taheri and Jalili (2016), an effective algorithm, called the enhanced biogeography-based optimization (EBBO), has been proposed to moderate early convergence problem of the basic BBO algorithm. In the projected algorithm, to improve the overall performance of the basic BBO algorithm, the new mutation and migration operators are proposed. These two new operators improve the ability to converge the ordinary BBO algorithm and further discharge premature and stagnation convergence.

Kamal et al. (2010), presented a new optimization method biogeography-based optimization (BBO) for antenna array combination. With the help of BBO algorithm, an ideal set of amplitudes of antenna elements provides a pattern of radiation with supreme side part level decrease. At last, they compared and contrasted this with other optimization methods. Antenna array combination makes goals at finding a physical arrangement whose radiation design is near to the desired.

Farswan and Bansal in 2018 proposed a mixture algorithm acquired by combining the concept of fireworks explosion of Fireworks Procedure and biogeography-based optimization and named as firework-inspired biogeography-based optimization (FBBO). The main feature in the projected FBBO algorithm is the combination of two dissimilar abilities to advance solution worth. The projected algorithm is verified on CEC 2014 standard problems.

Mathur in 2014 proposed a Biogeography-Based Optimization (BBO) algorithm to explain many types of Economic Load Dispatch (ELD) problems of the thermal power plants. The presented algorithm can manage monetary load report problems with limitations like communication losses, forbidden functioning regions, etc. The efficiency of the projected algorithm has been tested with few trial arrangements.

Wang and Song in 2016 made an elaboration in the simple migration equilibrium model of biogeography concept. The algorithm process is formed on the basis of population characterized by migration device of BBO algorithm. Description of seven linear or non-linear migration relation models has been made. Replication trials are tested on eight testing purposes to confirm the presented migration ratio models.

## 4 Emergency Vehicle Routing Problem

The motivation of this research is to solve emergency vehicle routing problem. Although in advance a catastrophe can be forecasted, there is always a possibility of shortcoming of warnings and a safe and robust method to reach the disaster venue and supply the necessary objects to the affected people and place. Many studies have been explored to evaluate the level of the effects of a natural catastrophe.

Our objective is to decide how to relegate vehicles from different regions (for example carports) to transport alleviation stuff and merchandise from different warehouse offices (terminals, docks, airports, railway station and depot) to the essential catastrophe points. Figure 2 shows the total framework included within the emergency VRP's with stuff in unexpected catastrophe.

As visible in the figure, the crisis help fabric diffusion framework can be a three-layer design. Transporting stuff is basically a vehicle steering issue involving development from various parking lots (first layer) to multiple capacity area.

Layer first covers the stopping areas; the material to be taken out for removal of all vehicles starts with stop parts. The number of vehicle types can change, but the number is known. Layer 2 fabrics is made up of save capacity location, which counts all produces stations, air terminals and comparable offices. We accept that at every saving point, the supply of supplies is shifted, which is not the supply of certain types of materials between periods of crisis.

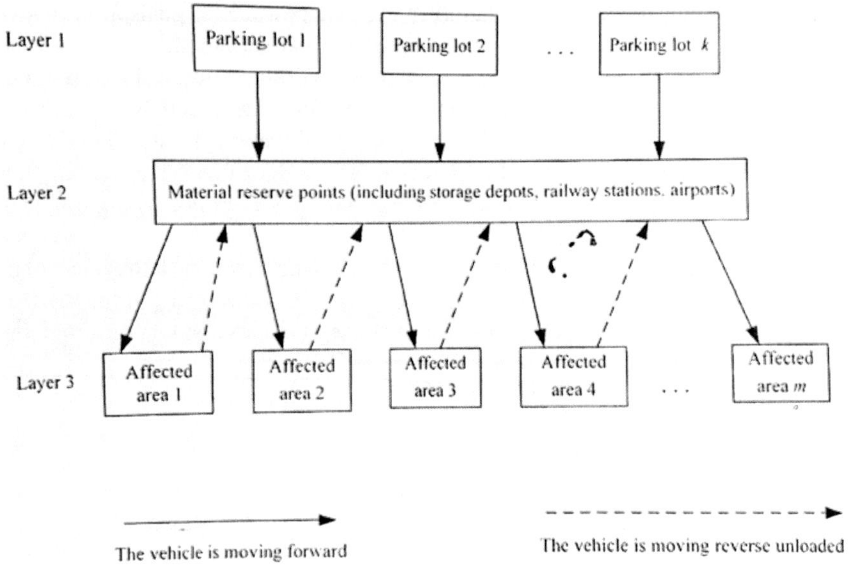


**Fig. 2** Illustration of a layered system of vehicle routing

Layer 3, the lower layer comprises the influenced regions that are the goals for alleviation dispersion. After a catastrophe, request for all sorts of crisis supplies will be created at all the effect focuses concurring to the catastrophe appraisal.

The vehicles from layer first to second; however, they may be now no longer stacked. Besides, this stream is one way, with the vehicles moving as those were from the stopping ranges to the different fabric reserve sites. They do not return to the stopping zones amid the crisis. Be that as it may, both vehicle stream and fabric stream, happen between Layers second and third. The material moves in a one-way unidirectional stream from layers second to third, while the vehicles make a two-way stream between these two layers. After a vehicle reaches the point of mayhem, it does now no longer right away go back to the parking range, however, stays within the entry space as a standby.

Both stuff and cars go with the drift among layers second and third. In any case, there may be no relation among every factor of the identical layer because FISCO suddenly is very notable after a request for supply, which usually exceeds the typical customer request for small clumps in commerce coordination. This contrast excess is reflected by the characteristics of the crisis aid supply requested by the affected focus. The imperative objective of post-disaster dispersal is to make the most predictable about all fiasco destinations while reducing the running time of a fully dispersed structure. In this way, the misfortune is decreased at every catastrophe point, and the depletion of vehicle operations also decreases. All vehicles can be reused through the dispersion framework, that is, the vehicle stream is reoriented among the moment and the layer third until all the desires of the catastrophe are met.

# 5 Mathematical Model of the EmVRP

With the relief material in sudden catastrophe under the guidance of Emergency VRP's conviction framework, the boundary conditions of the vehicle delivery system are defined as follows:

(1) The network has many demand and supply points. The demand and supply for crisis materials are known.
(2) There are numerous stopping zones, and each stopping range may have an assortment of vehicles. Each parked vehicle incorporates a number and parked vehicles are adequate to meet the requirement for material conveyance.
(3) There are assortments of crisis supplies to be conveyed. Each is distinctive in volume and weight, and each has distinctive stacking effectiveness. It is accepted that the stacking effectiveness of the fabric at the supply location is the same as the emptying effectiveness at the catastrophe location.
(4) One kind of material is delivered at each task.
(5) Multiple vehicles can serve at each catastrophe location.
(6) After the delivery work is completed, if the vehicle is still in supply, at that point, it will proceed to total the delivery task; something else, it will return to the beginning point at the stopping range to hold up for the following expedite.
(7) Between the points of each layer, there is neither material nor vehicle flow.

The objective function, model description and limitations conditions are defined as follows:

## 5.1 Objective Function

$$Min\ T = \sum_{l\in L}\sum_{k\in K}\sum_{i\in S} * d_{ki}/v_1 * y_{lki}$$
$$+ \sum_{l\in L}\sum_{i\in S}\sum_{j\in D}\sum_{g\in G} * t_g * x_{lijg} * y_{li} * Z_{lijg}$$
$$+ 2 * \sum_{l\in L}\sum_{i\in S}\sum_{j\in D} * d_{ij}/v_1 * y_{lij}$$
$$+ \sum_{l\in L}\sum_{i\in S}\sum_{j\in D}\sum_{g\in G} * t_g * x_{lijg} * y_{li} * z_{lijg}$$

## 5.2 Constraint Condition

$$x_{lijg} * w_g \le cap1$$

$$x_{lijg} * c_g \leq v_1$$

$$\sum_{g \epsilon G} * z_{liig} = 1$$

$$\sum_{i \epsilon S} * y_{lki} = 1$$

$$y_{lki} \in \{0, 1\}, \; y_{lki} \in \{0, 1\}, \; z_{liig} \in \{0, 1\}$$

$$y_{li} \in \{0, 1\}, \; y_{lj} \in \{0, 1\}$$

$$1 \leq l \leq |L|, \; 1 \leq i \leq |S|, \; 1 \leq j \leq |D|, \; 1 \leq g \leq |G|$$

where variables are defined as Decision variables.

$x_{lijg}$: the amount of g that is conveyed from the supply site i to j by first vehicle:
$z_{lijg} = 1$ First vehicle delivers supply g from site i to j = 0 else lϵL, iϵS, jϵD, gϵG
$y_{lki}$= 1 First vehicle passes edge (k, i) = 0 else lϵL, kϵK, iϵS
$y_{lij}$ = 1 Vehicle I passes edge (i, j) = 0 else lϵL, iϵS, jϵD
The set of materials/goods : $G = \{G_1, G_2, ..., G_p\}$;
The set of supply sites: $S = \{S_1, S_2, \ldots, S_n\}$
The set of catastrophe sites: $D = \{D_1, D_2, \ldots, D_m\}$
The set of parking areas: $K = \{K_1, K_2, \ldots, K_k\}$
The set of vehicles: $L(L_1, L_2, ...., L_l\}$
The set of edges: $E = \{(k, i)(i, j)|k \in K, i \in S, j \in D\}$
$V_l$: max volume of first vehicle;
$Cap_l$: the maximum dead weight tonnage of the first vehicle;
$v_l$: The velocity of first vehicle.
$w_g$: Material 'g' unit weight;
$c_g$: Material 'g' unit volume;
$t_g$: Material 'g' loading and uploading time.
$d_{ki}$: the distance between parking area and material supply site;
$d_{ij}$: the distance between the supply site and the demand point.

## 6 Results and Discussion

In this paper, a novel attempt is made to apply BBO to the problem of sudden disaster relief material. The presented problem is a discrete, np hard problem, which is extremely difficult to solve. BBO is originally proposed for continuous variable optimization problems. A novel attempt is made in order to apply BBO to this difficult problem.

**Table 1** Obtained results and comparison of BBO with state-of-the-art algorithms

| | ABC | BA | Modified BBO | CS | DE | EMBO | PSO |
|---|---|---|---|---|---|---|---|
| Best | 158.31 | 154.10 | 152.3 | 143.43 | 142.52 | 109.06 | 157.19 |
| Mean | 162.96 | 161.27 | 160.0849 | 145.26 | 144.68 | 118.18 | 159.03 |
| Worst | 167.35 | 167.86 | 186.20 | 150.21 | 146.10 | 140.77 | 162.23 |
| Std | 2.2516 | 3.1246 | 8.573617 | 8.573617 | 0.8202 | 13.2842 | 1.4138 |

BBO used for this problem is a modified version of the original BBO. It is based on constant mutation rate and with elitist strategy. Elitist parameter is set equal to 2, which represents the best two solutions saved after each iteration and gets replaced with two worst solutions. This helps in retaining the best solution obtained. In the next generation, it will be updated as well using migration and mutation operators like all other solutions. Hence, exploration and exploitation balance are maintained as well.

In review of the art of the state and No free lunch theorem of David Wolpert and William Macready appears in the 1997 'No Free Lunch Theorems for Optimization', it is always desirable to test a nature-inspired optimization technique for a real-life problem. The results are obtained in form of the objective function value. Time represents the objective function value and the purpose of the application is to minimize the time. However, results obtained by modified version of BBO are given Table 1. The bold entries give the results obtained. However, in comparison to MBO, modified BBO has performed poorly.

## 7 Conclusion and Future Scope

A novel attempt is made to solve the presented problem by Biogeography-based optimization. The results are obtained and compared with the existing literature. The comparison shows that MBO is more suitable candidate for solving sudden relief material problem; however, there are many variants of BBO are presented in the literature that can be tested for solving the problem.

## References

1. An Introduction to Optimization—Fourth Edition—Edison K.P Chong and Stanislaw H. Zak—Wiley
2. Optimization Techniques an Introduction by L.R Foulds Published by Springer Verlag New York—Heidelberg—Berlin
3. Optimization Techniques by A.K Malik , S.K Yadav , S.R Yadav—I.K. International Publishing House Private Limited

4. Numerical Optimization Techniques for engineering Design-Published by Vanderplaats Research and Develo (2001)
5. Using Monarch Butterfly Optimization to Solve the Emergency Vehicle Routing Problem with Relief Materials in Sudden Disasters by Jiao-Hong Yi, Jian Wang and Gai-Ge Wang
6. Wallace TG (2005) Distribution of animals (Two Volumes). Adamant Media Corporation, Boston, MA
7. Darwin C (1995) The origin of species. Gramercy, New York
8. MacArthur R, Wilson E (1967) The theory of biogeography. Princeton University Press, Princeton, NJ
9. Hanski and M. Gilpin (1997) Metapopulation biology. Academic, New York
10. Wesche T, Goertler G, Hubert W (1987) Modified habitat suitability index model for brown trout in southeastern Wyoming. North Amer J Fisheries Manage 2:232–237
11. Hastings and K. Higgins (1994) Persistence of transients in spatially structured models. Science 263;1133–1136
12. Muhlenbein H, Schlierkamp-Voosen D (1993) Predictive models for the breeder genetic algorithm: I. Continuous parameter optimization. Evol Comput 1:25–49
13. Back T (1996) Evolutionary algorithms in theory and practice. Oxford University Press, Oxford, UK
14. Parker K, Melcher K (2004) Themodularaero-propulsionsystemssimulation (MAPSS) users' guide, NASA. Tech Memo, 2004–212968
15. Simon D, Simon DL (2006) Kalmanfilterconstraintswitchingforturbofan engine health estimation. Eur JControl 12:331–343
16. Simon D (2006) Optimal state estimation. Wiley, New York
17. Mushini R, Simon D (2005) On optimization of sensor selection for aircraft gas turbine engines. In: Proc Int Conf Syst Eng, Las Vegas, NV, 9–14
18. Chuan-Chong C, Khee-Meng K (1992) Principles and techniques in combinatorics. World Scientific, Singapore
19. Garg V, Deep K (2015) A state-of-the-art review of biogeography-based optimization. In: Proceedings, 4th International Conference on Soft Computing for Problem Solving, Dec 26–28, 2014, NIT Silchar, Assam. Advances in Intelligent Systems and Computing, Springer, 336, 533–549
20. Garg V, Deep K (2016) Performance of laplacian biogeography-based optimization algorithm on CEC 2014 continuous optimization benchmarks and camera calibration problem. Swarm Evol Comput Elsevier 27:132–144

# Chapter 10
# Shoring-Based Formwork Optimization Using SOMGA for Multi-Storey Buildings

**Shilpa Pal, Dipti Singh, and Piyush Vidyarthi**

## 1 Introduction

Slab formwork is a pre-planned basic framework utilized to convey vertical and horizontal burdens with no failure in that timeframe. The sheathing of traditional slab formworks is made from plywood sheets, while joists, stringers, and shores act like beams and columns. Sheathing retains concrete and joists, stringers, and shores, serving as the supporting member as it holds the sheathing in place. Joists and stringers act horizontally while shores act vertically. Stringers are supported on shores while joists are supported on stringers. Shores are used to transfer the pressures uniformly to the lower points.

Shores carry the weight of formwork, freshly placed concrete, construction loads, and transfer them to the lower interconnected floors, whereas reshores are the type of shores that is placed just below a concrete slab or structural member after the original forms. Thus, restricting the new slab or structural member to deflect and support its weight as well as existing construction loads applied before the installation of the reshores.

Paul [12] described every aspect of horizontal deck formwork with the standard limits of deflection, bending, and shear, and, choosing the plywood thickness of joist spacings to calculate loads on formwork to have safe concrete elements. Mosalla et al. [11] developed an analytical approach to check the adequacy of the slab during

S. Pal
Department of Civil Engineering, Delhi Technological University, Delhi, India

D. Singh
Department of Applied Sciences, Gautam Buddha University, Greater Noida, India
e-mail: diptipma@rediffmail.com

P. Vidyarthi (✉)
Department of Civil Engineering, Gautam Buddha University, Greater Noida, India
e-mail: vid.piyush@gmail.com

 
D. Singh et al. (eds.), *Proceedings of International Conference on Scientific and Natural Computing*, Algorithms for Intelligent Systems,
https://doi.org/10.1007/978-981-16-1528-3_10

construction. They first time included the effect of lateral load. Hanna et al. [8] developed the first computerized method to optimize concrete slab formwork design. In this method, the cost of each component was taken and the comparison with a numerical study using design optimization showed that there is 9.9% cost saving. Dongping et al. [6] investigated the slab load for 20 different construction shoring/reshoring schemes. The result showed that the traditional method underestimated or overestimated the maximum slab load and depended mainly on the shoring/reshoring scheme. An approach was developed in which stiffness of the shoring and reshoring was kept rigid assuming the assistance of construction engineers to assure safe construction procedures. Calderon et al. [4] calculated the actual load during shoring and striking using the clearing procedure. Clearing removes forms and a number of shores from the hardened lower slab before casting the next slab. That method is known as shoring, clearing, and striking (SCS). SCS was improved by considering the modulus of the slab's elasticity depending on time variation and average deflection of each slab equal to the shores' moderate deformation.

Minimization of the construction cost of the shoring-based system is one of the most important factors for civil engineers without making any change in formwork effectiveness [11]. Hyun C et al. developed a building information modeling-based automatic formwork design system to optimize the formwork design process [9]. Design optimization of various structures is a vast topic of research in the field of structural design. Although several techniques are available in the literature for optimization due to easy implementation, population-based soft computing techniques are gaining attention nowadays for solving such type of problems, and Many soft computing techniques have been used by various researchers [3, 7, 10, 13, 15] for proving a safe slab formwork with minimum cost for solving engineering problems. In this paper, a hybrid technique SOMGA has been used for finding the optimum combination of the level of shore and level of reshore for four types of buildings, i.e., 05, 20, 40, 50 storeys for two grades of concrete and varying days of construction. The main reason for using this algorithm is that it combines two techniques GA and SOMA, with both the algorithm's best features. SOMGA tries to maintain the search space's diversity during the entire search and has fewer chances to get stuck in the local optimal solution or infeasible solution [5]. Another advantage of using this algorithm is that due to hybridization with SOMA, parameters need not be fine-tuned much. It works well with a small population size of 20 [14]. GA and SOMA also obtain results but were found to be unsafe. In this study, the sequence of load distribution on interconnected slabs during the construction process is based on the simplified method specified in ACI-347.2R05 [2]. Optimized results obtained by SOMGA have been validated with analytical calculation as per ACI.

## 2 Analytical Mathematical Formulations as Per ACI Code

According to ACI committee 347 [2], in a typical construction cycle for a multi-storey cast-in-place concrete building where both shores and reshores are used, there are four construction phases as shown in Fig. 1. The load on the slabs for all these stages is calculated and then checked by the strength of that particular slab for a period of time as per ACI codal recommendations. The strength calculation formula is given by Eq. 1, according to ACI209R-82 [1].

$$f_{ct} = \frac{t}{(a + bt)} \times f_{ck} \tag{1}$$

$f_{ct}$ = Compressive strength of concrete at any time after casting, N/mm$^2$.

$f_{ck}$ = Characteristic compressive strength of concrete 28 days after casting, N/mm$^2$ t = time elapses after casting, in days a = constant, depends on type of cement and curing method. b = constant, depends on the type of cement and curing method.

The values of a and b can be obtained from Table 1

Formulation for calculation of load on the intermediate slab, analysis of load on particular levels of shore and reshore in construction of multi-storey buildings for all the phases of construction are given in Eqs. 2 to 4

$$\begin{aligned} Load\ on\ slab &= (dead\ load\ of\ that\ slab) \\ &+ ((Position\ factor) * (weight\ of\ slab + construction\ loads)) \end{aligned} \tag{2}$$

$$Load\ on\ Shore = (weight\ of\ above\ slabs)$$

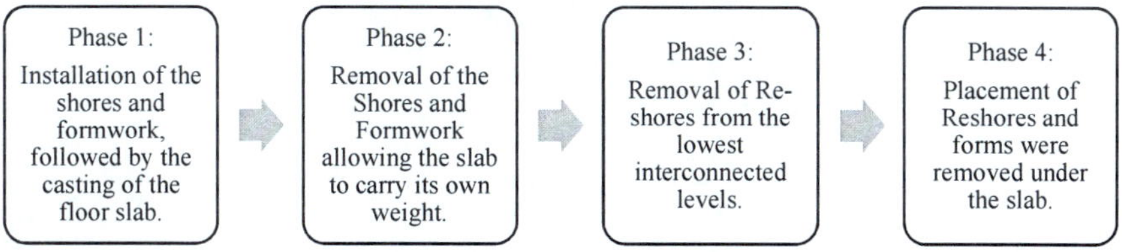


**Fig. 1** Different phases of construction by using shoring-based formwork

**Table 1** Values of constant a and b from ACI 209R-82

| Type of curing | Type of cement | a | b |
|---|---|---|---|
| Moist curing | Type I | 4.0 | 0.85 |
| | Type III | 2.3 | 0.92 |
| Stem curing | Type I | 1.0 | 0.95 |
| | Type III | 0.7 | 0.98 |

$$+ \ (weight\ of\ above\ shore\ level) + (construction\ load) - (weight\ of\ above\ slabs) \tag{3}$$

$$Load\ on\ reshore = (weight\ of\ above\ slabs) + (weight\ of\ above\ shore\ level) + (weight\ of\ above\ reshore\ level) + (construction\ load) - (weight\ of\ above\ slabs) \tag{4}$$

## 3 Problem Statement

The study has been carried out on four different storey buildings, i.e., 5, 20, 40, and 50 storeys for two different grades of concrete M20 and M25 and for two separate days of the construction cycle, i.e., 6 and 7 days. To find the optimum combination of shore and reshore, a population-based technique, SOMGA has been used to minimize cost and safe construction. The following objective function and constraints have been taken.

### *3.1 Objective Function and Constraints*

This study's objective function has four parameters: number of levels of shores, number of levels of reshores, grade of concrete, and grade of steel reinforcement.

The objective function is defined as:

$$cost(X) = x1 + x2 + x3 + x4 + x5 \tag{5}$$

$$\begin{aligned} x1 &= cost\ of\ shores \\ &= 30\%\ of\ total\ cost\ of\ (concrete + steel) \\ &\times\ no.\ of\ levels\ of\ shores\ (m) \end{aligned} \tag{6}$$

$$\begin{aligned} x2 &= cost\ of\ reshores \\ &= 10\%\ of\ total\ cost\ of\ (concrete + steel) \\ &\times\ no.\ of\ levels\ of\ reshores\ (n) \end{aligned} \tag{7}$$

$$\begin{aligned} x3 &= cost\ of\ concrete \\ &= constant\ cost\ coefficient\ (c) \\ &\times\ charactersistic\ compressive\ strength\ of\ concrete\ (d) \end{aligned} \tag{8}$$

$$x4 = cost\ of\ steel = constant\ cost\ coefficient\ (c1) \times strength\ of\ grade\ of\ steel\ (e) \quad (9)$$

$$x5 = Penalty\ cost = PC \quad (10)$$

In the objective function, cost of shores and reshores is 30% and 10% of the cost of concrete and steel, respectively, and is changeable without impacting the objective. The technique checks all the combinations of shores and reshores for different type of concrete grades. The stripping time is changed by changing the grade of concrete. The formula for stripping time calculation has been derived from the strength equation given by ACI code. The bounds for the number of levels of shore, reshores, stripping time, and cycle time are as follows:

$$1 \leq m \leq (z-1) \quad (11)$$

$$1 \leq n \leq (z-1) \quad (12)$$

$$2 \leq m + n \leq z \quad (13)$$

$$T = ts + tinstallation \quad (14)$$

Z = number of floors, ts = stripping time, T = cycle time $t_{installation}$ = installation time, to install the formwork per floor m = no. of levels of shores n = no. of levels of reshores.

The above-given objective function and constraint have been formulated taking following assumptions:

All floors have the same area, the same grade of concrete is used for all slabs, all slabs have an equal thickness, ground stiffness is infinite, axial stiffness of shores and reshores members is infinite, stiffness of interconnected slabs are equal, installation time is of 2 days, cost of each level of shore assumed to be 30% of the total cost of concrete and steel per floor, cost of each level of reshore is assumed to be 10% of the total cost of concrete and steel per floor and self-weight of concrete is assumed to be 1. The process of optimization using SOMGA technique is given in Fig. 2.

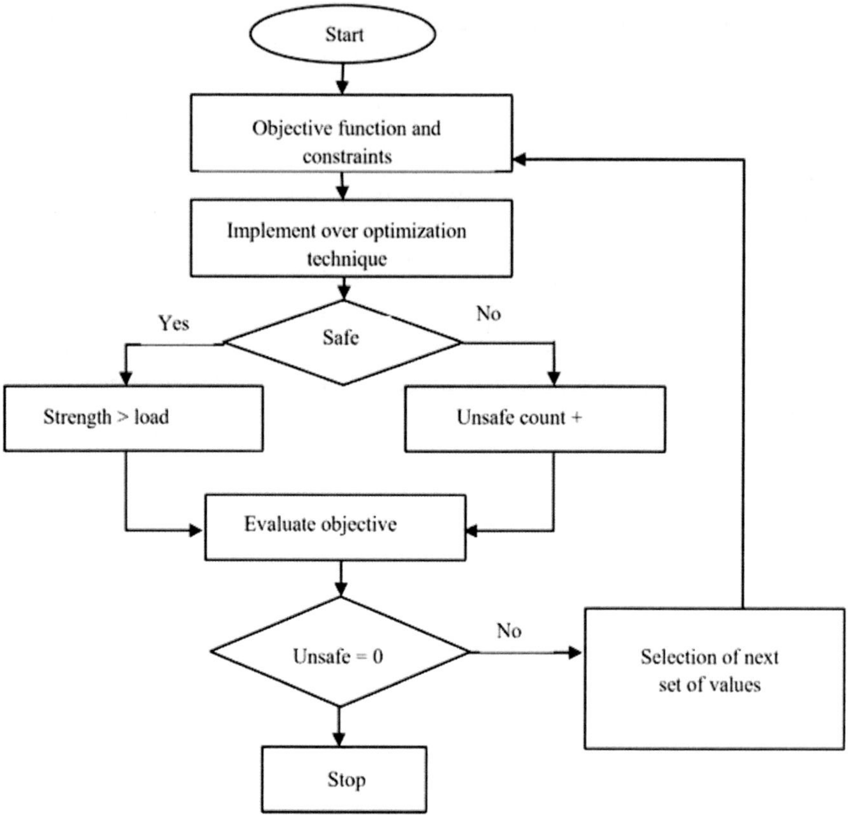


**Fig. 2** Optimization process of objective function

## 4 Self Organizing Migrating Genetic Algorithm

Self Organizing Migrating Genetic Algorithm (SOMGA) is a population-based stochastic search technique that is the hybridized variant of Self Organizing Migrating Algorithm (SOMA) and Genetic algorithm (GA) [5]. It is based on the competitive-cooperative behavior of a group of individuals looking for a food source. The working of this algorithm is easy to implement and has the features of two efficient algorithms SOMA and GA. Other algorithms use a large population size to start with this algorithm. It requires a small population size and has a low computational cost. In the working of SOMGA, first, a population of individuals is generated randomly. Then selection, crossover, and mutation operator are applied to find the new individuals in the search space. Based on their fitness value, the best individual is considered a leader and worst as an active individual. Then active individual moves towards the leader in steps of defined path length. The computational steps of this algorithm are as follows:

Step 1: Generate the initial population.

Step 2: Evaluate all individuals.

Step 3: Apply tournament selection on all individuals to select better individuals for the next generation.

Step 4: Apply the crossover operator on all individuals with crossover probability $P_c$ to produce new individuals.

Step 5: Evaluate new individuals.

Step 6: Apply mutation operator on every bit of every individual of the population with a mutation probability.

$P_m$.

Step 7: Evaluate the mutated individuals.

Step 8: Find leader (best-fitted individual) and active (worst fitted individual) of the population.

Step 9: For active individuals, a new population of size N is created, where N = (path length/step size). This population is nothing but the active individual's new positions towards the leader in n steps of the defined length. The movement of this individual is given by:

$$\begin{aligned} &x_{i,j}^{MLnew} = x_{i,j,start}^{ML} + (x_{L,j}^{ML} - x_{i,j,start}^{ML})tPRTVector_j \\ &where \quad t \in < 0, \ by\ step\ to,\ PathLength >, \\ &ML \ is\ actual\ migration\ loop. \\ &x_{i,j}^{MLnew} \ is\ the\ new\ positions\ of\ an\ individual. \\ &x_{i,j,start}^{ML} \ is\ the\ positions\ of\ active\ individual. \\ &x_{L,j}^{ML} \ is\ the\ positions\ of\ leader. \end{aligned} \tag{15}$$

*PRTVector*$_j$ (Perturbation vector) is created using the value of PRT parameter before an individual proceeds toward leader as follows: A random number *rnd* is generated (for each individual's coordinate) and then compared with PRT.

$$if\ rnd_j < PRT\ then\ PRTVector_j = 1;\ elsePRTVector_j = 0;$$

Step 9.1: Select the new population's best individual and replace the active individual with this best individual.

Step 10: Select the best individuals (in fitness) of the previous and current generation for the next generation via tournament selection.

Step11: If the termination criterion is satisfied, stop else go to step 3.

Step12: Report the best chromosome as the final optimal solution.

The parameters used are as follows: population size = 20, $P_c = 0.8$, $P_m = 0.001$ path length = 3, step size = 0.81, PRT = 1.

## 5 Results and Discussion

After analyzing the problem using SOMGA, four different cases, i.e., effect of different storey height keeping grade of steel as Fe 415 has been discussed in detail for following parameters:

(i) M20 grade concrete with 6 days of construction cycle
(ii) M20 grade concrete with 7 days of construction cycle
(iii) M25 grade concrete with 6 days of construction cycle
(iv) M25 grade concrete with 7 days of construction cycle

### *5.1 Case I: 05 Storey Building*

When the grade of concrete is 20 MPa, the grade of steel is Fe415 for construction of five-storey building with 6 days of the construction cycle. No safe combination of levels of shores and levels of reshores has been obtained, i.e., the strength of slab < load on the slab. With 7 days of the construction cycle, multiple reasonable safe combinations have been obtained with safety factor less than 1.5; but from practical construction, viewpoint for this factor of safety construction needs higher accuracy and monitoring. For this case, the minimum cost is obtained for a combination of 1 level of reshores with one level of shores, i.e., Rs.1289000.

When the grade of concrete is 25 MPa, the grade of steel is Fe415 for construction of five-storey building with 6 days of the construction cycle. The multiple reasonable safe combinations have been obtained with the factor of safety less than 1.5. Still, the minimum cost is received for a combination of 1 level of reshores with 1 level of shores, i.e., Rs.1299000. With 7 days of the construction cycle, the multiple reasonable safe combinations have been obtained with the factor of safety ranging from 1.5 to 1.1. For this case, the minimum cost is obtained for a combination of 1 level of reshores with 1 level of shores, i.e., Rs.1299000.

On increasing the grade of concrete from M20 to M25, the construction cost increases by 1% approximately. Because the cost of shores and reshores is dependent on the cost of concrete and steel, and the cost of the shore is more than the cost of steel. It is also observed that on decreasing the duration of the construction cycle, the factor of safety decreases from 1.5 to 1.1. Hence, it can be concluded that irrespective of the grade of concrete and construction cycle, minimum cost is obtained for a combination of 1 level of reshore with 1 level of shore for the construction of five-storey building, results have been summarized in Table 2.

### *5.2 Case II: 20 Storey Building*

When the grade of concrete is 20 MPa, the grade of steel is Fe 415, with 6 days of the construction cycle, no safe combination of shores and level of reshores has been

**Table 2** Combination of shores and reshores for five-storey building using SOMGA

| S. No | Grade of concrete | Grade of steel | Floor construction cycle | Combination of shores and reshores | Optimum combination of shores and reshores | Remarks |
|---|---|---|---|---|---|---|
| 1 | M 20 | Fe415 | 6 days | No safe combination | No safe combination | Irrespective of grade of concrete and steel, minimum cost is obtained for combination of 1 level of reshore with 1 level of shore |
| | | | 7 days | Multiple combinations | 1 level of shore with 1 level of reshore | |
| 2 | M25 | Fe415 | 6 days | Multiple combinations | 1 level of reshore with 1 level of shore | |
| | | | 7 days | Multiple combinations | 1 level of reshore with 1 level of shore | |

obtained, which concludes that slab strength is less than the load on slab in this case. And for 7 days of the construction cycle, for 1 level of shores till 9 levels of reshores, the condition is safe, but the factor of safety is less than 1.5, hence construction needs higher accuracy, which is considered as unsafe. The reason is that after the lowest reshore connected is removed the interconnected slabs do not have enough capacity to carry additional shared load, whereas for 1 level of shore and 12 levels of reshores, the condition is safe, but the cost is as high as Rs.13,00,000. The cost difference between 9 levels of reshores and 12 levels is only Rs. 3000 approximately. But the impact on the performance level is high.

When the grade of concrete is 25 MPa with 6 days of the construction cycle, multiple reasonable safe combinations have been obtained with the factor of safety less than 1.5. For this case, the minimum cost is obtained for the combination of 1 level of reshores with 1 level of shores, i.e., Rs.1299000. With 7 days of construction cycle, the multiple reasonable safe combinations have been obtained with the factor of safety ranging from 1.5 to 1.1. Minimum cost is obtained for a combination of 1 level of reshores with 1 level of shores, i.e., Rs.1299000.

It is also found that more safe combinations are always obtained when the number of levels of shores is less than levels of reshores. The reason behind this is that when removing shores is rapid, the load distribution is more and shared loads by the slabs are lesser, whereas, when the number of levels of shores is greater than number of levels of reshores, the condition is safe, but the impact of a higher number of levels of shore increases the overall cost. On increasing the grade of concrete to 25 MPa, 50% more number of combinations are obtained. This is because on comparing to 20 MPa, the stripping time for 25 MPa gets reduced this means grade of concrete has more influence than stripping time on the capacity of slab.

When the construction cycles for M25 grade of concrete are reduced from 7 to 6 days with installation time maintained at 2 days, the safe combination of shore

**Table 3** Combination of shores and reshores for 20-storey building using SOMGA

| S. No | Grade of concrete | Grade of steel | Floor construction cycle | Combination of shores and reshores | Optimum combination of shores and reshores | Remarks |
|---|---|---|---|---|---|---|
| 1 | M 20 | Fe415 | 6 days | No safe combination | No safe combination | Irrespective of grade of concrete and steel, minimum cost is obtained for a combination of 1 level of reshore with 1 level of shore |
| | | | 7 days | Multiple combinations | 1 level of the shore with 12 levels of reshore | |
| 2 | M25 | Fe415 | 6 days | Multiple combinations | 1 level of reshore with 1 level of shore | |
| | | | 7 days | Multiple combinations | 1 level of reshore with 1 level of shore | |

levels with levels of reshores obtained is less. The reason behind this is when the striping time is reduced, the strength achieved by the slabs is lesser.

However, for 20-storey building, irrespective of grade of concrete and construction cycle, minimum cost is obtained for a combination of 1 level of reshore with 1 level of shore, summarized in Table 3.

## 5.3 *Case III: 40 Storey Building*

The same observation has been seen for 40-storey building with 6 days and 7 days of construction cycle and an increase in grade of concrete from 20 to 25 MPa. The results are summarized in Table 4.

With grade of concrete equal to 20 MPa with construction cycle taken as 7 days for 1 level of shores till 13 levels of reshore, the condition is safe, but the factor of safety is less than 1.5. For 1 level of shore and 17 levels of reshore, combination is acceptable and has the lowest cost with safe condition. The cost difference between 13 levels of reshore and 17 levels of reshore is only Rs.3000, which does not impact much on the overall cost. For 25 MPa concrete with 6 days of construction cycle, minimum cost is obtained for a combination of 1 level of reshores with 1 level of shores, Rs.1299000. With 6 days of the construction cycle, all the combinations from 4 levels of shores and various levels of reshores to (m-1) level of shores are unsafe. When the shores are removed due to the decrease in the striping time, sufficient strength is not achieved by slab to hold the shared load (strength < load shared by the slab after re-distribution of load). For 7 days of construction cycle, minimum cost is obtained for a combination of 1 level of reshores with 1 level of shores, i.e.,

**Table 4** Combination of shores and reshores for 40-storey building using SOMGA

| S. No | Grade of concrete | Grade of steel | Floor construction cycle | Combination of shores and reshores | Optimum combination of shores and reshores | Remarks |
|---|---|---|---|---|---|---|
| 1 | M 20 | Fe415 | 6 days | No safe combination | No safe combination | Irrespective of grade of concrete and steel, minimum cost is obtained for a combination of 1 level of reshore with 1 level of shore |
| | | | 7 days | Multiple combinations | 1 level of shore with 17 levels of reshore | |
| 2 | M25 | Fe415 | 6 days | Multiple combinations | 1 level of reshore with 1 level of shore | |
| | | | 7 days | Multiple combinations | 1 level of reshore with 1 level of shore | |

Rs.1299000. Figures 3 and 4 show that when the grade of concrete is increased to 25 MPa with 7 days for construction cycle,

15% more number of combinations are obtained this is because on comparing to 20 MPa the stripping time for 25 MPa gets reduced this means, grade of concrete has more influence than stripping time on the capacity of slab.

When the construction cycle for M25 grade of concrete is reduced from 7 to 6 days with installation time maintained at 2 days, the combination of level of shore and level of reshores obtained is reduced almost by 70%. The reason behind this is when the striping time reduced, the strength achieved by the slabs is lesser. For this

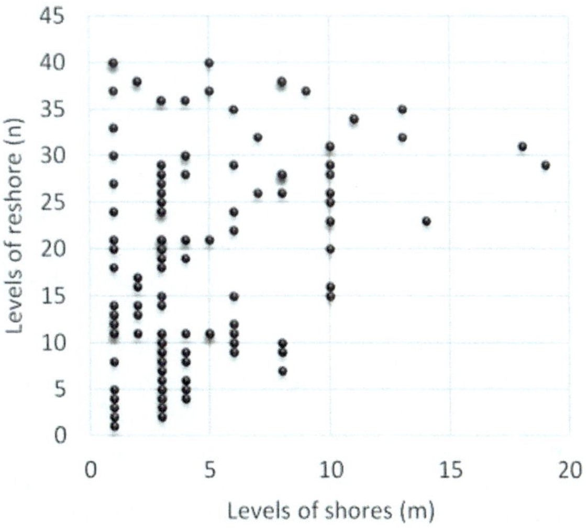


**Fig. 3** Levels of shores versus levels of reshores for M20

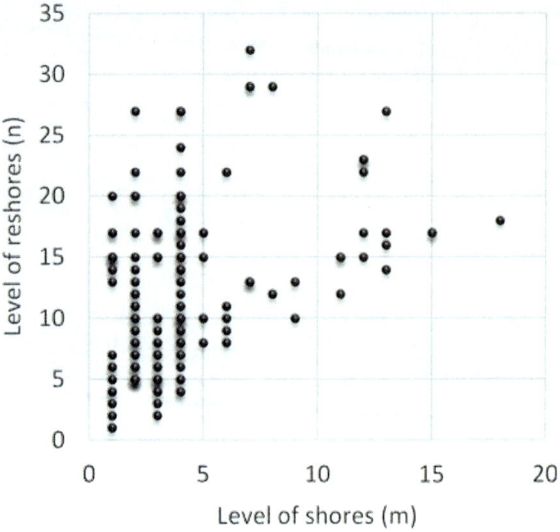


**Fig. 4** Level of shore versus level of reshores for M25

case also, irrespective of grade of concrete and construction cycle, minimum cost is obtained for a combination of 1 level of reshore with 1 level of shore.

## 5.4 *Case IV: 50 Storey Building*

Similar observations have been seen for 50-storey building with 6 days and 7 days of construction cycle and an increase in grade of concrete from 20 to 25 MPa. The results are summarized in Table 5. When grade of concrete is M25 with 6 days of construction cycle, the multiple reasonable safe combinations have been obtained with FOS less than 1.5, and the minimum cost of Rs. 12,99,000 is obtained for a combination of 1 level of reshores with 1 level of shores. From the analysis, it is observed that for 4 levels of shores and various levels of reshores, all combinations are unsafe. This occurs because when the shores are removed due to the decrease in stripping time, sufficient strength is not achieved by slab to hold the shared load, whereas, for 7 days of construction cycle, multiple reasonable safe combinations with FOS ranging between 1.5 and 1.1 have been obtained. However, the minimum cost of Rs. 1,299,000 is obtained for a combination of 1 level of reshore with 1 level of shores.

With increased grade of concrete to 25 MPa from 20 MPa, 15% more combinations are obtained. From Figs. 4 and 5, it can be observed that when the construction cycle for M25 grade of concrete is reduced from 7 to 6 days with installation time maintained at 2 days, the combination of the level of shore and level of reshores obtained is reduced almost by 70%. The reason behind this is when the striping time reduced, the strength achieved by the slabs is lesser.

**Table 5** Combination of shores and reshores for 50-storey building using SOMGA

| S. No | Grade of concrete | Grade of steel | Floor construction cycle | Combination of shores and reshores | Optimum combination of shores and reshores | Remarks |
|---|---|---|---|---|---|---|
| 1 | M 20 | Fe415 | 6 days | No safe combination | No safe combination | Irrespective of grade of concrete and steel, minimum cost is obtained for a combination of 1 level of reshore with 1 level of shore |
| | | | 7 days | Multiple combinations | 1 level of shore with 37 level of reshore | |
| 2 | M25 | Fe415 | 6 days | Multiple combinations | 1 level of reshore with 1 level of shore | |
| | | | 7 days | Multiple combinations | 1 level of reshore with 1 level of shore | |

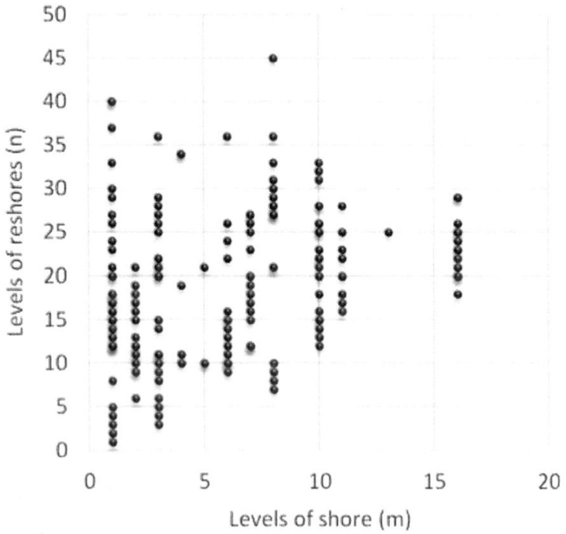


**Fig. 5** Levels of shores versus reshores for M25, 7 days of construction cycle

However, irrespective of grade of concrete and construction cycle, minimum cost is obtained for a combination of 1 level of reshore with 1 level of shore.

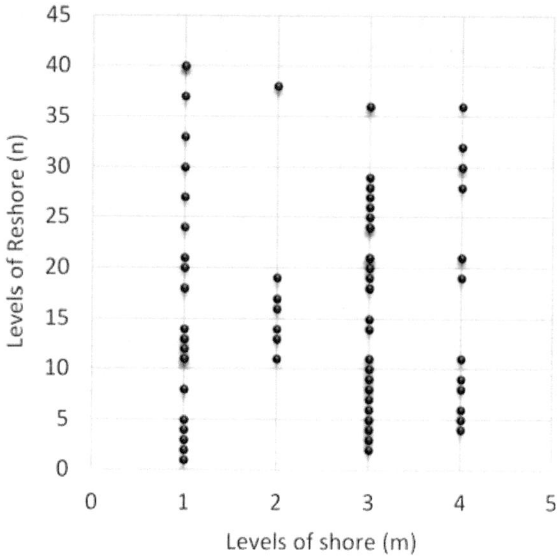


**Fig. 6** Level of shores versus reshores for M25, 6 days of the construction cycle

## 6 Conclusions

The objective is to find the best combination of shore and reshore to minimize the cost of shoring, reshoring and concrete for safe conditions during the construction process of high-rise buildings using SOMGA. From the study, it tends to be reasoned that regardless of the grade of concrete and the construction cycle and number of storey, the optimum combination is observed for 1 level of reshore with 1 level of shore used to construct a multi-storey building. Using the optimizing technique, results are found to be more accurate with less computational time than that of the simplified method according to the codal provisions of ACI347-2R:2005. When the grade of concrete increases, the number of multiple combinations increases as the stripping time reduces for higher concrete strength, which has been observed for building height 20 storey and above. With a decrease in construction cycle from 7 to 6 days, the number of combinations of shores and reshore reduces as the slab's strength reduces for higher grade of concrete. The optimization technique, namely, SOMGA, produces the optimum results and provides multiple outcomes that can be extended effectively to other engineering problems.

## References

1. ACI 209R-82, Prediction of Creep, Shrinkage and Temperature effects in Concrete Structures, Reported by ACI Committee 209
2. ACI 347–2R:2005, Guide for Shoring/Reshoring of Concrete Multi-storey Buildings, Reported by ACI Committee 347

3. AL-Tabtabai H, Alex AP, James R (1999) Slab formwork design using genetic algorithm. Durab Build Mater Compon 8:2407–2418
4. Calderon PA, Alvarado YA, Adam JM (2011) A new simplified procedure to estimate loads on slabs and shoring during the construction of multi-storey building. Eng Struct 33(5):1565–1575
5. Deep K, Dipti (2008) A self organizing migrating genetic algorithm for constrained optimization. Appl Math Comput, Elsevier 198(1):237–250
6. Dongping F, Haifeng XI, Xiaoming W, Chuanmin Z, Tingsheng Z (2009) Load distribution assessment of reinforced concrete buildings during construction with structural characteristic parameter approach. Tsinghua Sci Technol 14(6):746–755
7. Galeb AC, Zainab FA (2011) Optimum design of reinforced concrete waffle slabs. Int J Civ Struct Eng 1(4):862–880
8. Hanna SA, Ahmed BS (1995) Design optimization of slab formwork. J Const Eng Manag 121(2):215–221
9. Hyun C, Jin C, Shen Z, Kim H (2018) Automated optimization of form work design through spatial analysis in building information modeling. Autom Constr 95:193–205
10. Kamarthi SV, Sanvido VE, Kumara SRT (1992) Neuroform- neural network systems for vertical formwork selection. J Comput Civ Eng 6(2):178–199
11. Mosalla K, Wai-Fah C (1990) Design consideration for formwork in multistorey buildings. Eng Struct 12(3):163–172
12. Paul HS (1984) Charts aid in design of horizontal formwork. The Aberdeen Group, #C840648
13. Sahab MG, Toropov VV, Gandomi AH (2017) Optimum design of composite concrete floors using a hybrid genetic algorithm. Handbook of Neural Computation, Academic Press, Chapter 31:581–589
14. Singh D, Agrawal S (2016) Self organizing migrating algorithm with quadratic interpolation for solving large scale global optimization problems. Applied Soft Computing 38:1040–1048
15. Singh H, Rai HS, Singh J (2014) Discrete optimisation of one way slab using genetic algorithm. Int J Eng Bus Enterp Appl 9(2):116121

# Chapter 11
# Numerical Simulation of Heavy Rainfall Using Weather Research and Forecast (WRF) System

**Sushil Kumar, Bhanumati Panda, and P. V. S. Raju**

## 1 Introduction

The maximum rainfall was received in the east coast of India and central India due to summer monsoon depressions over the Bay of Bengal during June to September (JJAS). The monsoon depressions usually last for few days, which causes rainfall and sometimes very heavy rainfall due to their ability to produce phenomenal amounts of precipitation. Frequent and hazardous weather manifestations occurring over India cause landslide, flash floods and damage to crops and human beings in recent years. These incidences have major effects on the society, economy and environment. Therefore, improved and detailed prediction of the depression/cyclones and associated rainfall are the top priorities of interest for the meteorologist and researcher [1–4]. Due to the development of high-resolution mesoscale model, many researchers have carried out different modeling study for the improvement of high-impact weather prediction system including monsoon depression [5–8]. Particularly, over the last few decades, the high-resolution WRF (ARW) model is capable to predict the extreme weather condition over India [5, 6, 9–12]. Raju et al. [13] customized WRF (ARW) model using the combination of various physical parameterization schemes to study the tropical cyclone and recommended that the KF cumulus convection scheme along with the YSU PBL provides reasonable forecast over Indian region. Pattanaik et al. [14] also revealed that the YSU PBL, NMM (non-hydrostatic mesoscale model) of WRF

S. Kumar
Department of Applied Mathematics, Gautam Buddha University, Greater Noida, India

B. Panda
Department of Applied Science and Humanities, I.T.S Engineering College, Greater Noida, India

P. V. S. Raju (✉)
Centre for Ocean-Atmospheric Science and Technology, Amity University Rajasthan, Kant Kalwar, Jaipur, Rajasthan, India
e-mail: pvsraju@jpr.amity.edu; pemmani@gmail.com



D. Singh et al. (eds.), *Proceedings of International Conference on Scientific and Natural Computing*, Algorithms for Intelligent Systems,
https://doi.org/10.1007/978-981-16-1528-3_11

(ARW) scheme provide better track and intensity forecast. In this study, advanced WRF (ARW) core with suitable initial condition to simulate various characteristics associated with monsoon depression such as wind, sea level pressure, water vapor mixing ratios and heavy rainfall in Jharkhand and its neighborhood from 25th– to 27th July 2017. The synoptic description of the event is presented in Sect. 2 and the model WRF (ARW) description and the numerical experiments performed in the study are described in Sect. 3. Section 4 presents the results and discussion followed by conclusions in Sect. 5.

## 2 Synoptic Description of Land Depression

The study considered the depression formed over Gangetic West Bengal, south Bangladesh and north Bay of Bengal on 23rd July 2017 and further positioned as an organized system in early hours of 24th July. On 26th July 2017 at 0000 UTC, it strengthened into a depression over Jharkhand and its neighborhood with the center near Lat. 24.0 °N and Long. 85.0 °E. The depression moved northwesterly till 1200 UTC of 26th July 2017 and then moved westwards up to northeast Madhya Pradesh. The depression did not intensify further due to the absence of moisture and high vertical wind shear. It further lessened into a low-pressure area over northeast Madhya Pradesh at 0300 UTC of 27th July 2017 and moved westwards before weakening gradually. The remnant low-pressure area moved up to south of Rajasthan and its neighborhood.

## 3 Model

In the present study, the non-hydrostatic mesoscale WRF (ARW) model is used for the simulation of monsoon disturbances with a horizontal resolution at 9 km over the domain (15°N–30°N, 70°E–95°E) covering over northwest Jharkhand and its neighborhood (Fig. 1). The National Centers for Environmental Prediction (NCEP) FNL (FiNaL) data available a resolution of $1^\circ \times 1^\circ$ is used as initial and boundary conditions (ICs and BCs) for the model simulation. The time-varying lateral boundary conditions and sea surface temperatures (SST) were updated at every 6-hour interval from 00 UTC of 25th July 2017 to 18 UTC of 27th July 2017. The model was integrated up to 72 h starting from 00 UTC of 25th July 2017. The model domain consists of 15°N–30°N and 70°E–95°E. The land surface data with a horizontal grid spacing of 5 min are obtained from the US Geological Survey (USGS). Yonsei University (YSU) planetary boundary layer (PBL) scheme is used in the study because of its vertical diffusion compendium with a non-turbulent mixing in the PBL, which is relevant for boundary layer mixing [15]. Kain–Fritsch (KF) scheme for cumulus convection [16, 17] and Kessler microphysics are used. Rapid Radiation Transfer Model for longwave radiation [18] and Dudhia scheme for shortwave radiation [19]

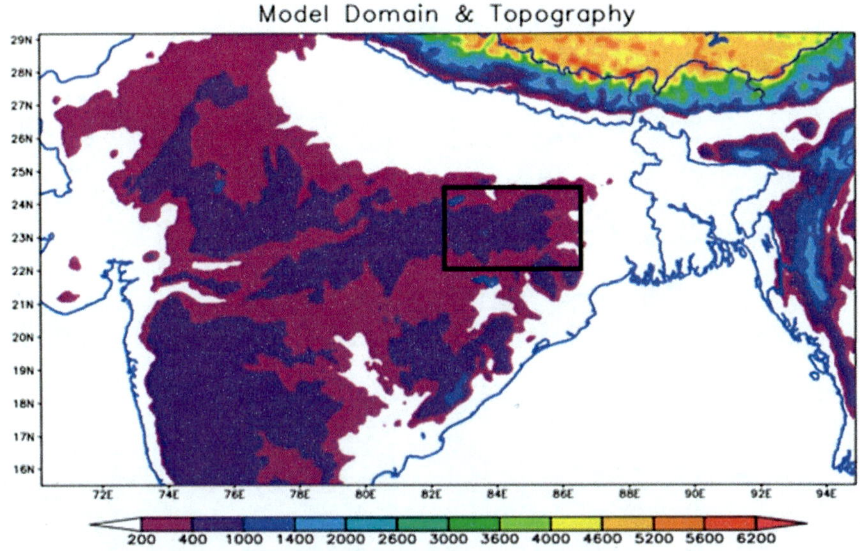


**Fig. 1** Model domain and topography used in this study

are also considered. Surface clay physics scheme MM5Monin-Obukhov scheme is considered in this study. The initial conditions at $1^0$ resolutions were interpolated from NCEPFNL (FiNaL) analysis and boundary conditions and SSTs were updated at every 6-hour interval from 00 UTC of 25th July 2017 from NCEP FNL analysis.

## 4 Results and Discussion

The WRF (ARW) model-simulated result consists of wind, radar reflectivity factor (unit: dBZ), sea-level pressure (SLP), water vapor mixing ratio (WVMR), the rainfall and the track of the depression are discussed in this section. The model-simulated result is observed over North-west Jharkhand and its neighborhood on 25th July 2017 during heavy rainfall caused by summer monsoon season. The model was integrated at 00 UTC on 25th July, 2017, which corresponds to the low-pressure area at the state of the observation. The model simulated wind field at 850 hPa level for the single model domain is shown in Fig. 2, which corresponds to 00 UTC of 25th July, 2017. It indicates clearly that the cyclonic flow over the northwest Jharkhand and its neighborhood are related with a depression. The circulation over northwestern region of Jharkhand, northerly flow was observed over Bay of Bengal to the south of 18 N with a strength of 15–20 m $sec^{-1}$ and southerly flow over central and northwest India. The model simulated results show a good agreement with the observation of IMD (15–28 m $sec^{-1}$). At 850 hPa level, the westerly flow observed over Jharkhand, Chhattisgarh, Bihar, Uttar Pradesh, Madhya Pradesh, etc. turning as

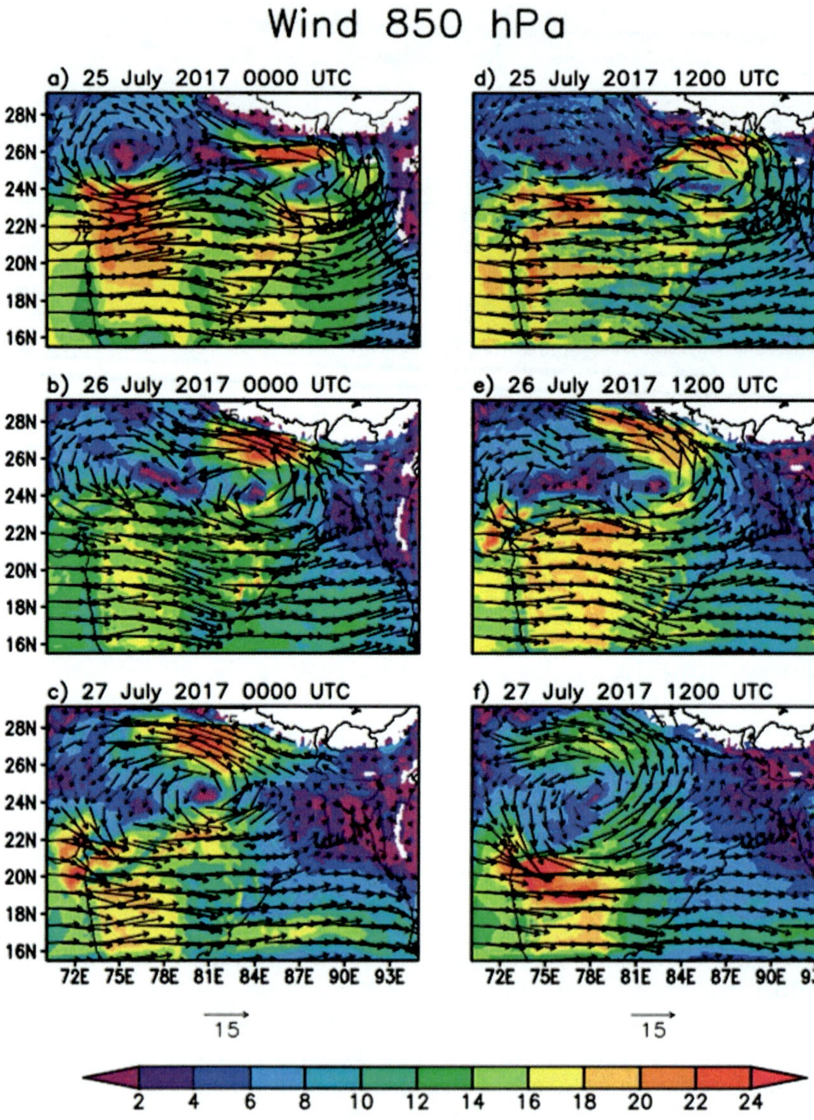


**Fig. 2** Spatial distribution of model simulation wind at 850 hPa during 25 July 2017 to 27 July 2017

north easterly and northerly forming a large anticlockwise circulation covering the northwest Jharkhand and neighborhood. The wind flow at 850 hPa shows the finer structure of a through formation over the Jharkhand and its neighborhood. Cyclonic disturbance is moving towards south east. The simulated wind intensity over the region is in good agreement with the observed daily weather observation of IMD.

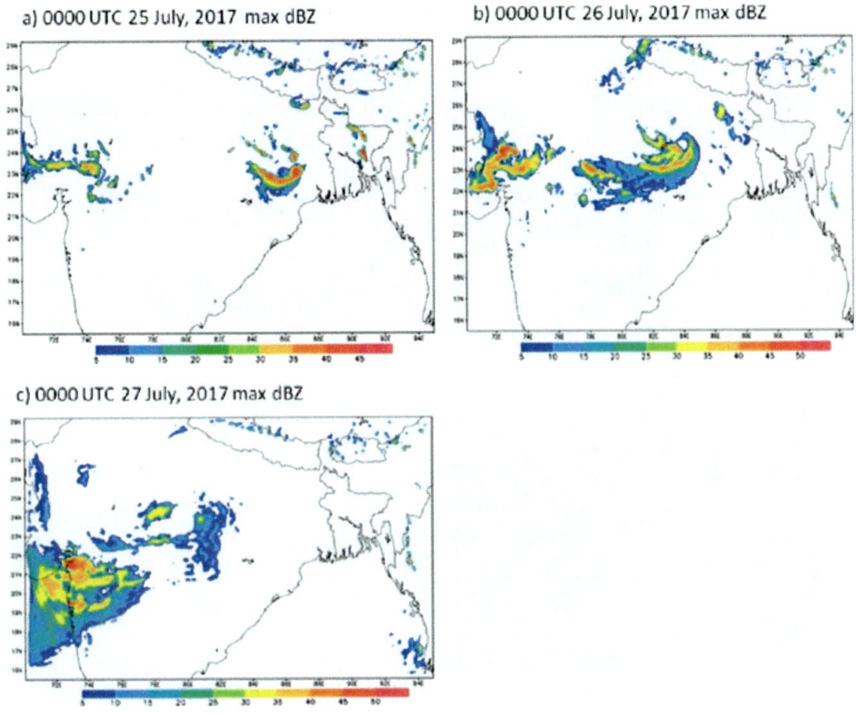


**Fig. 3** Maximum dBZ at 0000 UTC of **a** 25 July 2017, **b** 26 July 2017 and **c** 27 July 2017

The simulation results clearly denote circulations over Bay of Bengal at lower levels associated with moisture and dry flow from north in the middle levels.

Figure 3 shows the accumulated maximum reflectivity (unit: dBZ) measured by the weather radar at Jharkhand and its neighborhood. Max dBZ was found (>50 dBZ) over Jharkhand and its neighborhood. Simulation of cloud band structure depicts the maximum reflectance of clouds on 25th July over Jharkhand and in the gap of every 24 h reflectance varies over the central region of India covering Gujarat and Maharashtra.

The mean sea-level pressure at 00 UTC on 25th July, 2017 on Jharkhand and its neighborhood is depicted in Fig. 4. It was observed that on 25th July 2017, the sea-level pressure was 994 hPa, and after 24 h, there was a very slight increase in sea-level pressure, i.e. 994 hPa on 26th July. Finally, on 27th July, it was reached 999 hPa. The model-simulated sea-level pressure is closely matched with the data available by IMD at 00 UTC on 26th July, 2017. Figure 4a, b, c is simulated in interval of 24 h and sea-level pressure is shown accordingly in color.

For the prediction of depression, water vapor mixing ratio is very useful for tracing the properties of huge air masses in meteorology because they rise or fall in the atmosphere. The water vapor maxing ratio (WVMR) is recorded in various locations of Jharkhand, which is shown in Fig. 5. The maximum water vapor mixing

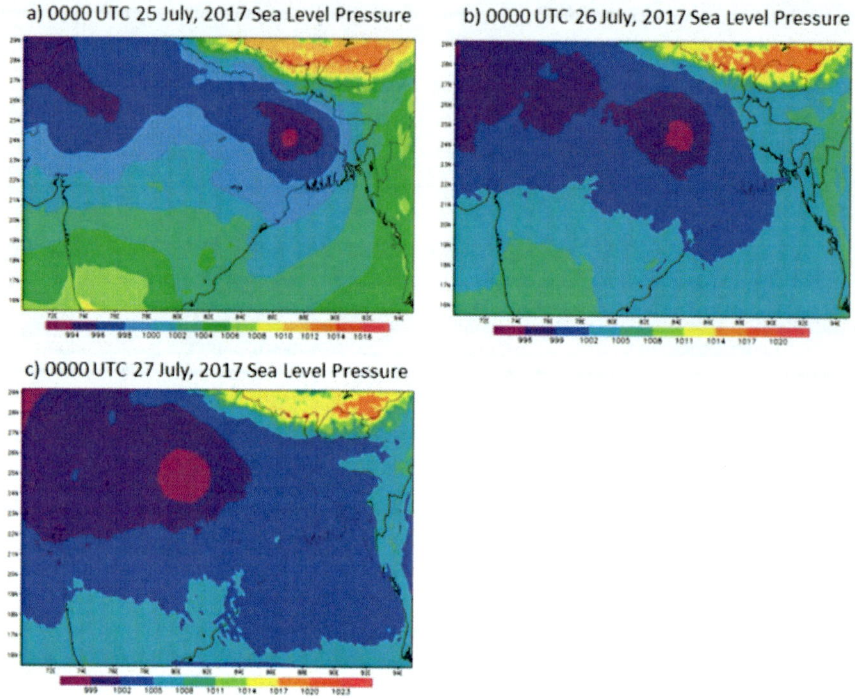


**Fig. 4** Sea-level pressure at 0000 UTC 0000 UTC of **a** 25 July 2017 **b** 26 July 2017 and **c** 27 July 2017

ratio is recorded as 19 g kg$^{-1}$ at an altitude of 950–1000 m at 12 UTC 25th July, 2017. The water vapor maxing ratio (WVMR) is recorded in Fig. 8 from 00 UTC of 25th July to 12 UTC 27th July at 23 N–24 N latitude and 84E-86E longitude. In Fig. 8a, the maximum water vapor mixing ratio is recorded 18 g kg$^{-1}$ at an altitude of 900–1000 hPa between 00 UTC of 25th July and 12 UTC of 27th July, whereas in Fig. 8b, the same amount is recorded at an altitude of 950–1000 hPa at 00 UTC of 26th July. In Fig. 8c, also the maximum water vapor mixing ratio is recorded 18 kg at an altitude of 950–1000 hPa at 00 UTC of 25th July. In Fig. 8e, the maximum water vapor mixing ratio is recorded 18 g kg$^{-1}$ at an altitude of 850–1000 hPa at 00 UTC of 25–26th July. Therefore, Fig. 6 represents the same amount of water vapor present at various altitudes from 25 to 27th July 2017.

The model simulated accumulated rainfall of the event on 25–27th July 2017 over north-west Jharkhand and neighborhood over the south east of India was 60 cm, whereas India Meteorological Department (IMD) rain-gauge recorded 64 mm rainfall. Such heavy rain events happened during the active phase of monsoon all over the north-west Jharkhand and neighborhood. An accumulated model-simulated rainfall in 24 h for 25–27th July is shown in Fig. 10. In Fig. 9, the observed data of IMD are shown and the model-simulated result is well matched with IMD observations.

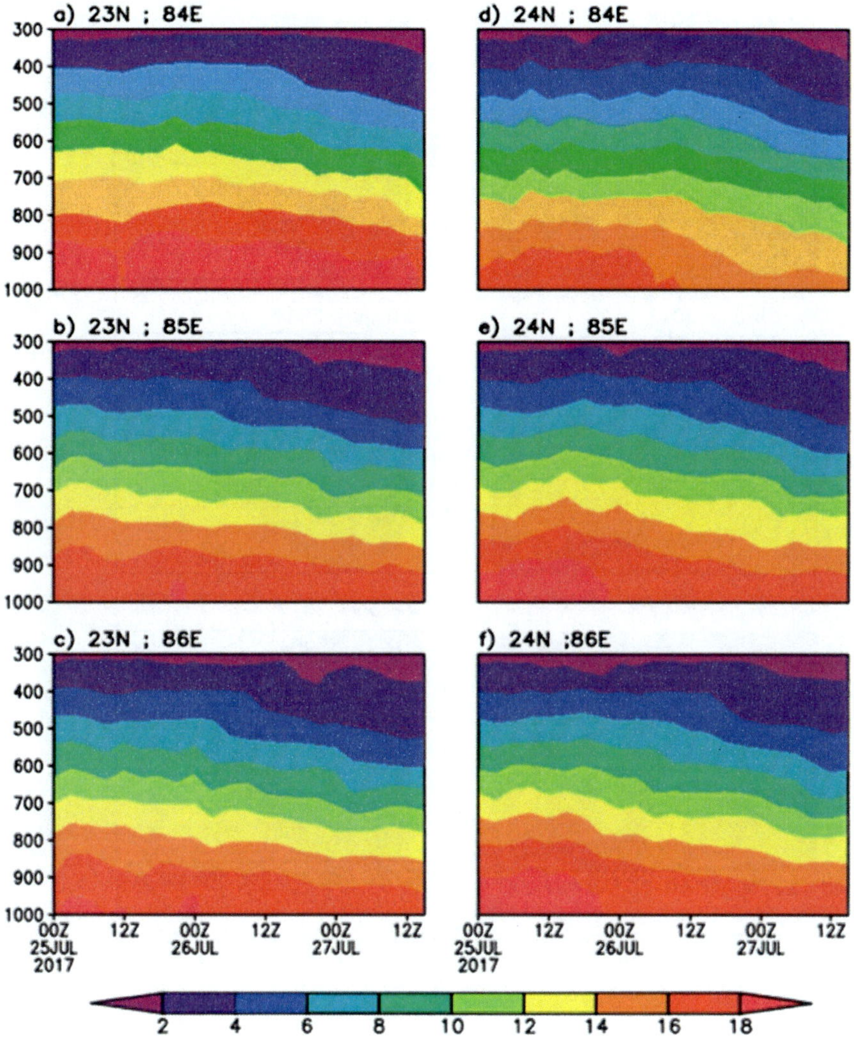


**Fig. 5** Model simulated water vapor mixing ratio at various lat/lon locations

Track of monsoon depressions produced by the WRF (AWR) model has been compared against the observed track of IMD is depicted in Fig. 7. For comparison of results, the WRF model is using Kain-Fritsch cumulus scheme and YSU planetary boundary layer (PBL) scheme considering the initial condition of 00 UTC of 25th July 2017. Integrating hours is taken 72 h for the forecast up to 00 UTC of 27 July 2017. The result reveals that the forecast track was almost in good agreement with the observed track and the model is capable of capturing the movement of monsoon depression for the forecast up to 72 h.

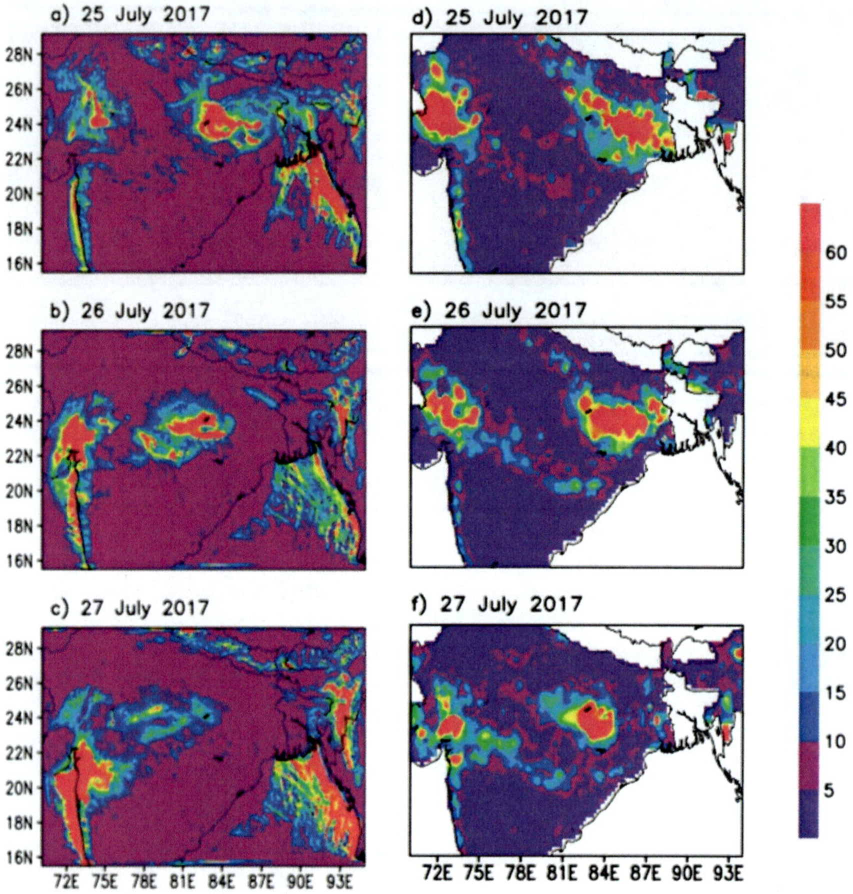


**Fig. 6** Accumulated 24 h rainfall of model simulation **a** 25 July 2017 **b** 26 July 2017 **c** 27 July 2017 and Observed Rainfall by IMD **d** 25 July 2017 **e** 26 July 2017 **f** 27 July 2017

## 5 Conclusions

The objective of this study is the incidence of heavy rainfall and characteristics related to the land depression using high-resolution non-hydrostatic WRF (ARW) mesoscale model. For the real-time forecast and exact prediction of the event, the initial condition plays a very important role and, in this study, we have considered the initial condition as 1° FNL with single domain. Simulation of the track and its intensity of cyclone is a very significant and challenging task for cyclonic disturbance mitigation in the coastal regions.

The observed rainfall recorded was 64 cm on 25th July 2017 over isolated localities of the Jharkhand and its neighborhoods, which shows a good agreement with

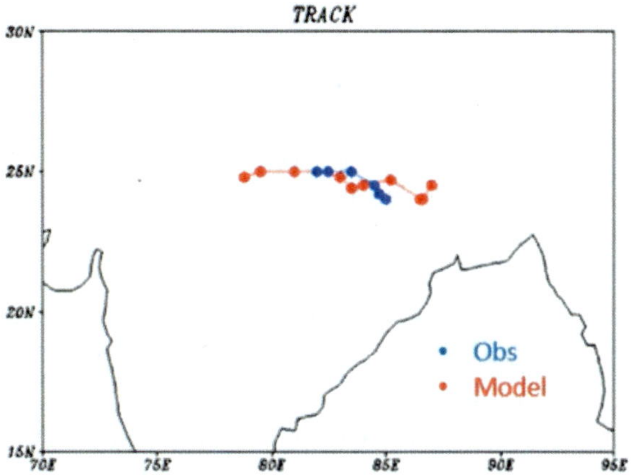


**Fig. 7** Observed and model-simulated track of the low-pressure system

the model-simulated result. The model produced maximum rainfall of 60 cm over Jharkhand and its neighborhoods.

The results indicated that the numerical experiment with the present configuration produced a good agreement of precipitation, in terms of low level wind at 850 hPa, max dBZ, sea-level pressure, water vapor mixing ratio and total accumulated rainfall at Jharkhand and its neighborhoods. It is clear that the numerical simulation of the depression with the WRF (ARW) model is well predicted with the suitable initial condition.

## References

1. Sikka DR, Rao PS (2008) The use and performance of mesoscale models over the Indian region for two high-impact events. Nat Hazards 44:353–372
2. Chang H-I, Niyogi D, Kumar A, Kishtawal CM, Dudhia J, Chen F, Mohanty UC, Shepherd M (2009) Possible relation between land surface feedback and the post-landfall structure of monsoon depressions. Geophys Res Lett 36:L15826
3. Raju PVS, Potty J, Mohanty UC (2012) Prediction of severe tropical cyclones over Bay of Bengal during 2007-2010 using high resolution meso-scale model. Nat Hazard 63(3):1361–1374
4. Sushil K, Routray A, Raju PVS, Panda B (2018) Intensity estimation associated with tropical cyclone 'Megh' by the spatial distribution of wind shear and CAPE. Int J Earth Atmos Sci 5(2):94–100
5. Routray A, Mohanty UC, Rizvi SRH, Niyogi D, Osuri KK, Pradhan D (2010) Impact of doppler weather radar data on numerical forecast of Indian monsoon depressions. Quart J Royal Meteorol Soc, Wiley Online Library
6. Kalra S, Kumar S, Routray A (2019) Simulation of heavy rainfall event along east coast of India using WRF modeling system: impact of 3DVAR data assimilation. Model Earth Syst Environ 5:245–256
7. Pattanaik DR, Kumar A, Rama Rao YV, Mukhopadhaya B (2011) Simulation of monsoon depression over India using high resolution WRF Model- Sensitivity to convective parametrization schemes. Mausam 62:305–320

8. Raju PVS, Mohanty UC, Potty J (2012) Simulations of tropical cyclone Nargis over Bay of Bengal using RIMES operational system. Pure Appl Geophy 169(10):1909–1920
9. Routray A, Kar SC, Mali P, Sowjanya K (2014) Simulation of monsoon depressions using WRF-VAR: impact of different background error statistics and lateral boundary conditions. Mon Weather Rev 142:3586
10. Sateesh M, Srinivas CV, Raju PVS (2017) Numerical simulation of tropical cyclone Thane: role of boundary layer and surface drag parameterization schemes. Nat Hazards 89(3):1255–1271
11. Routray A, Mohanty UC, Osuri KK, Kar SC, Niyogi D (2016) Impact of satellite radiance data on simulations of Bay of Bengal tropical cyclones using the WRF-3DVAR modeling system. IEEE Trans Geosci Remote Sens 4:2285–2303
12. Deb SK, Srivastava TP, Kishtawal CM (2008) The WRF model performance for the simulation of heavy precipitating events over Ahmedabad during August 2006. J Earth Syst Sci 117:589–602
13. Raju PVS, Mohanty UC, Potty J (2011) Sensitivity of Physical Parameterizations on prediction of tropical cyclone Nargis over the Bay of Bengal using WRF model. Meteorol Atmos Phys 113(3–4):125–137
14. Pattanaik S, Mohanty UC, Osuri KK (2012) Impact of parameterization of physical processes on simulation of track and intensity of tropical cyclone nargis (2008) with WRF-NMMModel. Sci World J 2012(671437)
15. Hong SY, Noh Y, Dudhia J (2006) A new vertical diffusion package with explicit treatment of entrainment processes. Mon Weather Rev 134:2318–2341
16. Kain JS, Fritsch JM (1993) Convective parameterization for mesoscale models: the Kain-Fritcsh scheme. In: Emanuel KA, Raymond DJ (eds) The representation of cumulus convection in numerical models. Am Meteorol Soc 246
17. Kain JS (2004) The Kain-Fritsch convective parameterization: an update. J Appl Meteorol 43:170–181
18. Mlawer EJ, Taubman SJ, Brown PD, Iacono MJ, Clough SA (1997) Radiative transfer for inhomogeneous atmosphere: RRTM, a validated correlated-k model for the longwave. J Geophys Res 102:16663–16682
19. Dudhia J (1989) Numerical study of convection observed during winter monsoon experiment using a mesoscale two-dimensional model. J Atmos Sci 46:3077–3107

# Chapter 12
# Parameter Estimation of VIC-RAPID Hydrological Model Using Self-adaptive Differential Evolution Algorithm

**Saswata Nandi and Manne Janga Reddy**

## 1 Introduction

Hydrological models are one of the most important tools in water resource management and hydrological studies for the representation of the hydrological processes of a catchment in mathematical form. They are used in various hydrological studies like flood forecasting, drought assessment, land-use change analysis, reservoir operation, water budget estimation, and many other purposes [6, 10]. With the advent of computational science, hydrological models have evolved from simple lumped variants to complex physics-based fully distributed hydrological model. However, with the increasing complexity, the parameters of hydrological models, which help in the implementation of physical laws within the model, have also significantly increased in their number in all hydrological models. In general, few parameters of these models cannot be precisely determined from the fields and must be estimated indirectly. This leads to the application of certain parameter estimation programs or calibration techniques in order to find suitable values for these unknown model parameters and deliver a reliable model performance. Earlier hydrological modelling studies have generally employed manual calibration techniques where model parameters are adjusted using some trial-and-error method for obtaining a satisfactory match between the model simulation and the observation data [2]. Although such manual calibration techniques are viable for estimating few parameters in simple hydrological models, their use is

S. Nandi (✉) · M. J. Reddy
Department of Civil Engineering, Indian Institute of Technology Bombay, Mumbai, India
e-mail: ce.saswata@gmail.com

M. J. Reddy
e-mail: mjreddy@civil.iitb.ac.in

M. J. Reddy
Interdisciplinary Programme in Climate Studies, Indian Institute of Technology Bombay, Mumbai, India



D. Singh et al. (eds.), *Proceedings of International Conference on Scientific and Natural Computing*, Algorithms for Intelligent Systems,
https://doi.org/10.1007/978-981-16-1528-3_12

impractical for calibration of the model with a large number of parameters, which is the case in current physics-based complex hydrological models.

In order to address such problems in the manual calibration, automatic hydrological model calibration procedures are implemented by recent studies. The population-based global optimization techniques have gained the most consideration for their use in the automatic calibration of hydrological models. The genetic algorithm (GA), particle swarm optimization (PSO), and shuffled complex evolution (SCE-UA) algorithms are some of the most commonly used population-based evolutionary algorithms that are used in hydrological modelling applications [2, 3, 8]. The differential evaluation (DE) [11] is another type of population-based evolutionary algorithm commonly used in the planning, design, and operation of water resource systems [7]. Despite the simplicity, efficacy, and various merits of DE, it has limited applications with hydrological modelling studies. However, similar to other evolutionary algorithms, the skilful performance of the DE greatly depends on the appropriate settings of its control parameters, namely, mutation factor (F) and crossover rate (CR). Fine-tuning these DE control parameters in case of its use in hydrological model calibration may create a substantial computational challenge. This study presents the application of self-adaptive differential evolution (SaDE) algorithm to handle such difficulty, which dynamically adapts the optimal values of these two control parameters of the traditional DE algorithm. The main purpose of this study is to demonstrate the application of the SaDE for the automatic calibration of the VIC-RAPID hydrological model for streamflow simulation. The proposed approach is demonstrated by estimating the 15 unknown VIC-RAPID model parameters for hourly streamflow estimation in the upper Krishna river sub-basin, India.

The remaining of this paper is arranged as follows: Sect. 2 presents the necessary steps in SaDE algorithm, a brief description of the VIC-RAPID hydrological model, selection of the calibration parameters, and chosen statistical performance measures for verification of model performance. Section 3 gives details about the study area and dataset used in the current study. Finally, the results and conclusions of the study are presented in Sect. 4 and Sect. 5, respectively.

## 2 Methodology

### *2.1 SaDE Algorithm*

The Differential Evolution (DE) algorithm proposed by Storn and Price (1996) has provided a faster and reliable solution for several optimization problems in engineering. Furthermore, the self-adaptive version of DE (called SaDE) offers more flexibility and avoids manual tuning of the parameters of the algorithm [6, 9]. The pseudo-code of SaDE-based procedure for calibration of the VIC-RAPID model is presented in Fig. 1.

**Pseudo-code for SaDE algorithm**

```
define problem dimension, and generate an initial population
set F(μF=0.5, σF=0.3) and CR (μCR=0.5, σCR=0.1) distribution statistics
get initial values of F and CR for each population from a normal distribution
calculate the fitness of each initial population
initialize generation counter g←0
while termination criterion not satisfied do
    for each population in current generation do
        select three solutions at random.
        generate mutation vector using Eq. (1)
        generate crossover vector using Eq. (2)
        calculate fitness of trial vectors
        if trial vector is more fit than parent then
            the parent is replaced with trial vector
        end if
    end for
    if g is divisible by 5 then
        update μF and μCR values to generate new F and CR values
    end if
    g←g+1
end while
```

**Fig. 1** Pseudo-code for SaDE algorithm

The algorithm shown in the above pseudo-code obtains the fitness of the population by running the VIC-RAPID model with the given population and computing the corresponding Nash–Sutcliffe Efficiency (NSE) as the objective function value. For generating offspring from parent population, mutation and crossover operation are performed. The mutation for the individual population is performed using the below relationship:

$$V_{i,G} = X_{r_1,G} + F_i(X_{r_2,G} - X_{r_3,G}) \tag{1}$$

where $V_{i,G}= [v_{1,i,G}, v_{2,i,G}, \ldots, v_{D,i,G}]$ is a mutant vector associated with population $i$ and generation $G$; $X_{r1,G}$, $X_{r2,G}$, and $X_{r3,G}$ are three randomly chosen population vectors from present generation; and $F_i$ is the mutation factor for population vector $i$.

The crossover operation is used to generate a trial vector by selecting certain features from the original vector and rest from the mutant vector. Crossover in SaDE is accomplished using the equation:

$$u_{j,i,G} = \begin{cases} v_{j,i,G}, & if\ (rand_j[0,1] \leq CR\ or\ (j = j_{rand})) \\ x_{j,i,G}, & otherwise \end{cases} \tag{2}$$

where $u_{j,i,G}$ is the *jth* decision variable for *ith* member at generation G; $U_{i,G} = [u_{1,i,G}, u_{2,i,G}, \ldots, u_{D,i,G}]$ is *ith* trial vector at generation *G*; $rand_j$ is a random value between 0 to 1 for each dimension of the vector; and $j_{rand}$ is a random number from 1 to *D*, the dimension of each vector.

The mean of mutation and crossover factor is updated to the mean of the corresponding factors leading to successful trial vectors after every five iterations. A new set of mutation and crossover factors are generated from a normal distribution using these new mean parameter values.

## 2.2 VIC-RAPID Model

Variable Infiltration Capacity (VIC) is a physically based, macroscale hydrologic model developed by Liang et al. [4]. This model divides the modelling domains into uniform grids and solves the water balance and/or surface energy budgets for each grid. Moreover, for better representation of the spatial heterogeneity, VIC can further divide each model grid into many vegetation tiles and then computes evapotranspiration, infiltration, runoff, and baseflow separately for each tile. This study employs a three-layer VIC model. The top two soil layers are responsible for generating the surface runoff and the bottom layer for the generation of baseflow. Evapotranspiration is computed by the Penman-Monteith method in the VIC model. The water balance in each VIC model grid is calculated as:

$$\frac{ds}{dt} = P - E - R \tag{3}$$

where dS/dt, P, E, and R are the change of soil water, precipitation, evapotranspiration, and runoff, respectively.

As the VIC model generates the runoff (both surface and sub-surface) in each model grid, a separate routing model is needed to transfer the generated runoff from the individual model grid to the basin's outlet. To achieve this purpose, the routing application for parallel computation of discharge (RAPID) model is utilized by this study. RAPID model route the river reaches flow through the river network into the basin outlet by solving a matrix-based Muskingum equation [1]. The input to the RAPID is generated surface and sub-surface runoff time series from the VIC model.

Total 15 parameters of the VIC-RAPID model are considered for the SaDE-based model calibration, which are (i) the infiltration parameter ($b_{inf}$) that regulates the separation of rainfall into direct runoff and infiltration, (ii) $W_s$ and $D_s$ are the parameters for controlling the baseflow (i.e., how fast the stored water in the third soil layer is departed as baseflow), (iii) $D_m$ represents the maximum velocity of baseflow, (iv) $Ex_1$, $Ex_2$, and $Ex_3$ are the exponents of the Brooks–Corey relationship for three soil layers and controls the vertical drainage between soil layers, (v) $K_{S1}$, $K_{S2}$, and$K_{S3}$ represent the saturated hydraulic conductivity of the three layers and has large impact on the runoff generation, (vi) $d_1$, $d_2$, and $d_3$ are the thickness of first, second, and third soil layer and have an effect on the available water for direct runoff, transpiration, and baseflow, respectively, (vii) $\lambda_k$ is a multiplier for Muskingum parameter that represents travel time of the flood wave in the river reach, and (viii) $x$ is another Muskingum parameters controlling the peak flow attenuation.

### 2.3 *Performance Measures*

Three statistical performance measures, namely, Nash–Sutcliffe efficiency (NSE), coefficient of determination ($R^2$), and percentage bias (PBIAS), are considered for evaluation of SaDE-based calibration of the VIC-RAPID model for hourly streamflow simulation. The equations used for these performance measures are given below:

$$NSE = 1 - \left[\frac{\sum_{i=1}^{n}(O_i - S_i.)^2}{\sum_{i=1}^{n}\left(O_i - \overline{O_i}\right)^2}\right] \tag{4}$$

$$R^2 = \frac{\left(n\sum_{i=1}^{n} O_i.S_i - \left(\sum_{i=1}^{n} O_i\right)\left(\sum_{i=1}^{n} S_i\right)\right)^2}{\left(n\sum\limits_{i=1}^{n}(O_i)^2 - \left(\sum\limits_{i=1}^{n} O_i\right)^2\right).\left(n\sum\limits_{i=1}^{n}(S_i)^2 - \left(\sum\limits_{i=1}^{n} S_i\right)^2\right)} \tag{5}$$

$$PBIAS = \left[\frac{\sum_{i=1}^{n}(O_i - S_i) \times 100}{\sum_{i=1}^{n} O_i}\right] \tag{6}$$

where $O_i$ and $S_i$ are observed and simulated flow values for the ith time step; $\overline{O}_i$ designate the means of observed flow values, respectively; n represents the total number of time steps.

## 3 Case Study

Brief details of the study area and data used are described below.

### 3.1 *Study Area*

The selected study area for this study is located in the upper Krishna river basin, Maharashtra, India and its location map is shown in Fig. 2. The study area falls in between latitude 16°1' N to 18°1' N and longitude77°3' E to 74°9' E and having an area of 15186 $Km^2$. The average annual temperature and rainfall are 25.8 °C and 947 mm, respectively. The average elevation is 590 meters from mean sea level. The

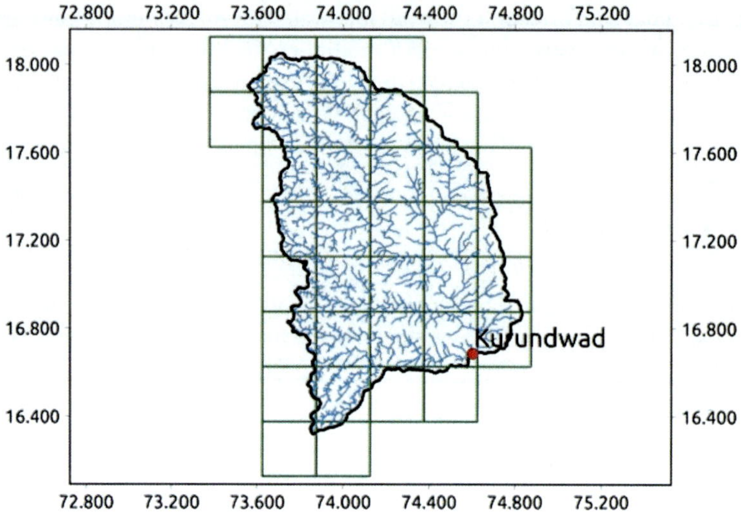


**Fig. 2** Location map of the study area

green boxes and blue lines in Fig. 2 represent the VIC model grids and RAPID river network, respectively. The outlet of the study area is located in the Kurundwad region and is shown in Fig. 2 with a red circle.

### *3.2 Data Used*

Hourly meteorological forcing (e.g., rainfall, temperature, wind speed, atmospheric pressure, incoming short and long-wave radiation, and vapour pressure) is collected from the IMDAA Regional Reanalysis product store at 12 km spatial resolution. SRTM (90 m) DEM is collected from National Remote Sensing Centre of India. The observed hourly streamflow at the basin outlet (Kurundwad) is obtained from the central water commission (CWC), India. Soil and land cover data are taken from the Food and Agricultural Organization's (FAO) harmonized world soil database and MODIS land cover database, respectively. All the river network information is collected from HydroRIVERS database of HydroSHEDS product.

## 4 Results

This study performs all the hydrological simulations at an hourly time scale for the period of 2000–2011. The SaDE-based calibration of the VIC-RAPID is employed for 7 years (2000–2006) and the remaining 5 years (2007–2011) is used for validation of the proposed methodology. Maximization of the NSE is considered for the estimation of the unknown model parameters in the SaDE algorithm. All the computations are performed using the Linux-based HPC cluster facility at the Department of Civil Engineering, IIT Bombay, which has 20 computing nodes with each holding 24Intel(R) Xeon(R) CPU E5-2650 v4 @ 2.20 GHz processors and 64 GBs of RAM.

Figure 3 shows the relationship between the average best objective function values (NSE) and the associated number of the model evaluation required to achieve those fitness values over 10 optimization trials. For the current model calibration, an NSE of 0.44 is obtained as the initial best value of the objective function value. It can be clearly seen from Fig. 3 that the calibration procedure is converged after attaining an objective function value of 0.80. A total of 1835 model evaluations are required to achieve this convergence with a population size of 72. A very low standard deviation of 0.07 is found as the objective function values of all populations at the end of the SaDE-based model calibration, which indicates a convergence of the whole population by the proposed calibration technique. Figure 3 also highlights the fact that a significant percentage of the overall improvement in the objective function value is achieved within 600 model evaluations. Such quick convergence can be a deciding factor for the selection of a calibration technique for complex hydrological models that require heavy computational demands.

The final estimated values of the 15 VIC-RAPID parameters using SaDE are presented in Table 1. The statistics of these 15-model parameters from the final generation of optimization is demonstrated in Fig. 4. The larger dispersion in the boxplots represents higher parameter equifinality (i.e., different parameter combinations delivering similar model performance) in the model parameters. From Fig. 4, it can be seen that the dispersions of boxplots for $D_S$, $D_m$, $W_S$, $d_2$, and $\lambda_k$ are very small, indicating their better identifiability using a model calibration approach. However, a large dispersion in boxplots for the other model parameters signified parameter equifinality in the current calibration problem.

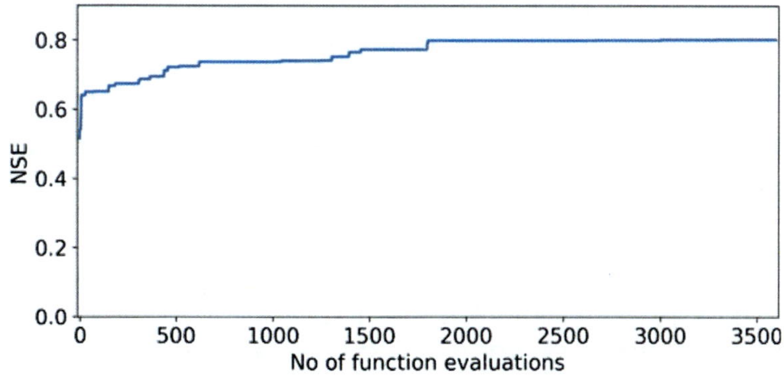


**Fig. 3** Convergence characteristics of SaDE-based hourly calibration of VIC-RAPID model during 2000–2006

**Table 1** Details of the optimal VIC-RAPID model parameters obtained from SaDE-based model calibration

| Parameter | Calibrated Value | Parameter | Calibrated Value |
|---|---|---|---|
| $b_{inf}$ | 0.3 | $K_{S1}$ | 6524.7 |
| $D_S$ | 0.0001 | $K_{S2}$ | 3500.9 |
| $D_m$ | 30 | $K_{S3}$ | 6612.5 |
| $W_S$ | 0.979 | $d_1$ | 0.49 |
| $Ex_1$ | 20.67 | $d_2$ | 0.51 |
| $Ex_2$ | 25.49 | $d_3$ | 0.75 |
| $Ex_3$ | 19.01 | $\lambda_k$ | 0.376 |
| x | 0.11 | | |

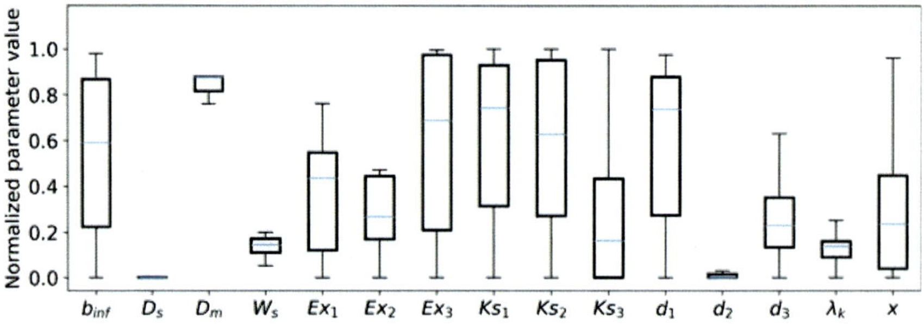


**Fig. 4** Box plot of normalized parameter values from the last iteration of the SaDE-based hourly calibration of the VIC-RAPID model

Comparison of simulated hourly flow from the VIC-RAPID using the optimal parameter values from the SaDE is made with hourly observed flow values, and the result is shown in Fig. 5. It is clearly noticeable from this figure that simulated streamflows closely match the pattern of the observed flow values for both calibration and

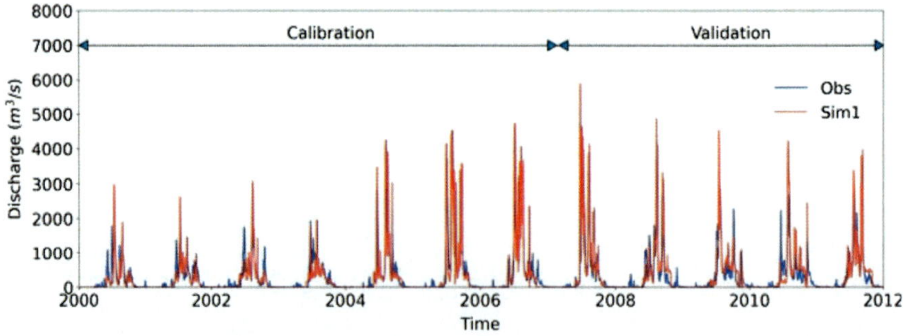


**Fig. 5** Comparison of VIC-RAPID simulated hourly streamflow with the observed flow for calibration and validation period

**Table 2** The result performance measures obtained for VIC-RAPID model calibration using SaDE

| | NSE | $R^2$ | PBIAS |
|---|---|---|---|
| Calibration | 0.80 | 0.89 | – 7.86 |
| Validation | 0.78 | 0.88 | 0.70 |

validation periods. The peak values of the observed data are also well captured in the VIC-RAPID streamflow simulation. To further evaluate the streamflow simulation, NSE, $R^2$, and PBIAS are computed for both calibration and validation period, and results are presented in Table 2.

Moriasi et al. [5] advocated a model performance can be labelled as satisfactory if NSE > 0.65, $R^2 > 0.5$, and PBIAS < ± 25%. All three considered performance measures satisfy such criteria during both calibration and validation periods, which is seen from Table 3 and further emphasizes the efficacy of SaDE-based calibration of the VIC-RAPID model in reliable streamflow simulation.

## 5 Conclusions

Parameter estimation of the hydrological models is one of the most important problems in the field of water resources planning and management. Past studies have shown that evolutionary algorithms can deliver the optimal parameter sets for hydrological modelling applications but may raise computational challenges in the case of physically based distributed hydrological models. Self-adaptive differential evolution (SaDE) is an evolutionary algorithm that incorporates a self-adaptation strategy to evolve the mutations and crossover factor of the traditional DE algorithm for improved accuracy and faster convergence. This study presented an automatic calibration scheme for the VIC-RAPID hydrological model using the SaDE algorithm. Results from the study showed that the SaDE methodology performed well, considering various statistical performance measures. The proposed approach is found to deliver a satisfactory calibration performance even with a small number of model evaluations. Furthermore, this study performed an identifiability analysis to identify complexity in the estimation of the model parameters and to check the presence of equifinality in the current model configuration. Overall, results from the study encourage further studies using SaDE for the automatic calibration of complex hydrological models.

## References

1. David CH, Maidment DR, Niu GY, Yang ZL, Habets F, Eijkhout V (2011) River network routing on the NHDPlus dataset. J Hydrometeorol 12(5):913–934
2. Duan Q, Sorooshian S, Gupta VK (1994) Optimal use of the SCE-UA global optimization method for calibrating watershed models. J Hydrol 158(3–4):265–284

3. Kamali B, Mousavi SJ, Abbaspour KC (2013) Automatic calibration of HEC-HMS using single-objective and multi-objective PSO algorithms. Hydrol Process 27(26):4028–4042
4. Liang X, Lettenmaier DP, Wood EF, Burges SJ (1994) A simple hydrologically ased model of land surface water and energy fluxes for general circulation models. J Geophy Res Atmos 99(D7):14415–14428
5. Moriasi DN, Arnold JG, Van Liew MW, Bingner RL, Harmel RD, Veith TL (2007) Model evaluation guidelines for systematic quantification of accuracy in watershed simulations. Trans ASABE 50(3):885–900
6. Nandi S, Reddy MJ (2020) Spatiotemporal analysis of water balance components and their projected changes in near-future under climate change over Sina Basin, India. Water Resour Manage 34:1–19
7. Reddy MJ, Kumar DN (2020) Evolutionary algorithms, swarm intelligence methods, and their applications in water resources engineering: a state-of-the-art review. H2 Open J 3(1):135–188
8. Shafii M, Smedt FD (2009) Multi-objective calibration of a distributed hydrological model (WetSpa) using a genetic algorithm. Hydrol Earth Syst Sci 13(11):2137–2149
9. Sirsant S, Reddy MJ (2018) Reliability-based design of water distribution networks using self-adaptive differential evolution algorithm. ISH J Hydraulic Eng 24(2):198–212
10. Sood A, Smakhtin V (2015) Global hydrological models: a review. Hydrol Sci J 60(4):549–565
11. Storn R, Price K (1997) Differential evolution—a simple and efficient heuristic for global optimization over continuous spaces. J Global Optim 11:341–359

# Chapter 13
# Numerical Simulation of Landfall Position and Intensity of Very Severe Cyclonic Storm 'Vardah' Over Bay of Bengal Using ARW Model

**Pushpendra Johari, Sushil Kumar, A. Routray, and Indu Jain**

## 1 Introduction

Tropical cyclones are a natural phenomenon that has caused disasters across the globe. These are also called Hurricanes and Typhoons in other parts of the world. Tropical cyclones are accompanied by strong winds, rainfall and storm surge that cause immense damage to life and property [14]. Of late, it has been noted that the frequency and severity of tropical cyclones are increasing. Along the Indian coast, the frequency is much higher on the eastern coast (Bay of Bengal – BoB) than on the western coast (Arabian Sea) about an average ratio of 4:1 [7]. About 7% of the world's cyclones happen in BoB.

Analysis of cyclonic activity in BoB has been the focus of researchers. Several statistical and synoptic methods have been applied for modelling cyclone behaviour in the North Indian Ocean (NIO) but the prediction of cyclone tracks and intensities have shown limitations [5, 13]. Out of these, the Advanced Research WRF (ARW) model has been giving encouraging results for forecasting weather phenomenon in NIO [1, 11, 12, 16–20, 24, 25]. The model is also being applied extensively world-wide for this purpose [18, 22, 23]. Though the forecasts in NIO now are much improved [15–17] but it is believed that further improvement is required in the modelled estimates of track and intensity from cyclone genesis to dissipation.

Most of the research with ARW has been focused on Odisha and West Bengal which are more to the north. The present study focuses on TC Vardah that made

P. Johari (✉) · S. Kumar
Department of Applied Mathematics, Gautam Buddha University, Greater Noida 201312, India
e-mail: Ppjohari@gmail.com

P. Johari · I. Jain
RMSI Pvt. Ltd, A-8, Sector 16, Noida 201301, India

A. Routray
NCMRWF, A-50, Sector 62, Noida 201309, India

 
D. Singh et al. (eds.), *Proceedings of International Conference on Scientific and Natural Computing*, Algorithms for Intelligent Systems,
https://doi.org/10.1007/978-981-16-1528-3_13

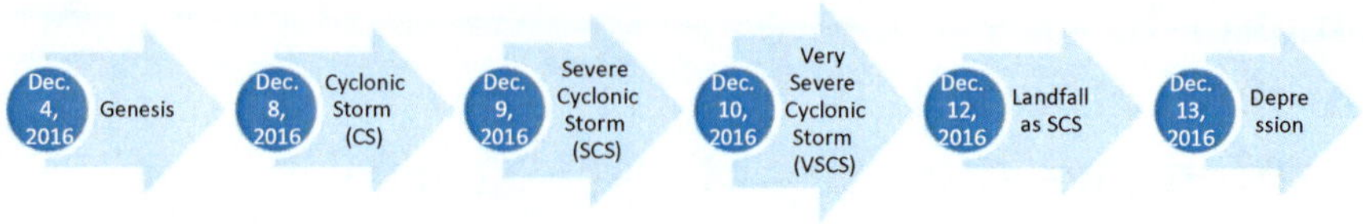


**Fig. 1** Life cycle of cyclone Vardah

**Table 1** Key parameters of cyclone Vardah

| Genesis | Landfall time (UTC) | Max CSLP | Max sustained wind speed |
|---|---|---|---|
| Over the south Andaman Sea and adjoining Sumatra | Crossed north Tamil Nadu coast near Chennai during 0950-1150 h UTC of **12 December 2016** | 975 hPa | 130 kmph |

landfall close to Chennai in the state of Tamil Nadu in December 2016 and attempts to simulate the various dynamic and physical processes associated with the storm by applying the ARW model.

## 2 Description of the Very Severe Cyclonic Storm—'Vardah'

Very Severe Cyclonic Strom, Vardah, was formed in the BoB on 4 December 2016 and dissipated as a Depression on 13 December 2016. Figure 1 shows the life cycle of the storm from genesis to dissipation.

The key parameters of the cyclone are specified in Table 1. The cyclone resulted in heavy crop loss in districts of Tiruvallur, Kanchipuram, Vellore and Tiruvannamalai in Tamil Nadu and caused about 24 deaths [7].

## 3 Brief Description of the ARW Model and Its Configuration

The National Center for Atmospheric Research (NCAR) and other Universities together have developed a mesoscale model called The Advanced Research Weather Research and Forecast (ARW) [28]. This study uses the ARW v3.8.1. The model implements non-hydrostatic, fully compressible equations, and a set of analytical variables including geo-potential, the wind in three dimensions, perturbation measures such as pressure, surface pressure, &turbulent kinetic energy. The model also includes prognostic scalars such as cloud water, water vapour mixing ratio etc. The model dynamics have the terrain following hydrostatic-pressure is the vertical coordinate, Arakawa C-grid is the horizontal coordinate, third-order Runge–Kutta scheme for time integration. Several options for specifying boundary requirements (starting conditions, Lateral, Top, Bottom) are available in the model. The model also supports options for nesting, spatial discretisation, and a various number of physical schemes.

To model the regional-scale circulations of cyclone Vardah, which are influenced by large scale flow, the regional mesoscale analysis system WRF (ARW) is implemented. ARW has the ability to capture the meaningful cyclonic characteristics from lesser resolution inputs provided by the GFS model [2, 27] as starting conditions and at various boundaries. The LAMs could generate a good estimation of regional climate [4, 9]. In modelling weather, regional-scale circulations are influenced by large scale flows, so the model domain was set from (96.45° E, 22.15° N) over Myanmar in the North East to (96.45 ° E, 5.37 ° N) further away from Andaman and Nicobar Islands in the South East and (73.26 ° E, 22.15 ° N) over Gujarat in West to (73.26 ° E, 5.37 ° N) further away from Lakshadweep Islands in the Arabian Sea in the South West with a horizontal resolution of 9 km giving 280 X 210 grids cells covering the BoB as depicted in Fig. 2.

In the present study, the model physics employed includes YSU scheme [6] for Planet Boundary Layer, Kain-Fritsch (new Eta) scheme [8] for cumulus parameterisation, Ferrier (new Eta) scheme for modelling cloud micro-physics, radiation through Rapid Radiative Transfer Model (RRTM) [10] for longwave and Dudhia

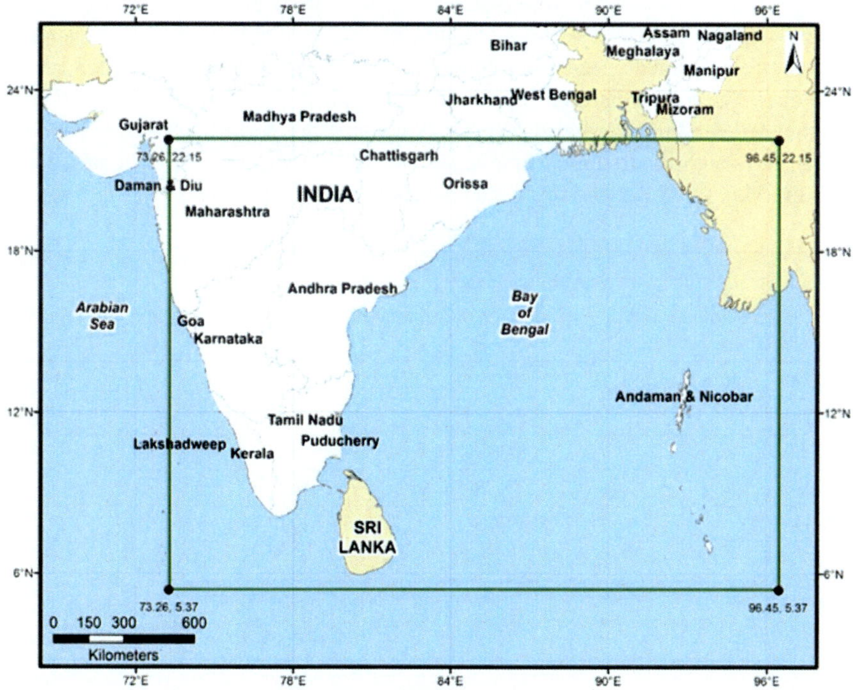


**Fig. 2** WRF model domain for Vardah cyclone study

Scheme [3] for shortwave and Unified Noah Land Surface Model for Land Surface modelling.

## 4 Numerical Experiments

The study involved carrying out five different sets of numerical simulations with the ARW model. For all the numerical simulations, the WRF model is initialised using 3-hourly forecasts from the National Centers for Environmental Prediction (NCEP) Global Forecast System (GFS) analysis data at 0.5° × 0.5° (latitude–longitude) grid. 4 analysis cycles are provided by GFS—00, 06, 12, 18 UTC, with forecasts for each analysis cycle being available up to 16 days. The United States Geological Survey (USGS) Global datasets [26] are used to build topography, land use and vegetation at 30 s horizontal resolution. These datasets are available for download at http://nomads.ncep.noaa.gov/pub/data/nccf/com/gfs/prod/.

All experiment simulations are made starting at 0000 UTC on the simulation start date using the initial conditions on that date. The observed storm track position and intensity are taken from the IMD report [7] for validating the simulation results. Table 2 shows the particulars of all the simulations carried out as part of this study.

The model outputs facilitate in investigating the Average Track Forecast Error, Average Along Track Error from different model initialised time (8–13 December 2016; total 5 runs). At every 30-minute interval, model outputs are saved to examine the event translational speed from genesis on 8 December to its dissipation on 13 December. The spatial distribution of wind at a height of 10 m, 850 hPa and 200 hPa, accumulated rainfall, were also investigated to test the performance of the model over BoB. In addition to these, some other parameters namely Vorticity, Latent Heat Flux and Maximum Reflectivity were also analysed.

**Table 2** Details of the experiments conducted

| Simulations start date and time in UTC (dd/hhhh) | Stage of Strom at IC | Forecast length valid at 13/0000 UTC |
|---|---|---|
| 08/0000 | CS | 120 h |
| 09/0000 | CS | 96 h |
| 10/0000 | SCS | 72 h |
| 11/0000 | VSCS | 48 h |
| 12/0000 | VSCS | 24 h |

# 5 Results and Discussion

The results from all the experiments more or less followed a similar trend, so the paper primarily discusses the results from the first and fourth experiment. The following sections exhibit a useful discussion of the results of the study.

## 5.1 *Simulation of the Cyclone Track*

Figure 3 shows the Vardah cyclone track simulated by the model for the five advisories issued by IMD from 8 to 13 December. The modelled tracks have been plotted against the observed cyclone track as published by IMD. The figure clearly shows that the model efficiency improves with time. The major shift in the observed track happened on 10 December. The model could not simulate this shift in simulated tracks for initial conditions of 8, 9 and 10 December 2016. These tracks closely followed the observed track for the initial 24 h but then deviated. For initial conditions 11 and 12 December, model simulations were very close to the observed track. This may be because of the use of the Kain-Fritsch scheme which more accurately simulates depth and intensity than the trajectory and landfall location [21].

Table 3 shows the error in the forecast of the landfall position as simulated by the ARW model for VSCS Vardah. IMD reported landfall at 1030 IST hrs of 12

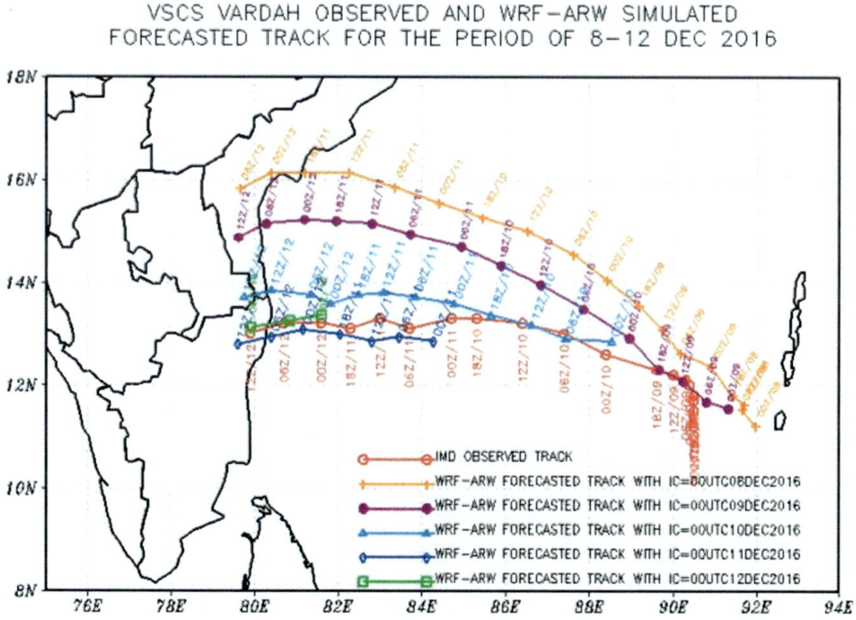


**Fig. 3** Observed and ARW simulated forecasted tracks for the period of 8 to 12 December 2016

**Table 3** Comparison of Observed and ARW simulated landfall location for the period 8–12 December 2016

| Initial conditions | Landfall Point (°N/°E) | | Landfall time (hours) | | Operational error | | ARW forecast in UTC |
|---|---|---|---|---|---|---|---|
| | ARW forecast | Actual IMD | ARW forecast | Actual IMD | LPE * (km) | LTE * (hrs) | |
| IC = 08122016 | 16.1232/81.2319 | 13.1/80.3 | 11/2330 | 12/1030 | 348 | -11 | 1800/11122016 |
| IC = 09122016 | 14.9798/79.9308 | 13.1/80.3 | 12/1430 | 12/1030 | 211 | +4 | 0900/12122016 |
| IC = 10122016 | 13.668/80.1402 | 13.1/80.3 | 12/2030 | 12/1030 | 65 | +10 | 1500/12122016 |
| IC = 11122016 | 12.9405/80.4131 | 13.1/80.3 | 12/1130 | 12/1030 | 21.5 | +1 | 0600/12122016 |
| IC = 12122016 | 13.2133/80.4677 | 13.1/80.3 | 12/1430 | 12/1030 | 22.6 | +4 | 0900/12122016 |

December 2016 at location 13.1°N/80.3°E. The modelled landfall location based on the advisories 11th and 12 December 2016 shows the least landfall position error of about 22 km away from the reported landfall location with landfall time errors of + 1 h and +4 h, respectively, which is a very good forecast.

* LPE stands for Landfall Point Error and LTE stands for Landfall Time Error. A '+' value indicates delayed landfall, and '–' indicates early landfall

Positive (negative) values of Along Track Error (ATE) indicate that the forecasted TC position lies ahead (behind) as compared to the observations whereas positive (negative) values of Cross Track Error (CTE) indicate that the forecasted track is to the right (left) of the observed track.

Figures 4–6 show the average, along and cross-track forecast error, respectively. The average CTE has been positive for all simulation durations, except for the simulation with 11 December conditions where the CTE is slightly negative. This implies that the average ARW forecast track lies to the right of the observed track for nearly all the simulations. This may be interpreted as a bias shown by the ARW model to

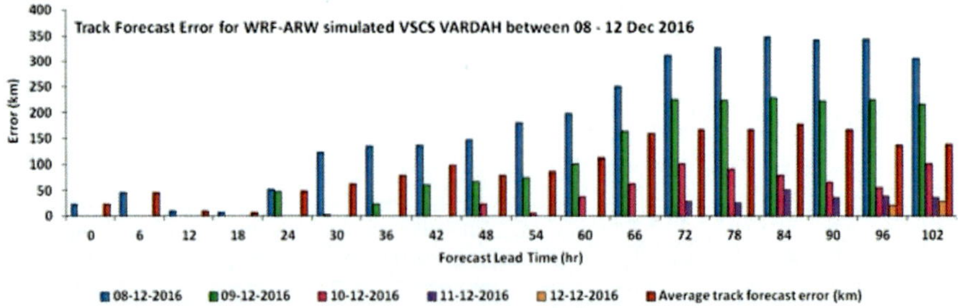


**Fig. 4** Track forecast error

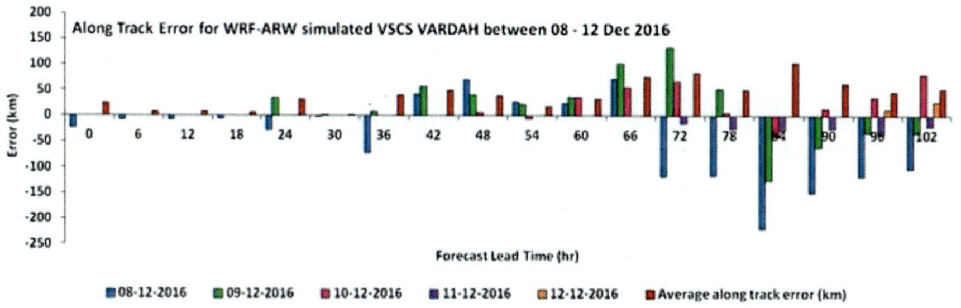


Fig. 5 Along track error

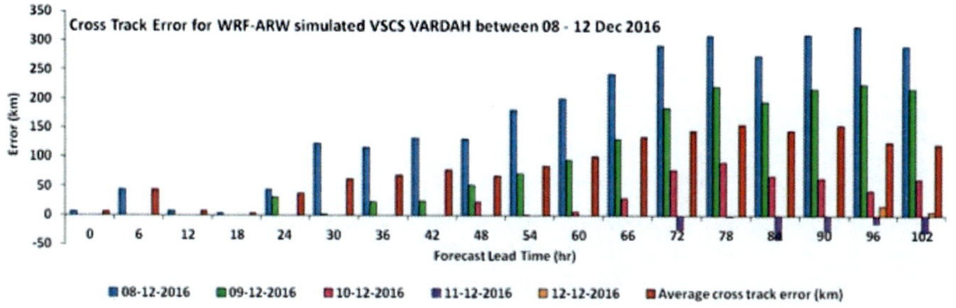


Fig. 6 Cross track error

predict the eastward movement of the TC over the domain considered. Similarly, analysing the ATE we observe that values are negative for the most part only for the simulation with 8 December conditions for all other simulation mostly it is positive. This indicates that the forecasted track positions are generally ahead of the observed track positions except for the simulation with 8 December conditions. Comparing the CTE and ATE values, it's observed that for all forecast lengths, except forecasts of 24 h or less, the ATE values are smaller as compared to CTE values. This indicates that the spread of the forecasted track relative to the observed track is more. Also mostly positive CTE values are also indicative of Vardah being more like a straight moving cyclone. Also, the average CTE has been reducing with lead time thereby indicating that the model forecasts improve with lead time and are very good over 48 h of simulations.

## 5.2 *Simulation of Winds*

The low-level wind (850 hPa & 10 m) flow distribution (in knots) simulated by the ARW model for Cyclone Vardah is depicted in Figs. 7, 8, 9 and 10, respectively. In all the figures one can observe

(a) the wind flows during formation (11 December 2016)

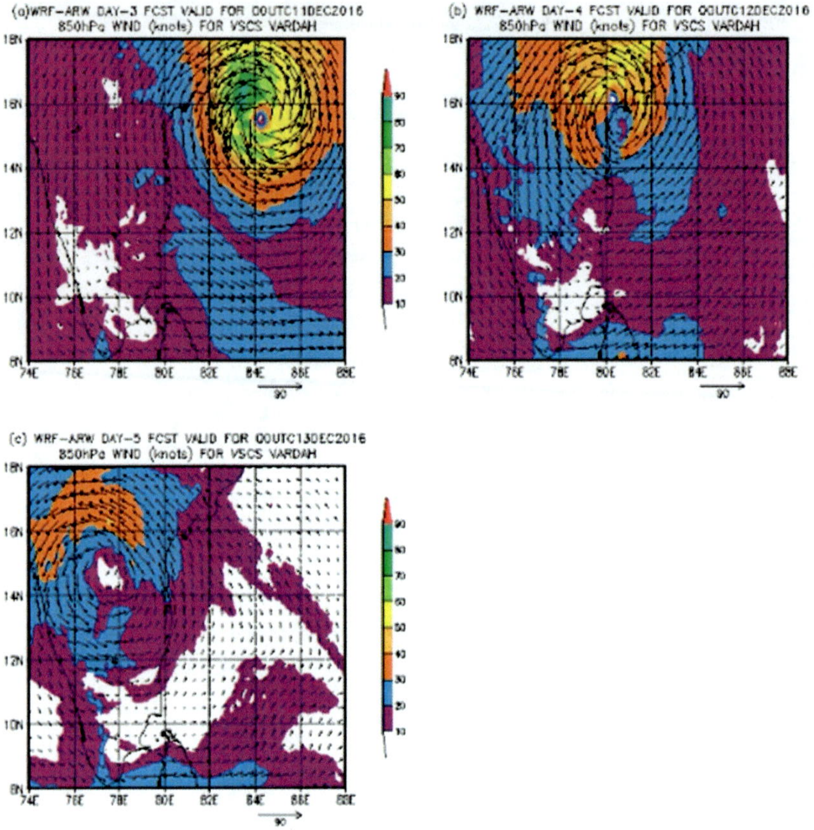


**Fig. 7** Distribution of low level (850 hPa) wind flow (knots) at various stages from pre-landfall (11 December 2016) to post-landfall (13 December 2016) with 8 December IC

(b) the wind flows at the mature stage just before landfall (12 December 2016)
(c) the wind flows at the mature stage just after landfall (13 December 2016)

Figures 7 and 8 present the distribution of winds simulated with 8 and 11 December ICs at various stages from pre-landfall (11 December 2016) to post-landfall (13 December 2016) at 850 hPa and Figs. 9 and 10 present the distribution of winds simulated with 8 and 11 December ICs at 10 m, respectively. A quick analysis of the figures reveals that the low-level (850 hPa) vorticity advection (Figs. 7 and 8) are simulated reasonably well by the model.

Comparing Fig. 8 (winds at 850 hPa) and Fig. 10 (winds at 10 m) reveals that the average wind speed at 850 hPa is about 10 knots faster than the average wind speed at 10 m before landfall and just at landfall at the same location. This is primarily because friction at the height of 10 m is more than at 850 hPa. Once again due to more friction winds at 10 m disperse much faster than at 850 hPa after landfall as depicted in Figs. 8c and 10c. A comparison of Figs. 9 and 10 reveals that the wind

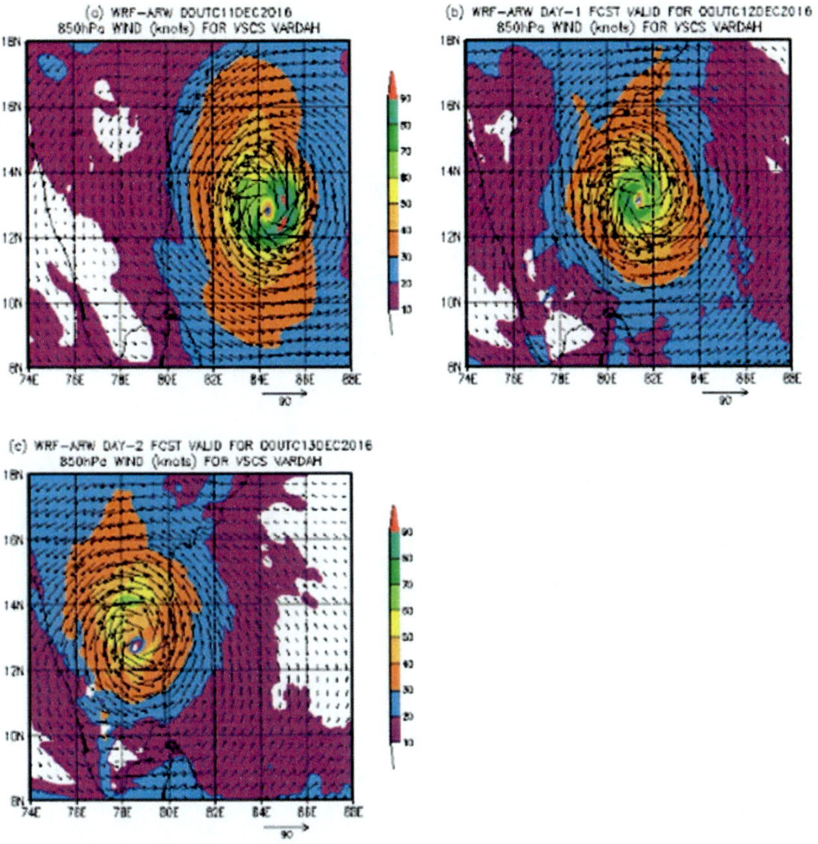


**Fig. 8** Distribution of low level (850 hPa) wind flow (knots) at various stages from pre-landfall (11 December 2016) to post-landfall (13 December 2016) with 11 December IC

simulation with 11 December IC is stronger, and resembles more with the observed winds patterns than with 8 December IC.

Figure 11 shows a bar graph comparing the 3 hourly maximum observed sustained surface winds and estimated central pressure with the 3 hourly modelled maximum sustained surface winds and estimated central pressure based on 8 December and 11 December initial conditions. A quick review of Fig. 11 reveals that the simulation with 8 December initial conditions shows a good match with observed winds only for 8 December and 13 December at 00 h. Also, with 8 December initial conditions the model fails to simulate the central pressure as close to that of observed. The analysis clearly indicates that simulated central pressure with an initial condition of 11th is in reasonably good agreement with IMD observations.

The performance of the model is much better with the 11 December initial conditions. Here also the estimation of the wind speeds is much better than central pressure. At several 3 h intervals, the model sustained surface winds nearly match that of the

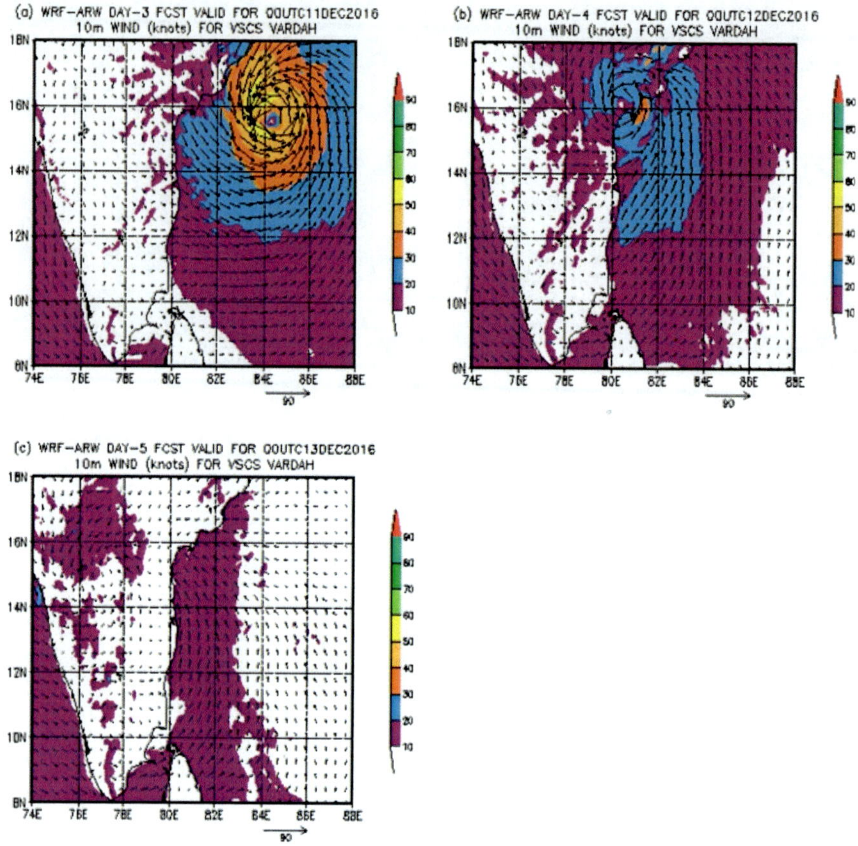


**Fig. 9** Temporal variation of wind at 10 m starting from pre-landfall (11 December 2016) to post-landfall (13 December 2016) with 8 December IC

observed ones (11/0000, 11/0600, 11/1500, 11/2100, 12/1800). At all other intervals, the variation is around 15% of the observed value. In the case of estimated central pressure, the model took about 24 h to stabilise (from 11/0000 to 12/0000), thereafter the model performance improved and at 12/0000, 12/1500 and 13/0000 the modelled central pressure matched exactly with the observed values.

## *5.3 Rainfall Analysis*

The model simulated 24-h accumulated rainfall based on 8th and 11 December initial conditions are shown in Figs. 12 and 13. The model simulated rainfall estimates are in the same range in both the figures with major high (16 to 32 cm) and very high (32 to 64 cms) rainfall bands and smaller extreme rainfall (greater than 64 cm) bands.

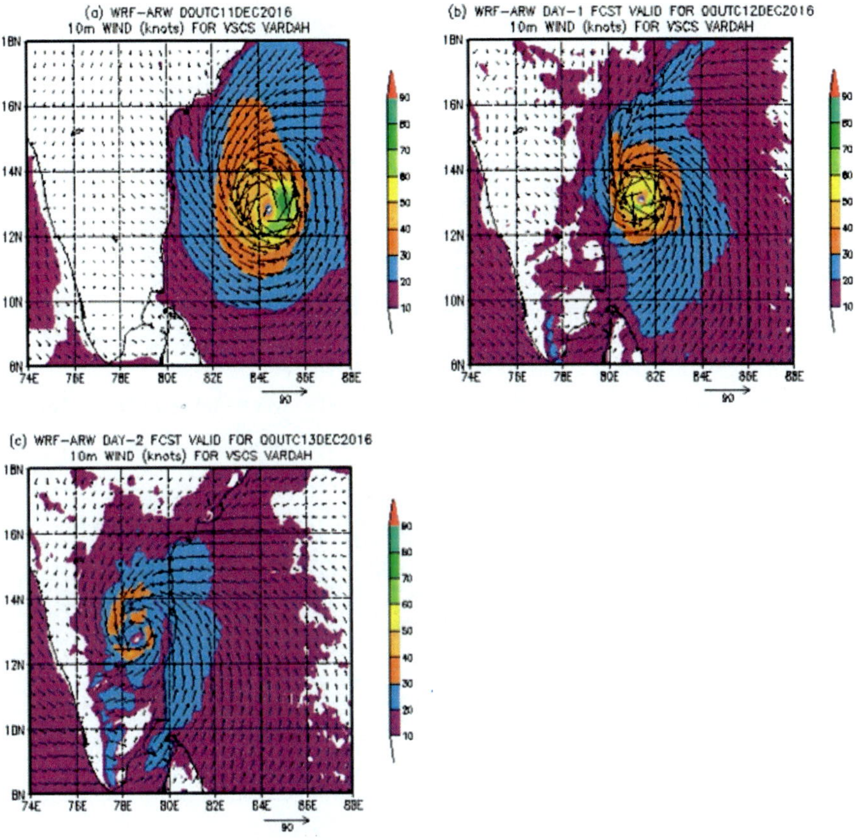


**Fig. 10** Temporal variation of wind at 10 m starting from pre-landfall (11 December 2016) to post-landfall (13 December 2016) with 11 December IC

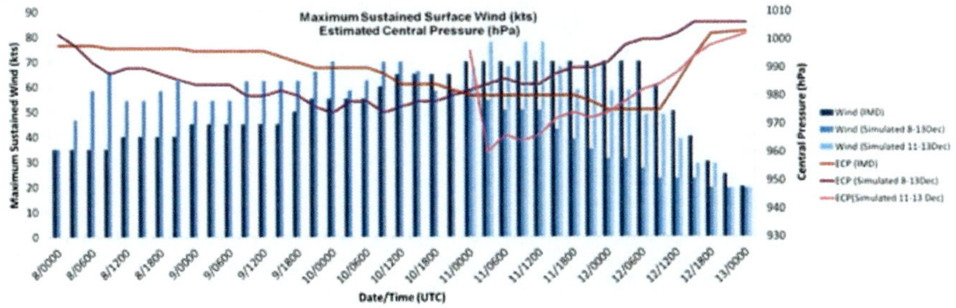


**Fig. 11** Comparison of observed and modelled maximum sustained surface winds and estimated central pressure

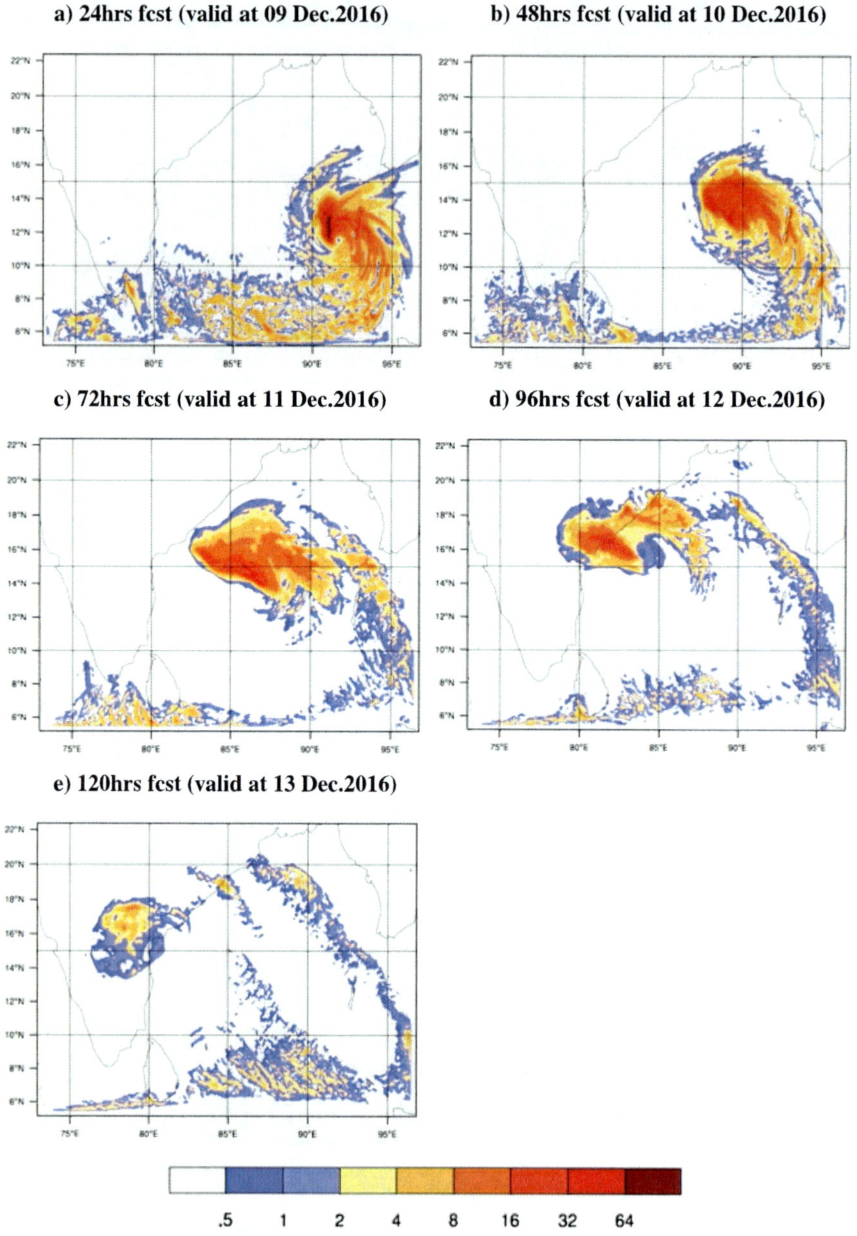


**Fig. 12** Simulated accumulated precipitation (cm) based on initial conditions of 8 December 2016

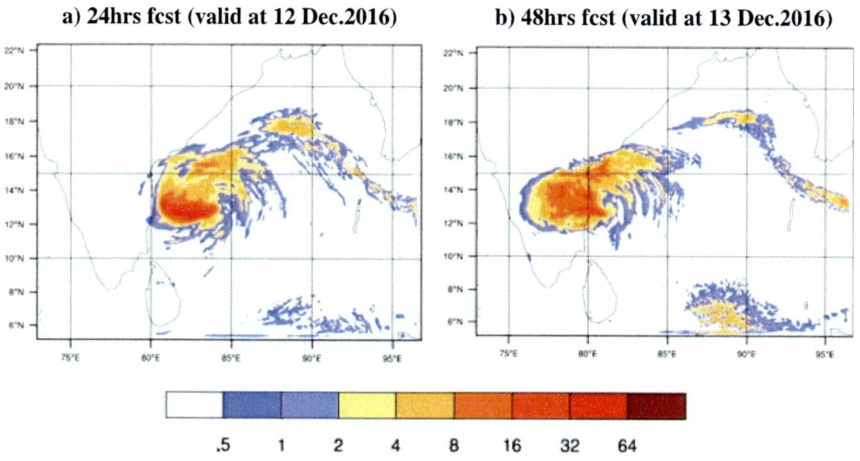


**Fig. 13** Simulated accumulated precipitation (cm) based on the initial condition of 11 December 2016

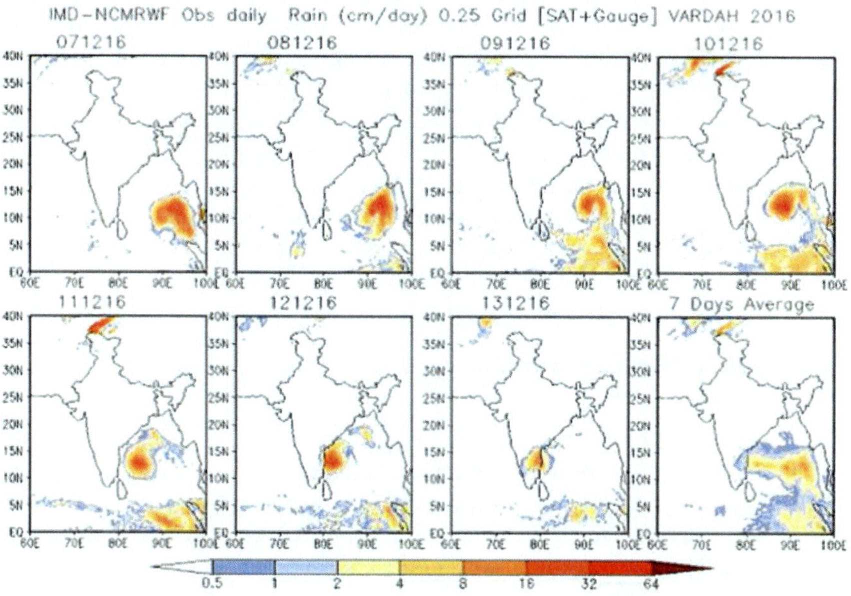


**Fig. 14** IMD-NCMRWF GPM merged gauge rainfall during 7–13th and 7 days average

The major difference between the two figures is in the location and spread of rainfall. The location difference directly correlates with the modelled cyclone tracks for 8 and 11 December advisories.

24 h accumulated observed rainfall, ending at 0830 h IST of date, associated with the system is depicted in Fig. 14 based on IMD-NCMRWF GPM merged gauge rainfall data [7].

Comparing Figs. 12 and 14, it is evident that ARW simulated rainfall is very closely following the observed pattern for all the days from 9 to 13 December. Both the minimum and maximum have been captured fairly well by the model. The areas, with rainfall between 16 and 64 cms, are very well represented across all days. On 9 December there is a slight overestimation of rainfall higher than 64 cms whereas on 10 December there is a slight underestimation of the rainfall higher than 64 cms. On 11–13 December, the model estimated rainfall is in the right range as compared to the observed rainfall but the area experiencing the rainfall is lower as compared to the observed rainfall as well as the location of the rainfall is more towards the north as compared to observed.

The model estimates of rainfall based on the 11 December (Fig. 13) initial conditions compare very well with the observed (Fig. 14) for 12 and 13 December. The spread of the rainfall is in the same area as well as the rainfall estimates are in the same range for 12 and 13 December. Also, areas experiencing high (16 to 32 cm), very high (32 to 64 cm) and extreme rainfall (greater than 64 cm) have been very well captured by the model close to the land as well as northwards away from the land. Even the moderate to low rainfall areas are also very well captured.

This comparison on the estimated rainfall based on 8 an 11 December initial conditions with respect to the observed rainfall clearly indicate that the variations are due to deviation in the track as compared to the observed track. The estimated cyclone track based on the 8 December initial conditions has much more deviation from the observed track as compared to the 11 December initial conditions based on track estimation.

## 5.4 *Relative Vorticity Analysis*

The model computed low level (850 hPa) relative vorticity on 11 December at 0000 UTC with ICs 8th, 9th, 10th and 11 December is shown in Fig. 15a–d. At 0000 UTC of 11th Dec, the low-level relative vorticity with IC 11 December was about 230 × $10^{-5}$ $S^{-1}$ which is only an 8% variation when compared with IMD observations of 250 × $10^{-5}$ $S^{-1}$ to the south-southwest of the system centre and thus in very good agreement with the IMD observation. While a stronger low-level relative vorticity of about 270 × $10^{-5}$ $s^{-1}$ is simulated by model with IC of 10 December Furthermore, it is interesting that vorticity remains fairly positive in the lower level with all ICs when the system was in the growing stage.

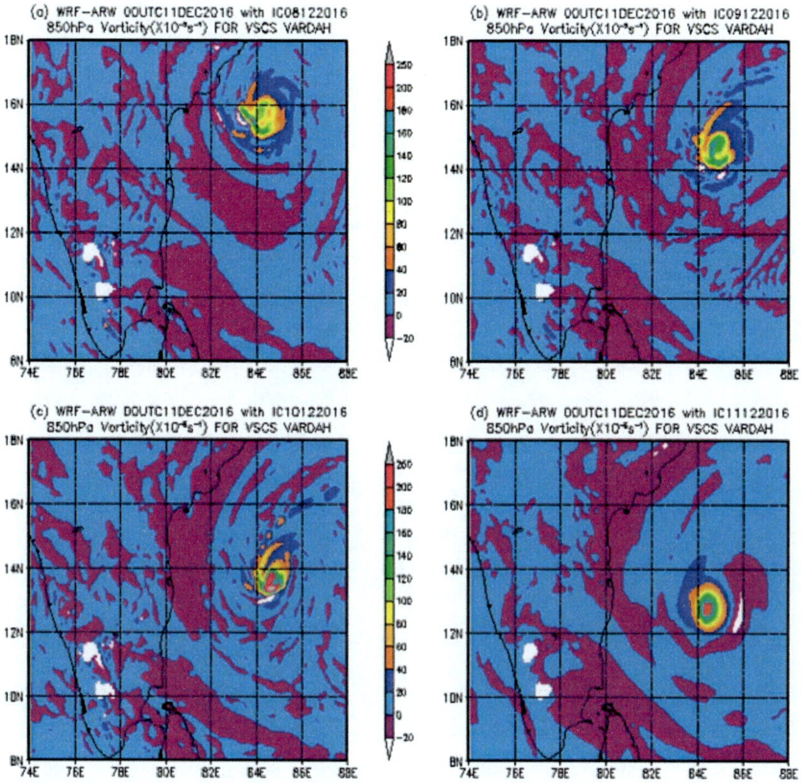


**Fig. 15** Low level relative Vorticity on 11 December 2016 at 0000UTC with all ICs

## 5.5 *Latent Heat Flux Analysis*

The latent heat flux with IC of 11 December 2016 for 48 h of model simulation at 03 h, 12 h, 24 h, 36 h and 48 h depicted in Fig. 16a–e respectively. It is clearly seen from the figure that high latent heat near the oceanic surface on 11 December at 12 h and on 12 December 00 h and it starts decreasing from 12 December at 12 h onwards when cyclone approaching towards the coast, it may be attributed to change in daily SST anomaly which may respond to the decrease in TC intensity. The intensification of tropical cyclones can be limited by some factors like cold sea surface temperature and dry air in the vicinity of the cyclone. Dry air limits the ascending motion because it is less buoyant than moist air at the same given temperature and it also decreases the net condensation and latent heat release. This decrease in latent heat is clearly visible in Fig. 16e which is the simulation for 13 December 2016 at 00 h, when a cyclone crossed the coast north of Tamil Nadu about 9 h ago.

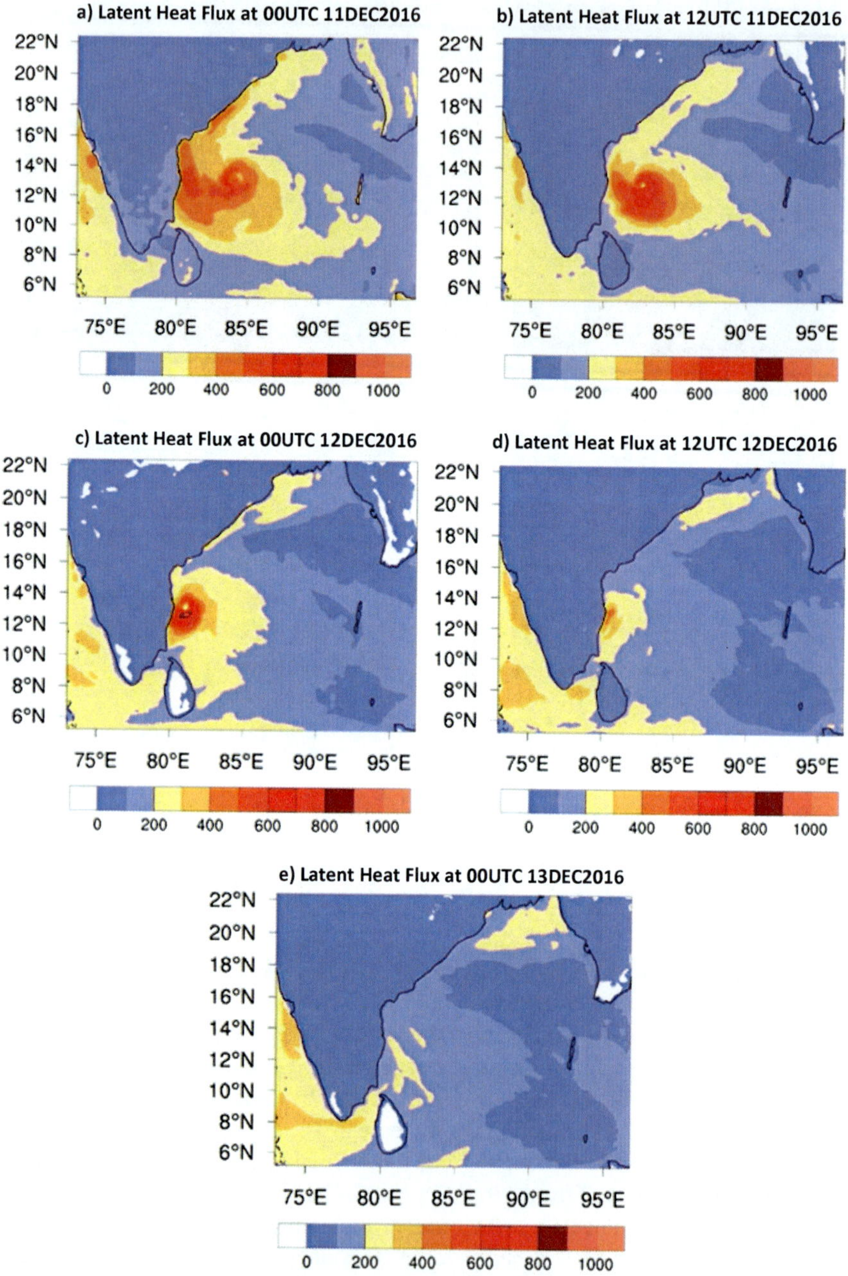


**Fig. 16** **a–e** Surface latent heat flux ($Wm^{-2}$) associated with the Vardah valid for the 11 December 2016 00 h to 13 December 2016 00 h

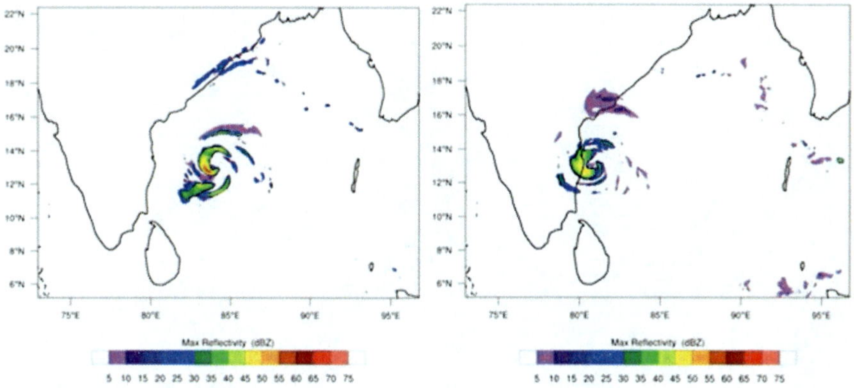

**Fig. 17** Model simulated maximum reflectivity (DBZ) valid for 0300 UTC 11 December 2016 (Left) and at 0300 UTC 12 December 2016 (Right)

## *5.6 Maximum Reflectivity Analysis*

Model simulated maximum reflectivity (DBZ) with IC of 11 December 2016 at 0300 UTC of 11 December and 12 December is given in Fig. 17. Larger values (45–75 DBZ) in the eyewall over the Bay of Bengal and lesser values (15-30 DBZ) near/over the East Indian coast supporting the explanation of rainfall distribution on this day (Fig. 17 left). The simulation of TC Vardah shows a well-captured comma cloud organisation around the cyclonic-eye. A GPM's DPR showed radar reflectivity values greater than 82 DBZ to the satellite at 12 December 0241 UTC (Fig. 18). GPM is a joint mission between NASA and the Japanese space agency JAXA. Thus the computed maximum reflectivity is found to be in reasonable agreement with the available observed value.

# 6 Conclusions

The ARW model captured the track and several features of the Vardah cyclone over the Bay of Bengal reasonably. Forecast variation/error is calculated for max wind, central pressure and relative vorticity. The analysis of the model outputs revealed:

- With the initial condition of 8 December, the model simulations of winds, central pressure, relative vorticity as well rainfall did not corroborate very well with the observed values.
- With the initial condition of 8 December, the estimates were more close to the observed in the initial two days of simulations but deviated a lot later

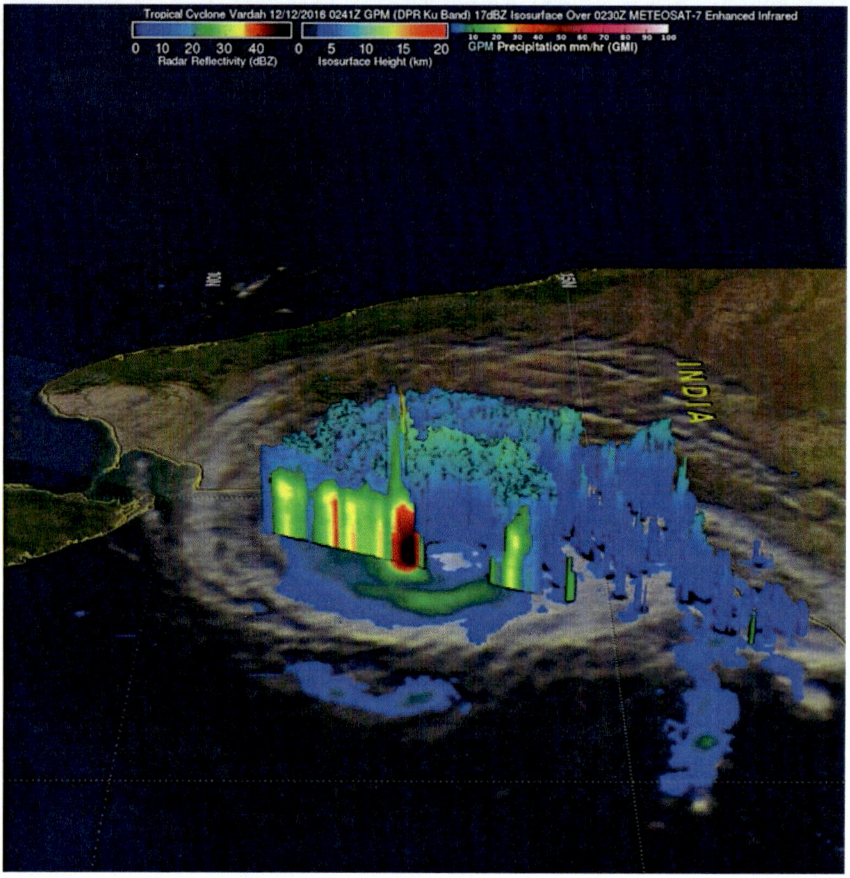


**Fig. 18** On 11 December at 9:41 p.m. EST (12 December at 0241 UTC) GPM reported maximum reflectivity values greater than 82 DBZ

- With the initial condition of 11 December, the model simulation of winds, rainfall, maximum reflexivity and relative vorticity were found very close to the observed values
- Even with the initial condition of 11 December, the model took 24 h to deliver a good estimation of central pressure

Based on the above it infers that overall the forecasts are found to be in reasonable agreement, which reveals the robustness of the model to use for forecast purpose for the cyclonic event. It was also observed that the model forecasts have high accuracy over a two day period but the accuracy reduces if the forecasts are longer than 2 days. This is reflected in the simulated Vardah cyclone track for 8–13 December where the track deviated from the observed more whereas the 11 December simulation had very low track error when compared to the observed track. This may be because of the use of the Kain-Fritsch scheme which more accurately simulates depth and intensity than the trajectory and landfall location ([21]). This is also reflected in the relative vorticity, winds and rainfall estimation for various days.

The performance of the model is much better with the 11 December initial conditions but the estimation of the wind speeds is much better than central pressure. At several 3 h intervals, the model sustained surface winds nearly match that of the observed ones and at other intervals, the variation is around 15% of the observed value at most. While the rainfall estimates were on the lower side with 8 December initial conditions, the estimates were very well captured by the simulation with 11 December initial conditions. The model not only captured the high, very high and extreme rainfall areas but moderate to low rainfall areas were also well captured. Similarly, the model estimates of vorticity were in good agreement (variation of about 8%) with 11 December initial conditions whereas estimates were higher with the previous day's initial conditions.

The model lags by a day in the estimation of central pressure, this is reflected in both 8 December as well as 11 December estimates of central pressure. This could also be interpreted as the simulated track approaches the coast it deviates from the observed track probably because the local conditions are not getting very well captured. There is a need for further analysis using more representative land use data than the default USGS dataset.

## References

1. Bhaskar Rao DV, Hari Prasad D (2006) Numerical prediction of the Odisha super-Tropical Cyclone (2006) sensitivity to the parameterization of convection, boundary layer and explicit moisture processes. Mausam 57:61–78
2. Camargo SJ, Barnston AG, Zebiak SE (2005) A statistical assessment of tropical cyclones in atmospheric general circulation models. Tellus 57A:589–604
3. Dudhia J (1989) Numerical study of convection observed during the Winter Monsoon Experiment using a mesoscale two dimensional model. J Atmos Sci 46:3077–3107
4. Elia R, Laprise R, Denis B (2006) Forecasting skill limits of nested, limited-area models—a perfect-model approach. Am Meteorol Soc Monthly Weather Rev 130
5. Gupta A (2006) Current status of tropical cyclone track prediction techniques and forecast errors. Mausam 57:151–158
6. Hong YS, Noh JY, Dudhia J (2006) A new vertical diffusion package with an explicit treatment of entrainment process. Mon Wea Rev 134:2318–2341
7. IMD (2016) Very severe cyclonic storm, 'Vardah' over the Bay of Bengal. IMD Report 06–13 (December 2016)
8. Kain JS (2004) The Kain-fritsch convective parameterization—an update. J Appl Meteor 43:170–181
9. Laprise R, Varma MR, Denis B, Caya D, Zawadzki I (2000) Predictability of a nested limited-area model. Am Meteorol Soc Notes Corresp 4149 (December 2000)
10. Mlawer EJ, Taubman SJ, Brown PD, Iacono MJ, Clough SA (1997) Radiative transfer for inhomogeneous atmospheres: RRTM, a validated correlated-k model for the long wave. J Geophys Res 102D:16663–16682
11. Mohanty UC, Mandal M, Raman S (2004) Simulation of Odisha super-tropical cyclone (1999) using PSU/NCAR mesoscale model. Nat Hazards 31:373–390

12. Mohanty UC, Osuri KK, Routray A, Mohapatra M, Pattanayak S (2010) Simulation of Bay of Bengal tropical cyclones with WRF model: Impact of initial and boundary conditions. Marine Geod 33:294–314
13. Mohanty UC, Gupta A (1997) Deterministic methods for prediction of Tropical Cyclone tracks. Mausam 48:257–272
14. Mohapatra M, Bandyopadhyay BK, Nayak DP (2013) Evaluation of operational tropical cyclone intensity forecasts over north Indian ocean issued by India meteorological department. Nat Hazards 68:589–601
15. Mohapatra M, Mandal GS, Bandyopadhyay BK, Tyagi A, Mohanty UC (2012) Classification of cyclone hazard prone districts of India. Nat Hazards 63(3)
16. Mohapatra M, Nayak DP, Sharma RP, Bandyopadhyay BK (2013) Evaluation of official tropical cyclone track forecast over north Indian Ocean issued by India meteorological department. J Earth Syst Sci 122:433–451
17. Mohapatra M, Sikka DR, Bandyopadhyay BK, Tyagi A (2013) Outcomes and challenges of Forecast Demonstration Project (FDP) on landfalling cyclones over Bay of Bengal. Mausam 61:1–15
18. Osuri KK, Mohanty UC, Routray A, Kulkarni MA, Mohapatra M (2012) Customization of WRF-ARW Model with physical parameterization schemes for the simulation of Tropical Cyclones over the North Indian Ocean. Nat Hazards 63:1337–1359
19. Osuri KK, Mohanty UC, Routray A, Mohapatra M (2011) The impact of satellite-derived wind data assimilation on track, intensity and structure of Tropical Cyclones over the North Indian Ocean. Int J Remote Sens 1–26
20. Panda J, Giri RK (2012) A comprehensive study of surface and upper air characteristics over two stations on the west coast of India during the occurrence of a cyclonic storm. Nat Hazards 64:1055–1078
21. Parker CL, Lynch AH, Mooney PA (2017) Factors affecting the simulated trajectory and intensification of Tropical Cyclone Yasi (2011). Atmos Res 194:27–42
22. Routray A, Mohanty UC, Osuri KK, Kar SC, Niyogi D (2016) Impact of satellite radiance data on simulations of Bay of Bengal tropical cyclones using the WRF-3DVAR modelling system. IEEE Trans Geosci Remote Sens 54(4):2285–2303
23. Routray A, Mohanty UC, Rizvi SRH, Niyogi D, Osuri KK, Pradhan D (2010) Impact of doppler weather radar data on simulation of Indian monsoon depressions. Quart J Royal Meteorol Soc 136:1836–1850
24. Routray A, Singh V, George JP, Mohandas S,1, Rajagopal EN (2016) Simulation of tropical cyclones over Bay of Bengal with NCMRWF regional unified model. https://doi.org/10.1007/s00024-016-1447-0
25. Srinivas CV, Venkatesan R, Vesubabu V, Nagaraju C (2010) Impact of assimilation of conventional and satellite meteorological observations on the numerical simulation of a Bay of Bengal tropical cyclone of Nov 2008 near Tamil Nadu using WRF model. Meteorol Atmos Phy 110:19–44
26. USGS data. Retrieved from http://www.mmm.ucar.edu/wrf/users/downloads.html
27. Walsh K, Watterson IG (1997) Tropical cyclone-like vortices in a limited area model: comparison with observed climatology. J Climate 10:2240–2259
28. WRF modeling system. Retrieved from http://www.mmm.ucar.edu/wrf/users/downloads.html

# Chapter 14
# Fuzzy-AHP-Based Indexing Model for Performance Assessment of Highways and Expressways

**Rohit Sharma, Prateek Roshan, and Shobha Ram**

## 1 Introduction

Highways and expressways are the most widely utilised mode of transportation. The highway management system connects large metropolitan cities and rural communities network across the country. Highway frameworks have made development conceivable, for travelling in simpler ways and provided the most convenient means of travel. Highway transportation is particularly significant and helpful with regard to the economy. The present study uses the fuzzy AHP technique to create a highway design and performance assessment index model. When applied to an existing highway, this model can give scores, which help compare two or more highways based on design and performance. The fuzzy AHP includes three main procedural elements: forming a hierarchy model, obtaining the priorities for each parameter or criteria, and checking the consistency of experts' judgment. First of all, all the factors affecting highway design and highway performance are identified, and these factors and parameters are formed into a hierarchical model. After constructing a hierarchical model, the weights of each parameter are calculated using the results obtained from the experts' opinion by applying the α-cut method, as proposed by Chang (2009) [1].

R. Sharma · S. Ram
Gautam Buddha University, Greater Noida, UP 201310, India
e-mail: rohitsharmagbu19@gmail.com

S. Ram
e-mail: shobharam@gbu.ac.in

P. Roshan (✉)
Delhi Technological University, Delhi 110042, India
e-mail: prateek_2k19phdce02@dtu.ac.in



D. Singh et al. (eds.), *Proceedings of International Conference on Scientific and Natural Computing*, Algorithms for Intelligent Systems,
https://doi.org/10.1007/978-981-16-1528-3_14

# 2 Literature Review

## 2.1 Review on Analytic Hierarchy Process

The analytical hierarchy process was first presented by Saaty [2]. It is a fundamental approach to decision making. It deals with complex decision making and helps the decision-maker to carry out simple pairwise comparisons based on expert judgments. AHP is a multi-criteria decision-making method representing a complex problem in the form of a hierarchical structure and develops a preference for alternatives based on expert opinion. The modification of conventional AHP by integrating fuzzy science into it was carried by Buckley [3], following which fuzzy AHP methods have found many applications in the past studies.

## 2.2 Review on Design of Highways

Sherman and Watkins [4] presented a summary of several reports on quality control of highway construction material. The authors reported that personnel training, reproducibility of tests, preservation of engineering judgment, and administration cost are some of the problems that need additional attention. They also discussed some possible solutions to these problems. It includes determining the precision of the test method and procedures for confirming that a laboratory is in parental control. The authors proposed using moving average and control charts as one of the procedures of quality control. They stated that doing so will give control without incurring any additional cost.

Gibreel et al. [5] presented a literature review on the geometric design consistency of two-lane rural highways in Europe and North America. High consistency in highway geometric design is very important in aspects of design and evaluation of highways to gain smooth and safe traffic operation. In previous decades the research work on highway design consistency is classified into speed considerations, safety considerations and performance considerations. The authors stated that the geometric design parameters affect the operating speed or design speed of highways.

Tsai and Chang [6] suggested a framework for developing sustainable items for highway design. The authors identified the highway-related sustainable items after the literature review. After that, specific sustainable items were selected by authors through interviews with practitioners, including designers, constructors, and maintenance owners. A checklist comprises 60 techniques, and sustainable material items were generated and tested on four highway design projects. The authors found that over 52% of the developed sustainable items were reviewed, and 50% included in the design.

Hallowell et al. [7] performed a research study to quantify how pairwise spatial and temporal interaction has an on-base level risk of 25 prevalent highway construction tasks in the region of United States. The authors quantified 600 risk interactions by obtaining and aggregating over 23,500 individual ratings with certified experts using the Delphi method. They reported that incompatible tasks might increase the base-level risk by up to 60%.

Hassan et al. [8] reviewed the practice work and research work related to design criterion on conformity with the 3D nature of the highway. The geometric design is a significant phase in highway design that is directly related to traffic safety. Sight distance, vehicle stability, driver comfort, drainage, and aesthetics are critical elements. Highways may risk their safety and economy when designed by using 2D standards of highway design. The authors explained that the modelling of vehicle dynamics in the current standards with the point mass formula might be inaccurate. Also for the drainage of surface water current standards have some recommendations. However, the coordination between the combined alignments lacked as reported by the authors.

Cedergren [9] studied the effective drainage system, describing the basic features of pavement drainage, inflow-outflow estimation, criteria for designing subsurface drains, and protecting sub-surface drainage system. They stated that damages caused to highways pavement by traffic impact and water trapped in the slow-drainage system are the most severe damage that affects highway safety and performance.

Al-Sobky and Mousa [10] devised a new smartphone application to find traffic density and speed. The suggested system consists of two smartphones and two cars and to count the vehicles between the two cars we need an observer. This count is used with tracking data to give determined the speed and density of traffic. The manual traffic counts and travel speed were used to find the density of traffic. The two densities are compared to each other, and a test makes a confirmation that the mean difference between the two densities is not significant at 5% level. The authors reported an average error of 8.2% when the founded flow is compared with the actual counts. This system was then applied to determine density on an elevated road in Egypt, and the level of service was measured on 15 sections of the road. This system's application is restricted to daytime, continuous flow condition and traffic streams with less percentage of heavy vehicles.

## *2.3 Review on Performance and Safety of Highways*

Anderson et al. [11] examined the relationship of safety measures of geometric design for two-lane rural highway alignments. Safety measures that were estimated consisted of speed reduction on a horizontal curve relative to the preceding tangent or curve, the average rate of vertical curvature, the ratio of individual curve radius to the average radius, average radius, and maximum radius ratio minimum radius. These safety measures were related to accident frequency and recommended as a candidate measure for the determination of geometric design consistency. With the help of Poisson, negative binomial and lognormal regression analysis relationship between consistency measure and accident frequency was derived by the authors.

Battikha [12] expressed that the quality management function is performed in highway construction and is showed as an interconnected system, which identifies the leading quality. The authors identified records and documents that are used in these activities. For the involvement of multilevel quality management, contractors, engineers, and managers, a model was defined. The quality management tasks, construction quality control-related scheme and quality assurance are all described in this model.

Kononov et al. [13] derived the connection between safety and congestion, examining the link between safety and the number of lanes on urban freeways. The connection between safety and obstruction on urban freeways was investigated using safety performance functions (SPFs) for dual-lane freeways in Colorado, California, and Texas. Authors utilised neural networks to determine the underlying relationship between safety and exposure. Relating safety to the degree of congestion suggested that safety deteriorates with the quality of service expressed through the amount of service. Comparison of SPFs of dual-lane freeways suggests that adding lanes might at first lead to a brief safety improvement that disappears as congestion will increase.

Jalayer et al. [14] briefly stated that the application of the highway safety manual (HSM) at the state level has the potential to permit transportation agencies to address safety considerations proactively. However, the widespread use of HSM faces vital barriers as several departments of transportation (DOTs) do not have comfortable HSM- require highway inventory knowledge. DOTs and native agencies use several techniques to gather highway inventory knowledge for alternative functions. The authors evaluated existing highway inventory strategies through a nationwide survey and a field trial of known promising highway inventory knowledge assortment (HIDC) strategies on numerous sorts of highway segments.

## 3 Methodology

Saaty's fundamental scale of absolute numbers which uses the scale of 1, 3, 5, 7, and 9 for relative comparison of the importance of two criteria over one another (Saaty 1980) [2] is used in the current study for the arrangement of gathered data of the experts' judgement. Roshan et al. [15, 16] have used the AHP procedure to calculate weight matrices; the same concept and technique have been adopted in this study. Consistency of judgement data was also checked using the consistency ratio check used by previous researchers [15, 16]. The α-cut method is used where the opinions of the expert are treated as a triangular function. It considers the uncertainty (α) and risk tolerance (λ) factors of the experts' judgement. The value of α and λ both are taken as 0.5 in the present study. The expert judgment data of fifteen experts were included in the study.

## 4 Proposed Model for Performance Assessment Indexing of Highways and Expressways

An exhaustive literature review of past research, bylaws and guidelines was carried prior to identifying the parameters that influence the design and performance of highways and expressways. The flow chart of the present study is shown in Fig. 1.

The indexing model proposed in this study was subdivided into two sub-components: the design index and the safety and performance index. The design index is sub-divided into six subcategories:

- layout
- geometric design
- the topography of selected site for highway construction
- traffic capacity
- the material used in construction and
- drainage system

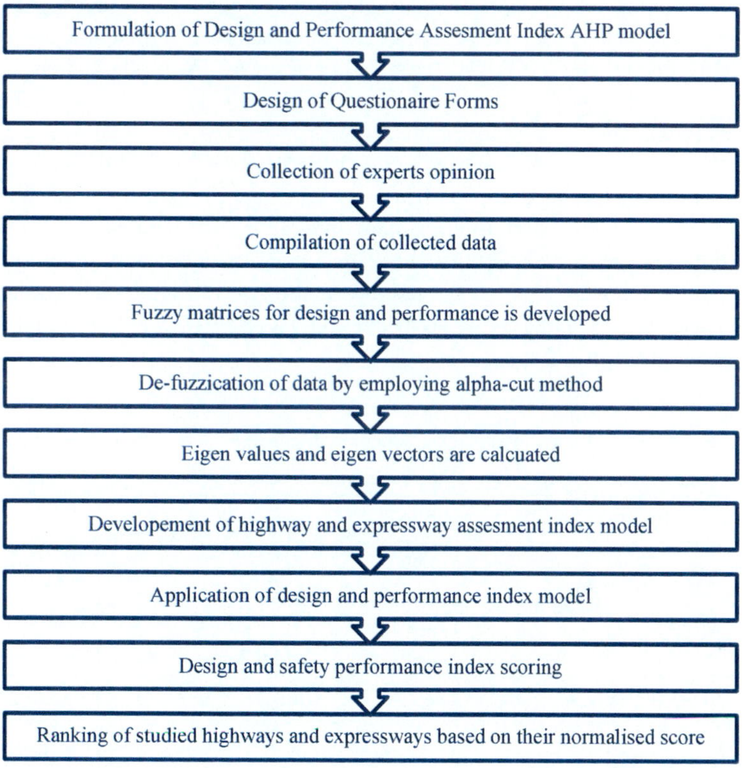


**Fig. 1** Flow chart of the proposed study

The safety and performance index is also sub-categorised into four criteria: quality control, traffic monitoring system, emergency services, and safety features.

The layout category is divided into two types (highways and expressways) and types of pavements (rigid and flexible). Geometric design under the design index consists of four sub-criteria, naming as follows: number of lanes, lane width, rotary, and U-turns. Topography in highways includes the type of terrain and the condition of the selected site for construction. Traffic volume, traffic composition, average daily traffic, and daily hourly traffic come under the proposed highway's traffic capacity. Concrete, bitumen, and WBM were sub-elements under the material category. The Drainage system is further sub-categorised into the open channel, screening, and proper timely maintenance.

There are four sub-criteria: quality control, traffic monitoring system, emergency services, and safety features that come under the category of safety and performance. Quality control consists of the construction process proper or improper, type of labour skilled or poorly skilled (i.e. unskilled), the material used in construction is good or poor, the drainage system provided in the highway is good or not. The Traffic monitoring system comprises a facility of emergency call box, proper toll collection, proper functioning CCTV cameras, variable message signboards, and night lights on the highway. Emergency services include police services, ambulance service in case of an accident, and fire brigade service in case of fire. Well-marked lanes, lane separators, sufficient proper shoulder, guard rails, and dividers are considered under safety features.

Questionnaire survey forms were sent to the experts for their judgment. The information about the highways and expressways taken up for the application of the proposed design and performance index model was obtained by surveying and visiting the site location and talking to the respective construction authorities to get all the relevant data for the study. Data and information collection at different lengths and toll points were obtained. The highways and expressways which were taken for the study are given in Table 1.

**Table 1** Highways and Expressways for the proposed design and performance index model

| S.no. | Highways and expressways | Starting point | Destination point | Length (km) |
|---|---|---|---|---|
| 1 | Eastern peripheral EW | Kundli | Palwal | 135 |
| 2 | Yamuna EW | Noida | Agra | 165 |
| 3 | NH-91(GT road) | Ghaziabad | Kanpur | 126 |
| 4 | NH-44 | Srinagar | Tirunelveli | 3745 |
| 5 | NH-24 | Delhi | Lucknow | 438 |

*NH - National Highway, EW - Expressway

# 5 Calculations Involved

## *5.1 Scoring for the Performance Indexing Model for Highways and Expressways*

The criteria related to highway design and safety and performance are placed in the first column, and in the next column, the relative weights of each criterion are placed at appropriate locations. The value of measurable and observable criteria in highways is placed in the column next to alternatives. The measurable items observed in the highways have a value equal to 1, and others have 0. The weighted sum score is the global score for any given option with respect to each criterion. Similarly, the total score at the topmost level is determined. The model's maximum score has obtained when all criteria for design and safety and performance meet with their maximum weights obtained. The maximum score found for the highway design and performance index is 0.8816. The Performance assessment Index hierarchical model for highways and expressways with the individual weights of each element at each level is shown in Fig. 2. The highways are listed in Table 1 for which the design index score is evaluated depending upon the types of attributes observed. The detailed calculation for each is not shown due to brevity.

## *5.2 Relative Performance Assessment Index and Normalised Score of Highways and Expressways*

Once the final scores of highways and expressways were obtained, they are normalized at a scale of 100. The scaling procedure taken up for the study is explained below.

Let WS be the weighted score of any highway taking under consideration and $WS_{max}$ be the maximum possible score taken in the analysis. The value of WSmax is 0.8816, as mentioned earlier. The design, safety and performance index of highways was normalized at the relative scale of 100, represented by "DSPAI 100" and is evaluated as shown in Eq. (1):

$$\text{DSPAI} = \frac{WS}{WS_{\max}} \times 100 \tag{1}$$

Hence, the relative DSPAI scores obtained with their corresponding ranking based on performance score are represented in Table 2.

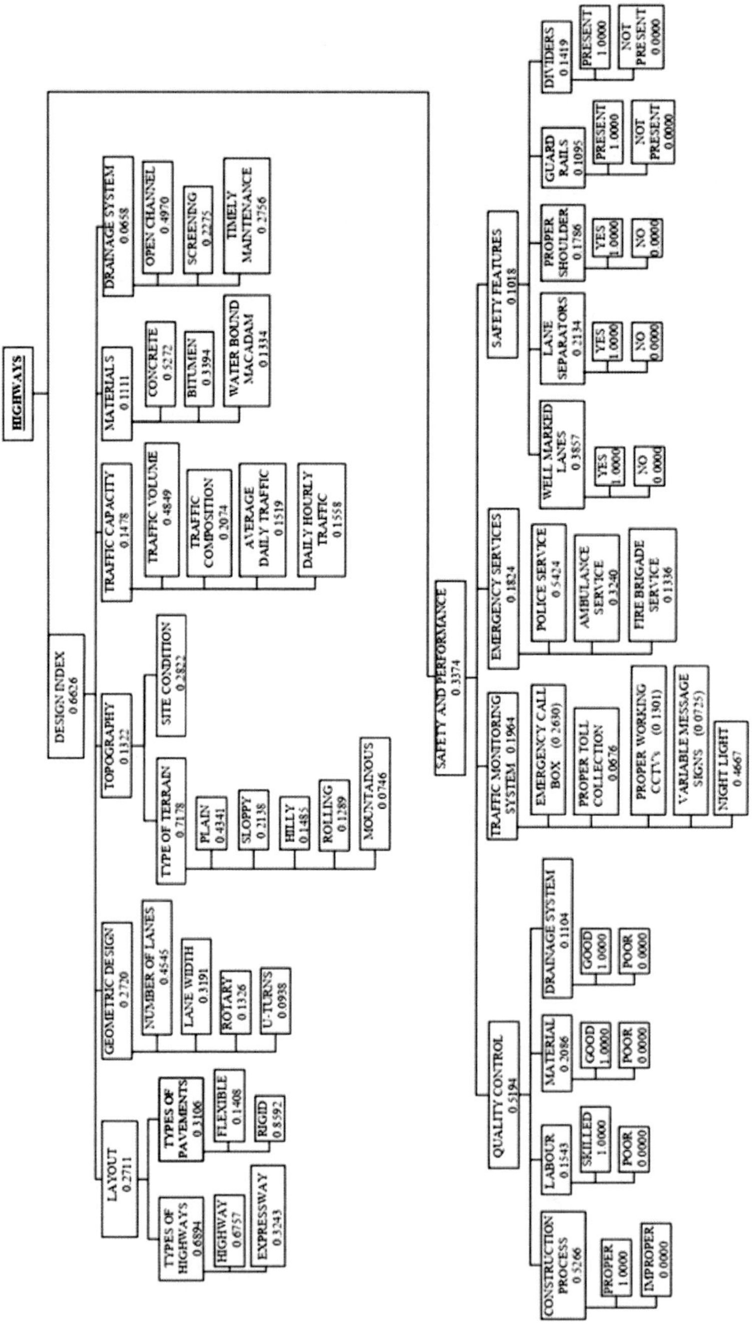


**Fig. 2** Performance assessment index hierarchical model for highways and expressways

**Table 2** Performance assessment index score of highways and expressways with obtained relative normalized scores (DSPAI) and their corresponding ranks

| S. no. | Highways and expressways | Index score | Normalized score (DSPAI) (DSPAI = $\frac{WS}{WS_{max}} \times 100$) | Rank |
|---|---|---|---|---|
| 1 | Eastern peripheral EW | 0.7497 | 85.038 | 1 |
| 2 | Yamuna EW | 0.6791 | 77.030 | 2 |
| 3 | NH-91(GT road) | 0.5348 | 60.662 | 5 |
| 4 | NH-44 | 0.6500 | 73.729 | 4 |
| 5 | NH-24 | 0.6548 | 74.274 | 3 |

# 6 Results and Discussion

The score of these highways and expressways are obtained and are normalised on a scale of 100. Ranking of these highways (HWs) and expressways (EWs) were also carried based on the obtained normalised scores. The DSPAI and ranking of Easter peripheral Expressway are the highest among all the sample highways taken in the study. The material used for making its pavement rigid, adequate number of lanes provided in the highway, site condition, topography, capacity of the highway for taking the certain limit of highway, the material used in construction, quality-controlled construction process, drainage system, and safety features were the factors that contribute for such a high design and performance score and accounted for first ranking. From Table 2 it can be seen that NH-91 is one of the highways among the samples highways taken for study, which have obtained the lowest DSPAI score and ranked last. The design factors of NH-91 somehow have weights near all other sample highways, but it lacks in its safety and performance owing to lack of good construction quality process, having lack of quality material and sound drainage system, etc., and also lacks in safety features as it does not have proper shoulders and lane separators. These factors contributed to the score of NH-91 being the lowest. Also, Table 2 shows that Yamuna expressways have the second-highest score among all highways chosen for this study. Yamuna Expressway has a score similar to the eastern peripheral expressways. They have some parameters in common which have similar weights. Some of them are, adequate lane width provided in the highway or expressway, the type of terrain on which they are constructed, etc.

## *6.1 General Findings*

Based on the proposed study, some critical findings which highlight the high and low scoring of the HWs and EWs considered in the study are discussed below:

- All the HWs and EWs have similar and adequate lane width provided during the construction. However, some of them have six lanes, and some have four lanes.

- None of the HWs and EWs was chosen for the study has rotary and U-turns except NH-24 which has U-turn but does not have rotary. This is very dangerous for commuters and due to which the score of all highways and expressways got reduced.
- Every Highway and Expressway taken up in the study is constructed on the plain terrain and has adequate site conditions in common.
- None of the studied HWs and EWs has an emergency call box through-out its length for any emergency case at any point on the highway.
- Some of the HWs and EWs have surveillance cameras along its length, and some of them have not. All the HWs have variable message signboards which are very useful for the one who is travelling through the highway.

### *6.2 Suggestions*

Some of the suggestion based on the above study was made to make highways and expressways safer, and for having long-life performance are discussed below:

- The dividers' width provided in the middle of the highways which separates the traffic from collision should be wider than the existing ones.
- To reduce the rate of accidents on highways and expressways during night time, night lights should be necessarily provided as they are crucial to make travel easier and safer for commuters.
- Emergency call boxes should all be provided with working facilities with proper instruction to be used by the traveller who is in an emergency. Emergency services like police service, ambulance service and fire brigade service have been present on all the studied highways and expressways at certain intervals. These services may facilitate immediate response, help and relief during an emergency.

## 7 Conclusions

Design and safety and performance assessment of highways is a critical and complicated task. A Fuzzy-AHP based design and safety and performance assessment index (DSPAI) for highways and expressways has been developed. $\alpha$-cut method was used in this study to handle the ambiguity and uncertainty of experts' judgments. For collecting the opinions of experts, a survey questionnaire is developed for assigning the relative weightage for each criterion at each level of the proposed model. The design and safety and performance model thus developed was implemented on sample highways. The design, safety and performance index model is applied on the five-sample HWs and EWs in the study to evaluate the DSPAI score, and it is found that the highway with the highest score has the best design, safety and performance aspect as per the model with respect to other HWs or EWs considered. The overall score of each sample highway and expressway was obtained. These were then ranked according to the final DSPAI score. The presence of good construction quality, better quality material, proper drainage system, adequate shoulders, adequate number of lanes, and proper lane separators are some factors that have led to higher DSPAI scores and, consequently, essential attributes for enhanced performance of HWs and EWs. Also, a well and effective traffic monitoring system and emergency services significantly contribute to enhanced safety and performance aspects of HWS and EWs as reflected in the study. The presented model is very easy to understand and implement as the value of measurable criteria for design and safety and performance index is easily determinable and quantifiable.

**Acknowledgements** The authors would like to thank Mr. Nishant Bhardwaj and Mr. Siraj Husain at BMSF Design Consultants, Gurugram, India for sharing their structural design expertise in refining the present model. The first author would like to thank the faculty and administration at Gautam Buddha University for their timely support during the research period.

## References

1. Chang C-W (2009) Applying fuzzy hierarchy multiple attributes to construct an expert decision making process. Expert System Appl 36:7363–7368
2. Saaty TL (1980) Analytical hierarchical process. McGraw hill, New York
3. Buckley JJ (1985) Fuzzy hierarchical analysis. Fuzzy Sets Syst
4. Sherman GB, Watkins RO (1968) Statistical quality control of highway construction material. Washington, DC, USA
5. Gibreel GM, Easa SM, Hassan Y et al (1999) State of the art of highway geometric design consistency. J Transp Eng Am Soc Civil Eng 125(4):305–313
6. Tsai CY, Chang AS (2012) Framework for developing construction sustainability items: the example of highway design. J Cleaner Prod 20(1):127–136, Science Direct
7. Hallowell M, Esmaeili B, Chinowsky P (2011) Safety risk interactions among highway construction work tasks. Construct Manage Econ 29(4):417–429
8. Hassan Y, Easa SM, Abd El Halim AO (1997) Design considerations for combined highway alignments. J Transp Eng 123(1), 60–68. ASCE. https://doi.org/10.1061/(ASCE)0733-947X(1997)123:1(60)
9. Cedergren HR (1974) Drainage of highway and airfield pavements. Robert E. Krieger Publishing Company, Incorporated Krieger drive Malabar, USA
10. Al-Sobky AA, Mousa RM (2016) Traffic density determination and its applications using smartphone. Alexandria Eng J 55(1):513–523
11. Anderson IB, Bauer KM, Harwood, et al (1999) Relationship to safety of geometric design consistency measures for rural two-lane highways. Transp Res Record J Transp Res Board Sage J 1658(1):43–51. https://doi.org/10.3141%2F1658-06
12. Battikha MG (2003) Quality management practice in highway construction. Int J Quality Reliabil Manage 20(5):532–550. https://doi.org/10.1108/02656710310476516
13. Kononov J, Bailey BA, Allery BK (2008) Relationships between safety and both congestion and number of lanes on urban freeways. Transp Res Record J Transp Res Board Sage J 2083(1):26–39
14. Jalayer M, Zhou H, Gong J, Hu S, Grinter M (2014) A comprehensive assessment of highway inventory data collection methods. J Transp Res Forum 53(2):73–92
15. Roshan P, Pal S, Kumar R (2018) Seismic damage assessment index of buildings using fuzzy-AHP approach. Int J Tech Innov Mod Eng Sci 4(08)

16. Roshan P, Pal S, Kumar R (2020) Performance assessment indexing of buildings through fuzzy AHP methodology. In: Ahmed S, Abbas S, Zia H (eds) Smart cities—opportunities and challenges. Lecture Notes in civil engineering, vol 58. Springer, Singapore

# Chapter 15
# Quality Factors Prioritization of Ready-Mix Concrete and Site-Mix Concrete: A Case Study in Indian Context

**Amartya Sinha, Nishant Singh, Girish Kumar, and Shilpa Pal**

## 1 Introduction

High demand for construction activities leads to a huge requirement and consumption of concrete mix. This huge demand has to be met in a shorter time for the timely completion of the projects. To meet such demand for concrete, in a shorter time span, the concept of Ready-Mix Concrete (RMC) plants is used. The usage of RMC eliminates the need for space required for the storage of cement, sand, aggregates, admixtures, etc., on the construction site. Moreover, the added advantage of RMC is the production of uniform concrete. Therefore, the quality of the concrete produced is much better and thus, the overall quality of the construction work is controlled effectively. However, the concept of using RMC is not feasible everywhere at all times and therefore, a lot of construction activities use Site-Mix Concrete (SMC), also known as In Situ Concrete, for concreting purposes.

Concrete Mix can be produced in two ways. One way is to store and maintain all the raw ingredients and machineries required at the construction site. Then the concrete can be produced at the construction site whenever required. Or, it can be produced at a separate location and then transported to the construction site. Former

A. Sinha (✉) · N. Singh · G. Kumar · S. Pal
Delhi Technological University, Delhi, India
e-mail: amartyasinha.dtu2022@gmail.com

N. Singh
e-mail: nishantsingh2539@gmail.com

G. Kumar
e-mail: girish.kumar154@gmail.com

S. Pal
e-mail: shilpapal@dtu.ac.in

 
D. Singh et al. (eds.), *Proceedings of International Conference on Scientific and Natural Computing*, Algorithms for Intelligent Systems,
https://doi.org/10.1007/978-981-16-1528-3_15

is known as In Situ Concrete, or Site-Mix Concrete and the latter is known as Ready-Mix Concrete and the location where RMC is produced is known as an RMC Plant [1].

Some studies were conducted for the quality improvement of RMC/SMC. Poovaragavan and Sekar [2] investigated continuous process improvement in Ready Mix Concrete Plants. Their study aimed at standardizing the quality of the concrete produced in the RMC plants using the Statistical Process Control (SPC) tools. In their study, they used the Relative Importance Index (RII) method for the ranking of quality factors. A review was conducted on factors affecting RMC Delivery Pattern by Afzal and Khan [3]. The study focused on the transportation process and related factors for RMC delivery. They adopted personal interviews and questionnaire surveys for their study. Alam et al. [4] studied the quality control of concrete Production in Dhaka City. They investigated the existing practices on the quality control of concrete production, by physical verification, and survey at various construction sites in Dhaka City.

From the above review, it is observed that SMC, though used widely in construction activities, does not gather much importance regarding its quality improvements. This study aims to identify the areas of vulnerability of the quality of the SMC which can be addressed to obtain higher quality and a more uniform mix. Also, here, it is intended to create a comprehensive list of different factors contributing to the variability of the quality of concrete mix for both RMC and SMC for a comparative study. Therefore, the following are the broad objectives of this study.

- To identify factors affecting the quality of RMC and SMC.
- Ranking of the quality factors of RMC/SMC using Relative Importance Index method.
- To analyze the process of mixing of RMC and SMC and identifying the quality factors ranked above, which require monitoring and further improvement.

Firstly, the quality factors for RMC/SMC are identified on the basis of extensive literature review and expert discussions. Thereafter, through a questionnaire survey and following the Multi-Criteria Decision Making (MCDM) Approach, the factors are ranked for both RMC and SMC. Lastly, the process of mixing of RMC and SMC are studied through relevant literature to determine which quality factors are properly addressed. The remaining factors are the critical areas where the high scope for improvement of quality of concrete is present.

The paper is organized into five sections. Section 2 deals with the details of the factors affecting the Quality of RMC and SMC which were selected for analysis in this study. Section 3 presents the methodology of the analysis. Section 4 gives a Case Illustration and the ranking of the factors. Lastly, Sect. 5 concludes this study.

## 2 Factors Affecting Quality of Ready-Mix Concrete and Site-Mix Concrete

Based on the extensive literature review, and expert discussions, the following quality factors for RMC and SMC are identified for the analysis of ranking.

### 2.1 *Quality Factors for Ready-Mix Concrete*

1. Quality of Ingredients used: A typical concrete mix comprises nearly 10% cement, 20% air and water, 30% sand, 40% gravel, and a certain number of admixtures. A typical construction aggregate, which includes sand and gravel, must have a rough surface texture and should be free from the coating and other dirt particles. Better quality ingredients will result in a better concrete mix [2].
2. Mix design adopted: Mix design means the quantity of suitable ingredients to produce concrete of desired properties in the most economical manner. Selecting the optimum mix design ensures the concrete with good workability in the plastic stage, desired minimum strength in the hardened stage and desired durability in the given environmental conditions [2].
3. Compaction and placement of concrete: Compaction means the removal of entrapped air bubbles from fresh concrete to make it dense in order to avoid honeycombing of concrete. Honeycombing may result in concrete cancer, if not treated early. Placement means deposition, distribution and consolidation of while casting. It must be placed, as nearly as practicable, in its final position to avoid re-handling. Moreover, it should be placed before the initial setting time starts so that it may not become plastic [4].
4. Skill of operators producing concrete at RMC plant: Skilled operators commit less mistakes in mixing correct proportions of ingredients while handling the machinery which, in turn, produces concrete of the required properties [2].
5. Calibration of RMC plant machinery: In a concrete batching plant, every ingredient is measured by weight with special scales, most of them being electronic. So, calibration must be done from time to time to maintain their accuracy [2].
6. Weather conditions and humidity at RMC plant and site location: Environmental Conditions like air temperature, wind velocity, relative humidity affect the quality of fresh and hardened concrete [5].
7. Age and condition of transit mixer: An ideal transit mixer maintains a homogeneous mix of cement, aggregate and water during its transportation. If transit mixers are old and not well maintained, problems of segregation and bleeding of concrete may arise [2].
8. Time of delivery of concrete: The time of delivery of concrete (Peak hour, Day, Night Time) also affects the quality of concrete as the transit time depends heavily on the time of delivery of concrete [3].

9. Traffic and Road conditions between RMC plant and site location: Delays and hauls caused by traffic and road conditions increases the time of delivery which changes the slump value of concrete and hence the workability [3].
10. Skills of labors in placing concrete: Skilled labors know different techniques of placing concrete depending on the site location. It can be done with buckets, hoppers, manual or motor-propelled buggies, conveyor belts, drop pipes and so on. The incorrect technique may result in re-handling of concrete and if the initial setting time is crossed, placement of concrete may get difficult [4].
11. Waiting time of transit mixer at the site: If the waiting time of the transit mixer increases, it can affect the quality of concrete. If the compressive strength of concrete is more, then Concrete may set rapidly in the mixer itself. Other environmental factors may also play a role in lowering the quality of concrete [6].
12. Coordination between RMC plant management and site staff: Effective Coordination between RMC Plant management and Site Staff may help in reducing any extra delay due to paperwork and counting of transit mixers.
13. Age and condition of machinery used in RMC plants: If machines in the plant are old and worn out, improper mixing of ingredients may be done which can lead to a poor and non-homogenous mix of concrete.
14. Degree of the overall site and RMC plant control: Without supervision, proper procedures of concrete production might not be followed by the crew members at the plant. Even at the site, supervision is required so that proper tests are conducted and the values are recorded before handling the concrete [4].

## 2.2 *Quality Factors for Site-Mix Concrete*

1. Quality of raw materials used: Whether the concrete is mixed in a plant, or in situ, the Quality of Raw Materials will affect the quality of the concrete mix.
2. Nominal mix ratio adopted: Usually, in the case of SMC, Nominal mix is used where mixing is done by volume and not by weight. This method generally leads to a non-uniform mix.
3. Skill of labors in producing Concrete mix: As discussed in RMC quality factors, skilled labors commit less mistakes in mixing the correct proportions of ingredients that produces concrete of the required properties.
4. Compaction and placement of concrete: Compaction is done to avoid honeycombing of concrete which may result in concrete cancer if not treated early. Placement of concrete means deposition, distribution and consolidation of concrete in the place where it is to harden. Concrete must be placed in its final position to avoid re-handling [4].
5. Skill of labors in placing concrete: As discussed in RMC quality factors, Workmanship is a significant factor affecting the quality of the concrete mix [4].

6. Weather conditions and humidity at the site: Environmental Conditions like Air temperature, wind velocity, relative humidity affect the quality of fresh and hardened concrete [5].
7. Uniformity and quality of concrete as compared to RMC plant: In an RMC plant, the proportion of every ingredient is measured with accuracy under a controlled environment which produces uniform and better-quality concrete. On the other hand, various conditions such as temperature and presence of impurities could not be controlled at the site due to which site mix concrete is less uniform and of poorer quality than RMC.
8. Age and condition of mixers: An ideal mixer combines all the ingredients homogeneously to form a concrete mix. If these are old and not maintained, mixing may not be done properly resulting in segregation of concrete mix. Mixer's performance also needs to be checked [4].
9. Record keeping of tests at the site: Record of every test conducted on the concrete at the site must be maintained by the crew members. This helps in determining the fault, if any, for the poor quality of concrete and rectifying it. If these records are not maintained properly, analysis and quality checks of SMC cannot be done as they are conducted for RMC.
10. Supply chain management at the site for raw materials: Delay in the supply of any raw material would affect the quality of concrete produced as other raw materials present in excess can suffer environmental degradation on prolonged storage.
11. Degree of overall site control: At the construction site, proper supervision is required so that every test to assess the quality of concrete is performed and their records are maintained which would help in rectifying any fault in the concrete production. The availability of trained and experienced personnel is a basic requirement of the success of a quality control plan [4].

## 3 Methodology

For comparing the quality of concrete produced in Ready-Mix Concrete Plant and that produced in situ, a survey was conducted in which responses were received according to multi-criteria decision-making approach (on a scale of 1 to 5) from respondents consisting of industry professionals, academicians and researchers with expertise in the domain of Civil Engineering. The survey was conducted to determine the Relative Importance Index of various factors discussed above which affect the quality of concrete mix in the case of RMC, as well as SMC.

The responses received from different people needed to be factored in appropriately as the level of expertise of respondents may vary. In this survey, to calculate the RII and rank these factors, a factor of 0.75, 1, and 1.25 was multiplied for responses from experts depending upon their experience and level of expertise in three categories.

For calculating the Relative Importance Index, (a score out of 1, with a higher value indicating higher importance), the Eq. (1) was used [7].

$$RII = \frac{\sum_{i=1}^{i=n} S_i f}{5n}. \quad (1)$$

where

$S_i$ is Score rated by *i*th respondent; f depicts Factor selected for the respondent's category (0.75 for junior researchers, 1 for academicians, 1.25 for industry professionals); n is the number of respondents = 22.

## 4 Case Study

If the Indian scenario is considered, especially in major cities and towns, RMC is replacing SMC gradually. While both RMC and SMC have their advantages and disadvantages, from the quality perspective, RMC holds a better position [1]. For analysis for the improvement of the quality of SMC, from the responses of the questionnaire survey and using the Eq. (1), the rankings were calculated.

For example, in the case of RMC, the RII value for the Quality of Ingredients used is calculated as:

$$RII = \frac{\sum_{i=1}^{i=n} S_i f}{5n} = \frac{0.75(48) + 1(4) + 1.25(39)}{5 \times 22} = \frac{90.25}{110} = 0.8068 \quad (2)$$

Similarly, for SMC, the RII value for Quality of Raw Materials Used is calculated as:

$$RII = \frac{\sum_{i=1}^{i=n} S_i f}{5n} = \frac{0.75(50) + 1(4) + 1.25(39)}{5 \times 22} = \frac{88.75}{110} = 0.8205 \quad (3)$$

The complete results for all the factors are shown in Table 1 and Table 2 for RMC and SMC, respectively.

In an RMC Plant, an overall strict and rigid control is generally practiced. Based on the mechanism of strict quality control being employed in the RMC Plants using the computerized technologies, the factors ranked in Table 1 can be analyzed. The Quality of Ingredients Used (Ranked 1 with RII 0.8068) is maintained in a uniform manner with the use of batching plants and strict monitoring. The Mix Design adopted (Ranked 2 with RII 0.7659) is implemented using computerized modules [2, 8] where quality control is easily achieved. The Machineries, and their calibration, and Conditions of Transit Mixers, etc., are given a lot of importance in the quality control process and therefore, the sources of poor quality from those areas can also be neglected for our study purposes. However, the Process of Delivery of Concrete,

**Table 1** Ranking of factors affecting quality of RMC

| Factors | Relative importance index | Rank |
|---|---|---|
| Quality of ingredients used | 0.8068 | 1 |
| Mix design adopted | 0.7659 | 2 |
| Compaction and placement of concrete | 0.7523 | 3 |
| Skills of operators producing the concrete at RMC plant | 0.7477 | 4 |
| Degree of overall site and RMC plant control | 0.7250 | 5 |
| Coordination between RMC plant management and site staff | 0.7023 | 6 |
| Calibration of RMC plant machineries | 0.7000 | 7 |
| Weather conditions and humidity at RMC plant and site locations | 0.6841 | 8 |
| Time of delivery of concrete (Early Morning, Late Night, etc.) | 0.6641 | 9 |
| Age and condition of transit mixer used | 0.6568 | 10 |
| Traffic and road conditions between RMC plant and site location | 0.6432 | 11 |
| Skills of labors in placing the concrete | 0.6205 | 12 |
| Waiting time of transit mixer at site | 0.6068 | 13 |
| Age and condition of machinery used in the RMC | 0.5864 | 14 |

**Table 2** Ranking of factors affecting quality of SMC

| Factors | Relative importance index | Rank |
|---|---|---|
| Quality of raw material used | 0.8205 | 1 |
| Nominal mix ratio adopted | 0.8114 | 2 |
| Skills of labors in producing the concrete mix | 0.7568 | 3 |
| Compaction and placement of concrete | 0.7523 | 4 |
| Skills of labors in placing the concrete | 0.7182 | 5 |
| Degree of overall site control | 0.7136 | 6 |
| Weather conditions and humidity at site locations | 0.6977 | 7 |
| Uniformity and quality of concrete as compared to RMC plants | 0.6705 | 8 |
| Age and condition of mixers | 0.6295 | 9 |
| Supply chain management at the site for raw materials | 0.6273 | 10 |
| Record keeping for tests at site | 0.5841 | 11 |

during the time of travel, and Coordination between RMC Plant Management and the Site Staff, which can also lead to the increased waiting time of the Transit Mixer at the site, remains the major concern along with the degree of Overall Site Control where possibilities of neglect in placing the concrete can happen. Other factors like weather conditions and humidity at the site control can be dealt easily with the correct calibration of the machineries.

In the case of Site-Mix Concrete, flexible Site Control is generally seen. The factors ranked in Table 2 can be analyzed in a similar fashion. Proper quality control for the Raw Materials (Ranked 1 with RII 0.8205) is very difficult to achieve. There is no computerized process of manufacturing the concrete nominal mix (Ranked 2 with RII 0.8114), and a lot of errors can happen during the mixing of the concrete which may lead to unwanted qualities in the concrete mix and defects in the overall structure. Moreover, usually, volume mixing is observed in the Site-Mix Concrete instead of Weigh Mixing which also results in a deteriorated and sometimes, uneconomical construction. Also, the quality of the Site-Mix Concrete depends a lot on the skills of labors producing and placing the concrete (Ranked 3, 4, and 5), which may, or may not be properly checked and controlled.

## 5 Conclusion

In this study, a thorough review of the Ready-Mix Concrete Production, as well as Site-Mix Concrete Production, is done in the perspective of the quality of the fresh concrete mix obtained in both cases. The ranking of the quality factors selected based on the literature review was done with the help of a questionnaire survey.

Few factors which can lead to the quality of Ready-Mix Concrete becoming poor are Overall Site Control, RMC Management and Site Staff Coordination, high waiting time of RMC after reaching the site locations, and the process of transit of the concrete and placement of the concrete mix at the site locations. These areas need to be addressed to further improve the quality of the RMC produced. The case of Site-Mix concrete is, however, a bit different. In SMC, few factors which significantly affect the quality of the concrete mix are Quality of the Raw Ingredients, Relatively Inaccurate Nominal Mix Ratio, and high dependence on the skills of labors producing and placing the concrete. The best way to address these factors is to develop and follow a similar framework as followed in the RMC Plants. Although it is not feasible to make the process of Site-Mix Concrete production computerized as in the case of smaller sites, this can lead to high economic losses, but in case of the site being larger in size, and being more significant and important in terms of quality maintenance, some fraction of the budget can be utilized to provide RMC like environment.

If the Concrete Consumption pattern in India is considered, it is highly dominated by Site-Mix Concrete production. Only major metro cities and tier 1–2 cities have the facility of the controlled environment of RMC where mechanical production with the help of computers and technologies is possible [1]. According to data given in this study, out of 450-475 Million m3 of concrete consumption in India as of Feb 2015, about 350 Million $m^3$ (about 73%) of concrete is produced in situ [1, 9]. In future, RMC production might become more economic and the SMC production might become more cumbersome and tedious. Although the future of the concrete construction industry may depend more on the RMCs, the Site-Mix Concreting cannot and should not be replaced completely by the former. Therefore, the need for replicating the controlled environment in the Site-Mix Concrete production is important, which can be done gradually by addressing the factors affecting the quality of the RMC and SMC as ranked above.

## References

1. Manjunatha LR, Sandhya Anvekar R (2015) History of readymix concrete industry in India. ABBS Manage Bus Entrepr Rev 6(1):20–41
2. Poovaragavan E, Chandra Sekar K (2016) Continuous process improvement in ready mix concrete plants. Int J Sci Eng Res 7(4):154–159
3. Afzal S, Khan ZR (2018) A review paper on factors affecting ready-mix concrete delivery pattern. Int J Construct Eng Manage 7(3):97–100
4. Alam MA, Habib MZ, Sheikh MR, Hasan A (2016) A study on the quality control of concrete production in Dhaka city. IOSR J Mech Civil Eng 13(3) Ver. III, 89–98
5. Almusallam AA (2001) Effect of environmental conditions on the properties of fresh and hardened concrete. Cement Concrete Compos 23:353–361
6. Naiknavare HP, Deshpande SD, Padhye RD (2013) Model chart of quality control process for ready mixed concrete plants. In: Second international conference on emerging trends in engineering. IOSR J Mech Civil Eng, 50–54
7. Rajgor M, Chauhan P, Patel D, Panchal C, Bhavsar D (2016) RII & IMPI: effective techniques for finding delay in construction project. Int Res J Eng Technol 1173–1177
8. Kashwani G, Liu E, Atif A (2019) Safety review of the quality ready-mix concrete (RMC) and workmanship in the construction industry. J Safe Eng 8(1):1–8
9. Kurian J (2015) Seminar document. ICI-KBC concrete panorama & deminar. Quality assurance & innovations in ready-mixed concrete

# Chapter 16
# Study on Optimization of Unreliable Server Queueing Systems: A PSO Based Survey

**Radhika Agarwal, Divya Agarwal, and Shweta Upadhyaya**

## 1 Introduction

Queueing theory gives techniques to investigate complex service systems in computing systems, communication networks, transportation, manufacturing and many more real life situations. Authors working in this field are always motivated to search an algorithm that could give best optimal results in studying the queuing models. Nature inspired algorithms (NIA) are currently some of the most powerful tools to get optimal solutions. There are dozens of optimization algorithms available to solve different real life models. NIA have played a key role in solving complex queueing functions.

In 1995, Kennedy [1] introduced Particle swarm optimization (PSO) after inspecting the migratory behaviour of swarms and birds. Since the last decade, many authors have tried their hands on PSO in different fields of optimization. Some literature that helps in better understanding of PSO, modification and hybridization done in PSO along with its industrial application can be seen [2–7] for better understanding.

Unreliable queuing system implies when a server may breakdown at some point of time and as a result may partially or completely stop working. Sometimes a server may intentionally take a break upon seeing no customer and that is termed as vacation. Thus in both the conditions the server stops working either partially or completely. Queueing model that incorporate unreliable server and server vacations depict the real world scenario more precisely. Many authors have worked on unreliable queueing systems considering server on different types of vacation too. But only few of them has used PSO algorithm to optimize their models.

R. Agarwal (✉) · D. Agarwal · S. Upadhyaya
Amity Institute of Applied Sciences, Amity University, Sector 125, Noida, UP, India
e-mail: radhika6696@gmail.com

D. Agarwal
e-mail: dagarwal1@amity.edu

S. Upadhyaya
e-mail: supadhyay@amity.edu



D. Singh et al. (eds.), *Proceedings of International Conference on Scientific and Natural Computing*, Algorithms for Intelligent Systems,
https://doi.org/10.1007/978-981-16-1528-3_16

Liou [8] investigated an infinite Markovian queue where the server which is unreliable may breakdown while working. Impatient customers could balk or renege depending on the choice of the customer. PSO is implemented to solve a minimization problem and obtain minimum cost. Later epsilon constraint method and GA (genetic algorithm) are used to get the Pareto optimal solution. Recently, Yang and Wu [9] investigated the M/M/1 queueing system under N-policy where the server are subjected to breakdowns and may go for vacations while working. They used PSO for the optimization of the cost function that included the threshold N and mean serving rates. Yang et al. [10] presented a M/M/2 queueing system and its steady-state analysis. The model had heterogeneous (two) servers where Server 1 can be relied upon which leaves for vacation only if the system became empty. Sever 2 can't be relied upon and breaks down while giving service to customers. Canonical PSO was used by the authors to obtain approximate numerical solutions for the optimal serving rates of both the servers. Shekhar et al. [11] investigated a randomized arrival control policy in the finite capacity queueing model with server working while on vacation and may face vacation interruption. They also considered the impatient behavior of the customers. They implemented PSO to get the minimal cost for their system.

Retrial queueing system is an upcoming area of interest for many authors. Once a customer enter a queue, he may find the server busy. As the server seem to be busy at a particular point of time, the customer may join a virtual space called orbit and may retry getting his service from there by revisiting again and again. Many authors have used PSO in retrial queuing models to optimize the cost function. Wang et al. [12] considered the customer's behavior which is strategic in an M/M/1 queueing system with constant retrial times where they follow N-policy. PSO is used to get the best solution for the social welfare problem. Zhang et al. [13] considered a queue that has a single server with retrial times in which they control the serving rate using state-dependent service policy. Using Canonical PSO algorithm they specified that most benefit could be obtained if the managers apply this state-dependent service policy rather then the classic service model. Yang and Wu [14] analysed a steady-state analysis of an M/M/1 queueing model with retrial times which may go on vacations while working, where the server may fail in the starting itself. They employed CPSO algorithm to deal with the cost optimization problem.

The following article includes the survey on unreliable queueing models where PSO or CPSO algorithms have been used to solve the cost optimization problem. The rest of the article is as follows. Section 2 includes the basic definition of PSO and its application in unreliable queuing models and some real life queueing models. Section 3 incorporates some famous optimization techniques including ACO, FA and Bat algorithm with their working and few queuing papers involving them. Finally, we put together our work by terminating in Sect. 4 as a concluding remark.

# 2 Particle Swarm Optimization (PSO)

Swarm and birds migrate from one place to other depending on the atmospheric conditions or their feeding habits. This migratory nature of birds and swarms was studied by Kennedy [1]. He observed that the interaction between the flocks can help in getting a precise location. He thought of its novelty in solving complex nonlinear problems and it gave birth to one among the popular optimization algorithm called PSO.

## *2.1 Implementation of the Algorithm in Queueing Models*

Molecule swarm improvement (PSO) is a sort of computerized reasoning recreation strategy. It is a stochastic optimizing calculation dependent on swarm intelligence theory. PSO accomplishes the development of particulars by moving to the ideal individual and the ideal position. The algorithm takes the situation of every molecule as a solution vector, with the goal that every molecule has a chronicled optimal position. As a resultant the entire particle swarm gets an optimal position. Hence refreshes the position and speed of every molecule. Along these lines, the entire multitude tends to the optimal position of the group. In most cases, the calculation can cause the system to merge rapidly. As a result of this feature, the algorithm turns into a successful analyzer. PSO has played keen role in optimizing different models of queueing systems. Many authors have applied this technique to get optimal cost and pricing. Major work done in optimizing unreliable server queueing systems are summarized in Table 1.

## *2.2 Application of PSO in Real Life Queueing Models*

Application of PSO was intially limited to machine repairman problem. Wang et al. [22] analyzed M/M/R machine repair problem with the service pressure condition involving imperfect coverage. In case a machine under operation fails, it may immediately be detected,located and would be replaced. They first applied the direct search method to the profit function following which they performed Quasi-Newton method to search the optimal solution. Further PSO algorithm was carried out to get the optimal parameters for maximum profit. Later [16], they analysed a M/M/1 queueing system with multi-vacation where the repairman is unreliable. It is a warm-standby machine repair problem. They compared the search results of the PSO algorithm to those of the commonly used search method to ensure that the optimizing quality of the PSO is not compromised.

Recently, authors have started applying PSO in many other sectors like network at interchange stations, beds allocation, designing of airport security screening and

**Table 1** Particle Swarm Optimization used in unreliable server queueing models

| Algorithms | Author | Short description | Year |
|---|---|---|---|
| PSO & Pareto front solution | Liou [8] | An infinite markovian queue where the server is unreliable which may face working breakdowns and impatient customers | 2013 |
| PSO and Quadratic fit | Laxmi et al. [15] | A finite buffer queue with multiple working vacations where customers may balk or renege while vacation interruption occurs under N-policy | 2013 |
| PSO | Wang et al. [16] | A multiple-vacation machine repair problem where the repairman is unreliable | 2014 |
| PSO | Yang and Wu [9] | M/M/1 queueing model under N-policy where server may breakdown and may go on working vacation | 2015 |
| GA & PSO | Zandieh et al. [17] | A non-homogeneous unreliable production lines with buffer allocation problem and preventive maintenance planning | 2017 |
| Canonical PSO | Wang et al. [18] | A retrial machine repair problem where server may breakdown while working under N-policy | 2019 |
| Canonical PSO | Yang and Wu [14] | M/M/1 retrial queueing model such that the server may fail initially and may work during vacations | 2019 |
| Canonical PSO | Yang et al. [10] | M/M/2 queueing model with heterogeneous (two) service provider where server may breakdown while working and may go on vacation for more than one time | 2020 |
| PSO | Ma and Zhang [19] | A queueing system with N-policy where the server is unreliable | 2020 |
| PSO | Upadhyaya [20] | Geo/G/1 queueing system with retrial times, following Bernoulli feedback where the server may fail initially in the starting itself | 2020 |
| PSO | Shekhar et al. [21] | Multi-server queueing system considering retention of reneging customers with Bernoulli's scheduled modified vacation policy | 2021 |

**Table 2** Application of PSO in real life queueing models

| Algorithms | Author | Short description | Year |
|---|---|---|---|
| PSO and Quasi-Newton | Wang et al. [22] | The service pressure condition involving imperfect coverage of warm-standby M/M/R machine repair problem | 2013 |
| PSO | Wang et al. [16] | An unreliable repairman with multiple-vacation warm-standby machine repair problem in M/M/1 queueing system | 2014 |
| PSO | Narasimhamu et al. [24] | Buffer allocation problem with single server in tandem closed queueing network | 2014 |
| PSO & GA | Liou [25] | M/M/1 queueing system with machine repair problem where server may go on multiple vacations and may breakdown while working | 2015 |
| Hybrid simulation PSO | Khattak [26] | Designing of staircase according to its breadth and pathway network at an interchanging station | 2019 |
| PSO | Hou et al. [27] | Beds allocation in hospitals based on queuing model | 2019 |
| MO-PSO | Souza et al. [28] | Multiple-objective optimization of single server general finite queueing networks | 2020 |
| QT-PSO | Naji et al. [23] | Designing of security screen at an airport using PSO augmented with queueing theory | 2020 |

many more. Naji et al. [23] proposed augmented along with PSO a method based on queueing system (QT-PSO) to presume passengers waiting time in a screening process at airport security. Their model could be reliable as it is able to comprehend disparity in the amount of present passengers, security machines at the time of service and security officers. To evaluate their proposed method, they have gathered real-life data at security guarding from an airport in Australia for two concurrent years i.e. 2016 and 2017. This made it more trustworthy. Some of the real life application of PSO in queuing models studied in the literature are listed in Table 2.

## 3 Other Nature Inspired Algorithms

Nature Inspired Optimization Algorithm has got a huge notification for their strategies in critical problem solving. A few genuine issues are accomplished utilizing nature inspired calculations and furthermore advancement with these calculations place a critical part in taking care of the issues. There are different strategies for characterizing and depicting meta-heuristic estimations. Some of them used in solving queueing problems are listed below along with their proceeding process.

## 3.1 Ant Colony Optimization (ACO)

By analyzing the ant's behaviour i.e. rummaging for food we achieve the shortest path in this optimization technique. In 1991, Dorigo [29] introduced The Ant Colony Optimization(ACO). The algorithm tells the way in which the ant access their actual path in real life. The algorithm considers the course of the ant as the genuine way. More the pheromone deposition much favorable is the path, whereas the pheromone vanishing guarantees that the non-ideal course found in the prior hunting stage isn't concurrent quickly.

Initially this algorithm was applied to solve the traveling salesman problem (TSP) [30] were we found that this system was much effective in solving optimization problems. With the advances in the algorithm's development, ACO has widely being applied to many fields including queueing and reliability modelling. Lu [31] presented a fuzzy location allocation model which is applicable for emergencies on large-scale basis. They have developed ACO algorithm to solve and test the model through a large-scale emergency. Laxmi [32] analyzed a finite buffer queue with multiple working vacations. She also considered balking and reneging of the customers along with Bernoulli scheduled vacation interruption in the queue where cost analysis is done using ant colony optimization.

## 3.2 Firefly Algorithm (FA)

Studying the bio-luminescence in the fireflies, Yang [33] (a British scholar) in 2008 proposed the Firefly Algorithm(FA). In this algorithm, the resultant vector is the position of a firefly, and the destination model decides the illumination of the light and the magnetization to other fireflies. The brightness of light gives the position of the pleasant or unpleasant. As thing are more attracted to short distance, so stronger the light with clearer atmosphere, better is the position of the firefly. The entire firefly multitude can be consequently partitioned into subgroups. Every subgroup will gather collectively around the global model of the greatest pinnacle and the base valley of the models. Therefore, this algorithm is very effective in finding multiple optimal solutions simultaneously.

Kwiecien [34] discussed two continuous time queueing systems with losses. The models incorporate finite capacity where the customers are impatient. He successfully used FA in optimizing these systems. He Later [35], presented Markovian systems with service times to be exponential distributed and arrival follows a poisson process with losses. In this model also he incorporated finite capacity where the customers are impatient along with finite number of jobs in closed queueing system. Verma [36] has given a detailed review on Firefly Algorithm including its application in engineering and scheduling problems.

### 3.3 Bat Algorithm

Bats' power to produced ultrasonic waves helps in echolocation to fly. In 2010, Yang [37] proposed an algorithm named Bat Algorithm by studying this bat's echolocation technique. The ultrasonic waves tells the distance from the nearby impediments along with getting the bugs. The bats can get the information about the dimension and heading of the bug by echo, such that bats can get their target more precisely. Every bat in the group shows a usable solution in the solution space and every solution corresponds to a fitness value dictated by the issue to be improved. It is very much like the PSO algorithm. Each and every bat follows the bat which is currently optimal in the space specified for search, by adjusting the frequency, pulse rate and loudness.

Laxmi [38] considered a finite buffer $M/M/1$ queueing system with N policy, multiple vacations where the server is still in working mode and Bernoulli schedule with vacation interruption. Under the two customers' impatient situations, balking and reneging, the optimum service rate of a queue problem is studied. Most recently Shekhar [39] studied the impact of new introduced emergency vacation in service system of a single service provider through queueing-theoretic approach. For optimal control they have carried out a comparative analysis of some meta-heuristic and heuristic optimization techniques.

## 4 Conclusion

The key objective of this survey is to provide a unified framework for using particle swarm optimization algorithm in queueing models. PSO has found wide application in optimizing the day-to-day life models along with several engineering based systems. It is found that earlier the use of PSO algorithm was limited to salesman repair problem. However, the diversified use of nature inspired algorithms for optimizing the problem of queues have also captivated the attention of many scholars and researchers as reported in this survey.

We have also seen in this article that since last decade, how researchers are now highly motivated to use these algorithms in optimising queuing systems. So in future definitely researchers will explore more, to get optimal results in queueing models by using these swarm based optimization algorithms. Some recently developed algorithms include Spider Monkey optimization (SMO), Grey Wolf Optimizer (GWO) and many more which can be used in the queueing models to get the optimal values.

## References


1. Kennedy J, Eberhart R (1995) Particle swarm optimization. In: Proceedings of ICNN'95—international conference on neural networks, Perth, WA, Australia, vol 4, pp 1942–1948 (1995). https://doi.org/10.1109/ICNN.1995.488968
2. Gonzalo E, Fernández-Martínez JL (2012) A brief historical review of particle swarm optimization (PSO). J Bioinf Intell Control 1:3–16
3. Shafiq A, Dobbie G, Koh YS, Riddle P, Rehman S (2014) Research on particle swarm optimization based clustering: a systematic review of literature and techniques. Swarm Evol Comput 17:1–13. https://doi.org/10.1016/j.swevo.2014.02.001
4. Zhang YD, Wang S, Ji G (2015) A comprehensive survey on particle swarm optimization algorithm and its applications. Math Prob Eng 2015:1–38. https://doi.org/10.1155/2015/931256
5. Benuwa BB, Ghansah B, Wornyo DK, Adabunu SA (2016) A comprehensive review of particle swarm optimization. Int J Eng Res Africa 23:141–161. https://doi.org/10.4028/www.scientific.net/JERA.23.141
6. Slowik A, Kwasnicka H (2018) Nature inspired methods and their industry applications-swarm intelligence algorithms. IEEE Trans Indus Inf 14(3):1004–1015. https://doi.org/10.1109/TII.2017.2786782
7. Sengupta S, Basak S, Peters R II (2019) Particle swarm optimization: a survey of historical and recent developments with hybridization perspectives. Mach Learn Knowl Extract 1:157–191. https://doi.org/10.3390/make1010010
8. Liou CD (2013) Markovian queue optimisation analysis with an unreliable server subject to working breakdowns and impatient customers. Int J Syst Sci 46(12):2165–2182. https://doi.org/10.1080/00207721.2013.859326
9. Yang DY, Wu CH (2015) Cost-minimization analysis of a working vacation queue with N-policy and server breakdowns. Comput Indus Eng 82:151–158. https://doi.org/10.1016/j.cie.2015.01.017
10. Yang DY, Chen YH, Wu CH (2020) Modelling and optimisation of a two-server queue with multiple vacations and working breakdowns. Int J Prod Res 58(10):3036–3048. https://doi.org/10.1080/00207543.2019.1624856
11. Shekhar C, Varshney S, Kumar A (2020) Optimal and sensitivity analysis of vacation queueing system with F-policy and vacation interruption. Arab J Sci Eng 45:7091–7107. https://doi.org/10.1007/s13369-020-04690-5
12. Wang J, Zhang X, Huang P (2017) Strategic behavior and social optimization in a constant retrial queue with the N -policy. Eur J Oper Res 256(3):841–849. https://doi.org/10.1016/j.ejor.2016.06.034
13. Zhang X, Wang J, Ma Q (2017) Optimal design for a retrial queue system with state-dependent service rate. J Syst Sci Complex 30(2):883–900. https://doi.org/10.1007/s11424-017-5097-9
14. Yang DY, Wu CH (2019) Performance analysis and optimization of a retrial queue with working vacations and starting failures. Math Comput Model Dyn Syst 25(5):463–481. https://doi.org/10.1080/13873954.2019.1660378
15. Laxmi PV, Goswami V, Jyothsna K (2013) Optimization of balking and reneging queue with vacation interruption under N-policy. J Optim. https://doi.org/10.1155/2013/683708
16. Wang KH, Liou CD, Wang YL (2014) Profit optimisation of the multiple-vacation machine repair problem using particle swarm optimisation. Int J Syst Sci 45(8):1769–1780. https://doi.org/10.1080/00207721.2012.757378
17. Zandieh M, Joreir-Ahmadi MN, Fadaei-Rafsanjani A (2017) Buffer allocation problem and preventive maintenance planning in non-homogenous unreliable production lines. Int J Adv Manuf Technol 91:2581–2593. https://doi.org/10.1007/s00170-016-9744-4
18. Wang KH, Wang J, Liou CD, Zhang X (2019) Particle swarm optimization to the retrial machine repair problem with working breakdowns under the N-policy. Queue Models Serv Manage 2(1):061–081
19. Ma Q, Zhang X (2020) Analysis and comparison of queue with N-policy and unreliable server. Discrete Dyn Nat Soc 2020:1–10. https://doi.org/10.1155/2020/6195080

20. Upadhyaya S (2020) Cost optimisation of a discrete-time retrial queue with Bernoulli feedback and starting failure. Int J Indus Syst Eng 36(2):165–196. https://doi.org/10.1504/IJISE.2020.110245
21. Shekhar C, Varshney S, Kumar A (2021) Matrix-geometric solution of multi-server queueing systems with Bernoulli scheduled modified vacation and retention of reneged customers: a meta-heuristic approach. Quality Technol Quantitat Manage 18(1):39–66. https://doi.org/10.1080/16843703.2020.1755088
22. Wang KH, Liou CD, Lin YH (2013) Comparative analysis of the machine repair Problem with imperfect coverage and service pressure condition. Appl Math Modell 37(5):2870–2880. https://doi.org/10.1016/j.apm.2012.06.024
23. Naji M, Braytee A, Al-Ani A et al (2020) Design of airport security screening using queueing theory augmented with particle swarm optimisation. Serv Orient Comput Appl 14:119–133. https://doi.org/10.1007/s11761-020-00291-0
24. Narasimhamu K, Reddy V, Rao C (2014) Optimal buffer allocation in tandem closed queuing network with single server using PSO. Proc Mater Sci 5:2084–2089. https://doi.org/10.1016/j.mspro.2014.07.543
25. Liou CD (2015) Optimization analysis of the machine repair problem with multiple vacations and working breakdowns. J Indus Manage Optim 11(1):83–104. https://doi.org/10.3934/jimo.2015.11.83
26. Khattak A (2019) Simulation-based particle swarm optimization technique for the breadth design of stairs and passageway network at interchange stations. MATEC Web Conf 272:01036. https://doi.org/10.1051/matecconf/201927201036
27. Hou Y, Zeng H, Wang Y, Wang X, Gao Q (2019) Optimization of beds allocation based on queuing model and the particle swarm optimization algorithm. Int J Circuits Syst Signal Process 13
28. Souza GL, Duarte AR, Moreira GJP, Cruz FRB (2020) A novel formulation for multi-objective optimization of general finite single-server queueing networks. In: 2020 IEEE congress on evolutionary computation (CEC), pp 1–8, Glasgow, UK. https://doi.org/10.1109/CEC48606.2020.9185827
29. Dorigo M, Maniezzo V, Colorni A (1991) Positive feedback as a search strategy. Technical Report 91-016, Dipartimento di Elettronica, Politecnico di Milano, IT
30. Yang QW, Ruan SN, Chen JF, Li B (2016) Application survey on swarm intelligence in the traveling salesman problem. Techniq Autom Appl 8:1–6
31. Lu X, Hou Y, Qiang S (2010) A fuzzy queuing facility location model with ant colony optimization algorithm for large-scale emergencies. In: 7th international conference on service systems and service management, pp 1–6, Tokyo. https://doi.org/10.1109/ICSSSM.2010.5530267
32. Laxmi PV, Singuluri I, Jyothsna K (2016) Ant colony optimization for optimum service times in a Bernoulli schedule vacation interruption queue with balking and reneging. J Indus Manage Optim 12(4):1199–1214. https://doi.org/10.3934/jimo.2016.12.1199
33. Yang XS (2008) Nature-inspired metaheuristic algorithms. Luniver Press
34. Kwiecień J, Filipowicz B (2012) Firefly algorithm in optimization of queueing systems. Bull Polish Acad Sci Tech Sci 60(2):363–368. https://doi.org/10.2478/v10175-012-0049-y
35. Kwiecień J, Filipowicz B (2014) Optimization of queueing structures by firefly algorithm. 516:331–346
36. Verma H (2020) A systematic review on firefly algorithm: past, present, and future. https://doi.org/10.36227/techrxiv.12122748
37. Yang XS (2010) A new metaheuristic bat-inspired algorithm. In: González JR et al. (eds) Nature inspired cooperative strategies for optimization (NICSO 2010). Studies in computational intelligence, vol 284, pp 65–74. Springer, Berlin, Heidelber. https://doi.org/10.1007/978-3-642-12538-6_6
38. Laxmi PV (2019) Bat algorithm for optimal service parameters in an impatient customer N-policy vacation queue. Int J Math Model Comput 9(2):113–122
39. Shekhar C, Varshney S, Kumar A (2020) Optimal control of a service system with emergency vacation using bat algorithm. J Comput Appl Math 364. https://doi.org/10.1016/j.cam.2019.06.048

# Chapter 17
# Brain Tumor Detection Through MRI Using Image Thresholding and GUI

**Aditi Verma, M. A. Ansari, Rajat Mehrotra, Pragati Tripathi, and Shadan Alam**

## 1 Introduction

In recent years medical imaging has enhanced a lot which has led to better planning and accuracy in surgical procedures as it also plays an important role in medical researches. Even though the cases of brain tumor are not that common but the rate at which malignant tumor spreads in the brain cells is very high. Thus, the detection of brain tumors at their early stage an important task. To detect brain tumors accurately we need a medical image by which we can precisely identify the presence of the tumor. For this purpose, we use CT scans and MRI scans. Both are preferred for the identification of the tumor. A tumor can be diagnosed via imaging and pathology. The brain tumor is imaged with MRI (Magnetic Resonance Imaging), giving a cross-section image of the brain [1]. Image processing is widely used by medical researchers for finding out better ways to identify and analyze a particular disease. Because of the complex structure of the human brain and also due to the complex shape, size, structure, and texture of the tumor it has become very difficult to cure brain tumor perfectly.

Therefore, advancing in the field of image processing where various methods have been introduced to detect the tumor efficiently and easily. Our goal used to detect brain tumors accurately with the help of image processing. Extraction of a brain tumor can be accomplished by performing various operations on MATLAB

A. Verma (✉) · M. A. Ansari · P. Tripathi · S. Alam
Department of Electrical Engineering, Gautam Buddha University, Greater Noida, India
e-mail: vermaaditi881998@gmail.com

M. A. Ansari
e-mail: ma.ansari@ieee.org

R. Mehrotra
Department of Electrical and Electronics Engineering, GL Bajaj Institute of Technology and Management, Greater Noida, India

 
D. Singh et al. (eds.), *Proceedings of International Conference on Scientific and Natural Computing*, Algorithms for Intelligent Systems,
https://doi.org/10.1007/978-981-16-1528-3_17

software. The image thresholding method is used to detect the tumor through MRI scans. MRI scan is taken as input and it is converted into a grayscale image. Then for the detection of tumor steps such as preprocessing, image segmentation, feature extraction, and image classification are performed. This project image thresholding technique is used to detect the tumor from an MRI image and then various features of that image are also extracted. Several segmentation techniques can be implemented to detect the tumor one of them is image thresholding, these techniques help in the detection of the tumor by considering the solidity of an image. As the solidity of the tumor is greater than the normal brain cells. But before the image is segmented it goes through various morphological operations and the image is also preprocessed to remove the noise. Brain tumors can be benign which do not contain cancer cells or they can also be malignant which contains cancer cells. The brain tumor is classified into four grades. Grade I tumors are slowly growing and most unlikely to spread around them. Grade II tumor is also slow-growing but they looked abnormal under a microscope. Grade III tumors actively reproduce tumor cells. Grade IV tumors are the most malignant tumors.

These are the following main steps for brain tumor detection:

***Preprocessing***: Its task is to improve the quality of the MR images

***Image segmentation***: In this, the image is converted into a binary image, and morphological operations are performed.

***Feature extraction***: It is the process of collecting higher-level information of an image.

***Image classification***: Classification is the last step in the process, it simply groups the assigned pixel regions.

## 2 Requirement Analysis

System requirements include both the requirement of the user and the software. Both of them must be fulfilled for obtaining better results. There can be more than one user also. In this system of brain tumor detection, there are many users such as a doctor, patient, and researcher. For the patient, the software must provide an easy way to see the tumor and understand its size and position. For doctors, the result should be reliable and appropriate by which cure can easily be done. For researchers, the result and techniques should be properly defined which might help in further improvement of the work. These all were the requirement of the user which can also be modified with time.

The functional requirement includes the selection of MRI of the brain as an input. So for this, a separate folder of the database has to be collected which contains brain MRI with and without tumor. The government also includes the process of extracting the tumor region from the scanned image and finding the boundary of

**Table 1** Rating of components

| Operating system | Processors | Disk space | RAM |
|---|---|---|---|
| Windows 10 | Any Intel or AMD × 86 processor | 26 GB for MATLAB | 2 GB RAM |

the tumor. Creating a program for easy access is also included in the functional requirement.

For nonfunctional requirements, we need to install MATLAB, the reliability of the system, and the performance of the software (Table 1).

## 3 Methodology

In recent days, medical image processing plays a vital role in medical research. Image classification, filtration, and segmentation are the major processing techniques. The morphological operation of MRI brain image before performs the anisotropic diffusion filter, which reduces the contrast between consecutive pixels. This image is resized and converted into a greyscale image [2]. Figure 1 describes the workflow of the system.

The accuracy of detecting brain tumor location and size takes the most important role in the successful diagnosis and treatment of tumors. Filter operation was used to remove unwanted noises as much as possible to assist better segmentation [3]. So, the detection of brain tumors needs to be fast and accurate [4]. Magnetic Resonance

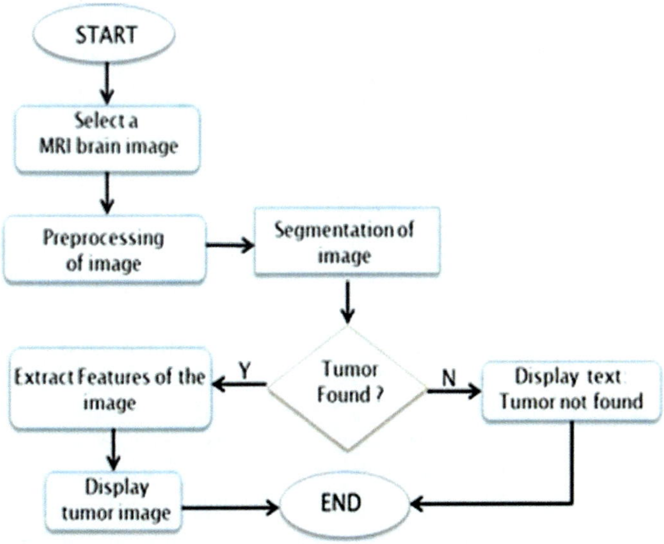


**Fig. 1** Flow chart of the system

Imaging (MRI) is a technique that is applied to display a detailed image of the attacked brain location. The medical imaging trick plays a significant behavior in the identification of the disease [5].

For accurate detection of a brain tumor, several steps are performed such as preprocessing, segmentation, feature extraction, and classification. Brain Tumor is a testing errand as symptoms show up in the progressive phases of the tumor. To provide accuracy in cerebrum tumor location MRI (Magnetic Resonance Image) is a widely preferred imaging modality. Giving better revelation of the tumor without affecting a normal tissue is an extremely difficult process [6]. All these processes are adopted for obtaining better and more reliable results.

- **Image Preprocessing**

T1-w MRI scans of the brain are taken from [7] as input, these images are grayscale and resized to 256 × 256 px for processing. Preprocessing technique is used to improve the detection of tumor region from the given MRI. The input image is a 2D grayscale image.

Processing of the input image is important because the image might be blurry or grainy which can affect the result of tumor detection. Anisotropic filtering and morphological operations are used to make the image smooth and remove the blur. In this work, a brain MRI taken as input with the help of anisotropic filtering blur is removed detail of the image is preserved. Anisotropic filtering makes the image smooth and preserves the sharpness of the edges. The input arguments for performing this function are the Gradient threshold, number of iterations, conduction method, and connectivity. The value of gradient threshold is a numeric scalar or numeric vector value. By increasing it the image becomes smoother. By default, the value of the gradient threshold is 0.01. The number of iterations determines the number of times filtering should be done. Connectivity describes the connection between the neighboring pixels which can be minimal or maxima. After anisotropic filtering, morphological operations are performed to make the image clearer and remove small grainy objects from the image. In this project, we are going to use morphological operations which are dilation and erosion. The relation makes the object more visible and fills small holes in the object. Erosion removes the islands and small objects so that substantive objects remain the same.

- **Image Segmentation**

In the segmentation of an image, the image is divided into subgroups of pixels. There are various techniques of image segmentation such as threshold method, edge-based segmentation, region-based segmentation, clustering-based segmentation, etc. In this paper, we are using the image thresholding method for segmentation as it is the most basic and powerful method to identify the tumor.

In brain MRI presence of a tumor is detected based on intensity. As the intensity of the tumor is greater than normal brain cells in an MRI. The brain MRI is converted to a binary image and is smoothed by anisotropic filtering as shown in Fig. 2. Then image thresholding method is applied to detect the presence of a tumor. The region

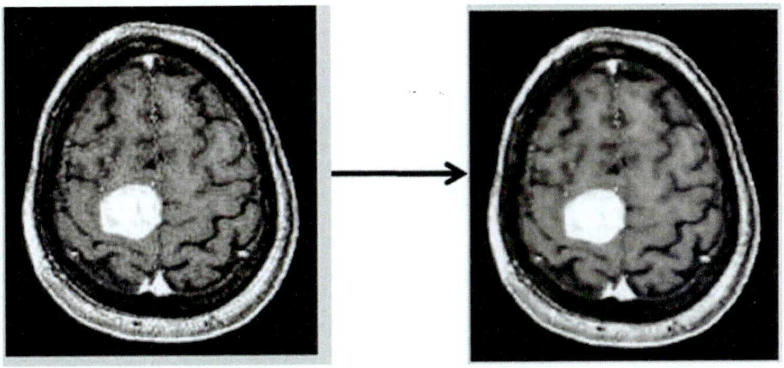

**Fig. 2** **a** Input image, **b** Processed image

which has solidity (intensity) higher than the given solidity is detected the presence of a tumor is confirmed in the given MRI.

Here we use two arguments solidity and area to detect the total area affected by the tumor. The threshold method is also subdivided into various types such as Simple Thresholding, Otsu's Binarization and Adaptive Thresholding. In simple thresholding, we replace the pixels either with black or white. If the intensity is greater than the threshold then it is converted to white otherwise to black. We use the region props command to get the solidity and area of the image. Then we define the specified value of solidity, if any area is present that has solidity above the defined value then it is considered as a tumor-affected area. The processing of magnetic resonance images (MRI) is one of the parts of image processing in the medical field, which is the most emerging field in the past few days [8].

- **Feature Extraction**

The accuracy of detecting brain tumor location and size takes the most important role in the successful diagnosis and treatment of tumors. So the detection of brain tumors needs to be fast and accurate. Brain tumor detection is considered a challenging mission in medical image processing [4]. Feature extraction is the process used to collect the information of the high level of the given MRI. Computer-aided detection applications have managed to make significant contributions to the medical world in today's technology [9]. This information includes the shape, texture, color, contrast, correlation, energy, entropy, autocorrelation, etc. This step is used to increase the accuracy of tumor detection. By using these prominent features we might gain some more information regarding the tumor. The textural finding and analysis can improve the diagnosis of the tumor. Features can be pixel intensity-based, texture-based, edge-based, etc. In feature extraction, we can also get the tumor alone image, tumor outline, and bounding box around the detected tumor. In the feature extraction step, we calculate the numerical value of various features. In this paper, we have the values of six features and they are contrast, correlation, autocorrelation, variance, energy, and entropy.

- **Contrast**: It is defined as the intensity of the pixel and its neighbor over the image.
- **Autocorrelation**: It is defined as the correlation of a signal with a delayed copy of itself also known as serial correlation.
- **Variance**: It is a measure of the square of average contrast.
- **Entropy**: It is calculated to determine the randomness of the textural image.
- **Energy**: Energy is defined as the amount of pixel repetition.

## 4 Implementation

In today's modern medical arena patients with brain tumors are increasing rapidly at a fast pace and above the par. Detection of brain tumors has become a challenging task to compete with [10].

The main objective of this work is to get some useful information regarding brain tumors. This paper aims to define an algorithm that gives the result of the tumor by extracting its features also from the input MRI. The resultant image obtained will give the information about the presence of tumor in the given MRI and it will also extract the features of the image. Automatic detection of brain tumors is a difficult task due to variations in type, size, location, and shape of tumors. In this paper, an abnormal brain MRI is chosen as input. Then it is converted into a binary image.

The image is then filtered with anisotropic filtering to remove the noise in the image. Two morphological operations are also performed which are erosion and dilation to make the image clearer. Anisotropic filtering and morphological operations are part of preprocessing of input MR image. These steps are important because they increase the accuracy of tumor detection thus helps us to get proper results. After preprocessing we next go to the segmentation step where the tumor is detected with the help of intensity (solidity). The area described is a highly dense area where the solidity is above the described value. After detecting the tumor, we create a bounding box around the tumor. The tumor alone image is also extracted by displaying the high dense area only. Then we will program to obtain a tumor outline to get the perfect image of the tumor. Input image, filtered image, bounding box image, tumor alone image, tumor outline image and detected tumor image all are obtained. Brain tumor segmentation is the most difficult task because it requires a huge amount of data.

There are many types of tumors present with different shapes and sizes. To detect those tumors automatically challenging task. Moreover, maintaining accuracy is also very difficult. The presence of tumors also affects the intensity of normal brain cells in MRI. In this paper, a total of five features are extracted which are contrast, correlation, autocorrelation, energy, and entropy. These are some of the measurable features which are extracted are computed from the brain MRI. All these five features are mainly intensity-based. They give high-level information about the image. Are very important for the study of tumor detection and also for the diagnosis is it increases the accuracy of the diagnosis system by selecting prominent features. The algorithm was carried out on MATLAB 2017R a. After preprocessing of the image, it is segmented and the tumor is also detected, feature extraction is also performed after the segmentation. After the completion of the whole programming, a user-friendly GUI is created which can be accessed by everyone with ease. For creating MATLAB GUI, we type the command "guide" on the command window. After that, another window will pop up. In which we can add axes, tables, pushbuttons, etc. The main aim of using this is to avoid the whole lengthy programming and for making it present in a better and user-friendly manner. In this, we will form 2 axes one of which is to take the input from the user and the other one to display the detected tumor. Three push buttons are used for input, detection, and feature extraction. A feature extraction table is also displayed.

## 5 Results and Discussion

Brain tumor detection in an early stage is a difficult task, as the imaging is quite unclear. The necessity of automated brain tumor segmentation and detection is high. Obtain an accurate MRI image of the brain tumor is challenging. An MRI image has high contrast images indicating regular and irregular tissues that help in differentiating the overlap margins [11, 12].

In this work, the tumor is detected by the image thresholding method and a GUI is also created to get the table of extracted features. An input MRI is taken from the user and then it is filtered and segmented and features of the image are also extracted. If an input image is taken which does not contain a tumor then a notification will pop up of no tumor.

Firstly, we have to ask the user to select an input image for tumor detection. Then the selected image is resized to 256 × 256 pixels. After that, the image is converted to binary. Then the tumor is detected and outlined with a dotted yellow line. The filtered image, tumor alone image and detected tumor image are displayed together. The detection of a tumor depends upon the value of intensity (solidity) we have selected. Thus, the proper value of intensity should be selected to obtain an accurate result. Accuracy plays an important role in the study of tumor detection as well as in the diagnosis. So proper value of the threshold should be selected. In the given result of (Fig. 3), there is no tumor present.

For the same above result (Fig. 4) of tumor detection, we have the GUI through which we can detect the presence of tumor easily (Fig. 5). Simply just push the button load image then select the input image from the dataset and click on detect. The image of the detected tumor will automatically be shown on the other axes.

For feature extraction a table has to be formed with it a pushbutton must also be added to get the features of the image. All five features must be included in the program. After the detection of the tumor is completed then the feature of the image is extracted, all these are numeric values that are important to obtain high-level information of the tumor.

In the above result of Fig. 6, the values of the feature extracted are numeric. There are some default values also set to the given features. But in the given result

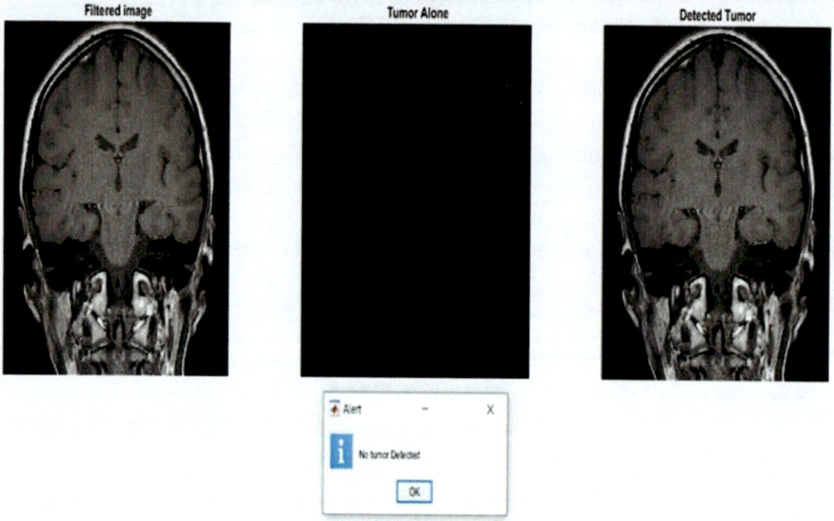


**Fig. 3** Result of MRI with no tumor present

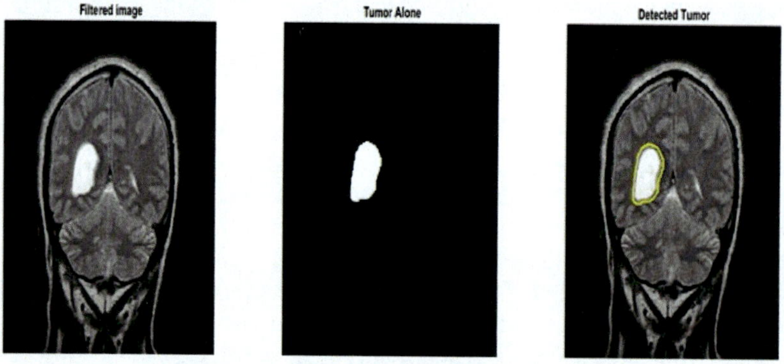


**Fig. 4** Result of MRI with detected tumor and tumor alone image

the values displayed are of the given image only. The value of contrast defines the intensity of pixels and their neighbors or we can also say that the difference between the darkest and the brightest region. The value of correlation varies between −1 and 1. Entropy describes the uncertainty or randomness of the MRI. Energy gives information regarding the uniformity of the pixels.

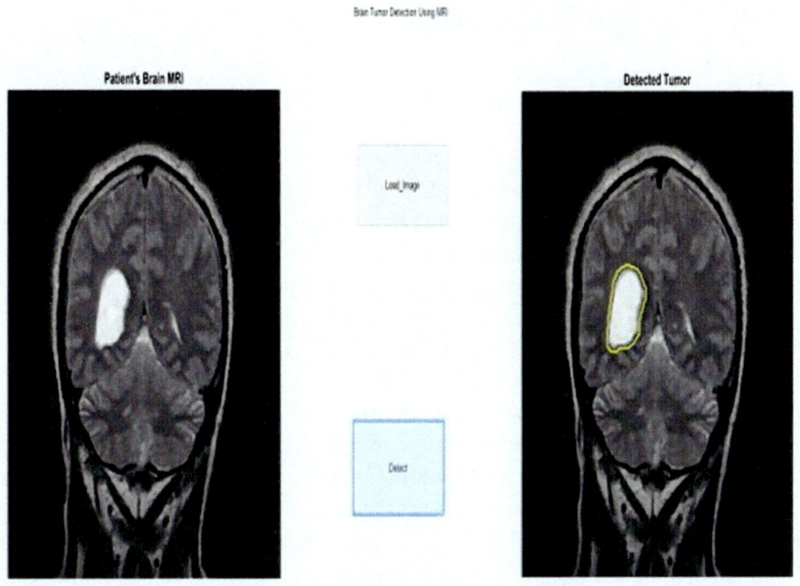


**Fig. 5** GUI result of MRI with detected tumor image

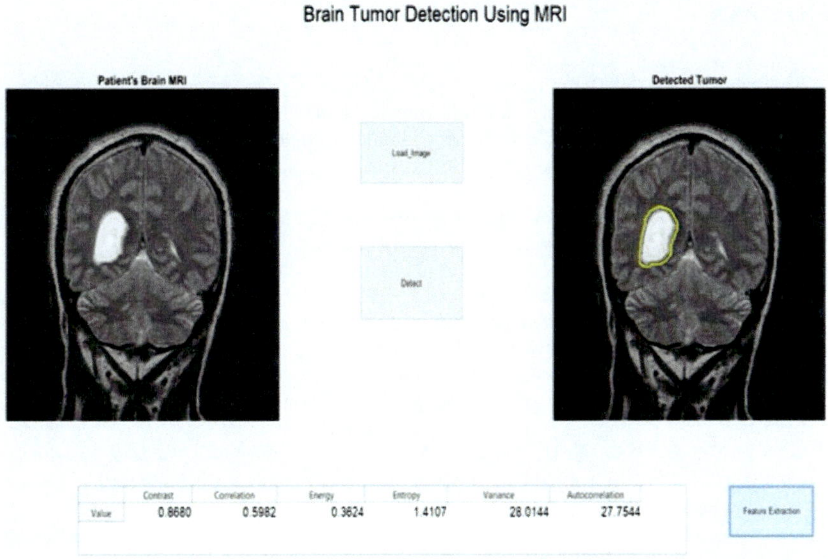


**Fig. 6** GUI result of MRI with detected tumor and extracted features

## 6 Conclusion and Future Scope

The proposed algorithm takes input from the user as a grayscale image that may or may not contain a tumor. The images are then converted to binary and processed by operations such as filtering, dilation, and erosion through MATLAB programming. The image thresholding method is used for segmentation. The detected tumor (if present) is then displayed and highlighted by a yellow outline. For feature extraction total of five features are extracted from the image. All these features are presented in form of a table. A user-friendly GUI is also created which makes it easy to use the program. The method of preprocessing segmentation and feature extraction is related to the intensity of the image. By making use of the fact that the intensity of the area where the tumor present is greater than the intensity of normal brain cells in an MRI. A possible extension of the presented work is that we can include more features that might give more information regarding the image. To get a more accurate result we can also program the input which is suitable for 3D images also. As 3D images more suitable for tumor detection than today's images. If we could connect the system to cloud storage then it would be beneficial for the patient as well as for the hospital. We can also increase the accessibility of this project by connecting it to the mobile phones.

## References

1. SuhibIrsheidat, RD (2020) Brain tumor detection using artificial convolutional neural networks. In: 11th international conference on information and communication systems (ICICS)
2. Mohanarathinam A (2020) Enhanced image filtration using threshold based anisotropic filter for brain tumor image segmentation. In: 3rd international conference on intelligent sustainable systems(ICISS)
3. Shahriar Sazzad TM, TanzibulAhmmed KM, Misbah Ul H, Mahmuda R (2019) Development of automated brain tumor identification using MRI images, 04 Apr 2019, 7–9 Feb 2019
4. Abdulbaqi HS, Mat M, Omar AF, Bin Mustafa IS, LoayKadomAbood (2015) Detecting brain tumor in Magnetic Resonance Images using hidden Markov random fields and threshold techniques. In: IEEE student conference on research and development
5. Rashid MHO, Mamun MA, Hossain MA, Uddin MP (2018) Brain tumor detection using anisotropic filtering, SVM classifier and morphological operation from MR images. In: International conference on computer, communication, chemical, material and electronic engineering (IC4ME2)
6. Akbar S, Nasim S, Wasi S, Ussama Zafar SM (2019) Image analysis for MRI based brain tumor detection. In: 4th international conference on emerging trends in engineering, sciences and technology (ICEEST)
7. Princeedey/brain-tumor-detection-and-segmentation-using-mri-images: https://github.com/princeedey/brain-tumor-detection-and-segmentation-using-mri-images
8. Santosh Hunnur S, Raut A, Kulkarni S (2017) Implementation of image processing for detection of brain tumors. In: International conference on computing methodologies and communication (ICCMC)
9. Kazdal S, Doğan B, YılmazÇamurcu A (2015) Computer-aided detection of brain tumors using image processing techniques. In: 23nd signal processing and communications applications conference (SIU)

10. Manisha, Radhakrishnan B, Padma Suresh L (2017) Tumor region extraction using edge detection method in brain MRI images. In: International conference on circuit, power and computing technologies (ICCPCT)
11. Samantaray M, Panigrahi M, Patra KC, Panda AS, Mahakud R (2020) An adaptive filtering technique for brain tumor analysis and detection. In: 10th international conference on intelligent systems and control (ISCO)
12. Mehrotra R, Ansari MA, Agrawal R, Anand RS (2020) A transfer learning approach for AI-based classification of brain tumors. Mach Learn Appl 2:

# Chapter 18
# A Study on Retrial G-Queues Under Different Scenarios: A Review

**Geetika Malik, Shweta Upadhyaya, and Richa Sharma**

## 1 Introduction

Retrial or recurrent queue is an integral part of queueing theory. A system in which the arriving customer keeps on retrying (at a constant retrial rate) to avail the same service on being rejected by the server is known as the retrial queueing system. In such cases, customers wait in the waiting area or '*orbit*' from where they can retry again and again to avail the service. Retrial queues are widely applicable in packet switching networks, production and management, telephone industry, banks, etc. A very basic example of a retrial queue in real life situation is making a phone call. Whenever a call is not connected due to some issue then we often tend to retry and make that call again. Thus, recurring calls and retrial queues are very much interrelated. This is why this concept is highly implied in call centres and cellular mobile networks. Another sector that can be considered as a multi-server retrial queueing system is hospitals where customers are categorized as a priority (patients with an emergency) and non-priority customers (routine check-up). We can say that wherever there is a need to minimize a queue length as well as the waiting time of retrying customers then the retrial queueing system plays its vital role. For further study of the applications one can refer to work done by Ohmura and Takahashi [32], Hui and Tao [22], Tran-Gia and Mandjes [51], Aguir et al. [1], Roszik et al. [46], Artalejo and Corral [8] and Do [18]. Ohmura and Takahashi [32], Hui and Tao [22] and Artalejo

G. Malik (✉) · S. Upadhyaya
Amity Institute of Applied Sciences, Amity University, Noida, India
e-mail: geetika.taluja@student.amity.edu

S. Upadhyaya
e-mail: supadhyay@amity.edu

R. Sharma
Department of Mathematics, J.K Laxmipath University, Jaipur, India
e-mail: richasharma@jklu.edu.in

 
D. Singh et al. (eds.), *Proceedings of International Conference on Scientific and Natural Computing*, Algorithms for Intelligent Systems,
https://doi.org/10.1007/978-981-16-1528-3_18

and Corral [8] have worked on retrial models and their impact on communication systems in different protocols. The application of retrial queues on cellular mobile networks has been explained by Tran-Gia and Mandjes [51], Roszik et al. [46] and Do [18].

Today when technology and communication system is on the highest pace of development, it is hard to find a queueing system which runs smoothly (without any interruption) at every instant. Not every customer entering the system gets served, some of them enter just to remove one or more positive customers (the one being in service), further causing an interruption or breakdown in service. We can broadly classify such arrivals in two types depending upon how many customers are removed or destroyed. The first type of negative customer puts one positive customer, one being at the head and in-service out on entering the system. The other is a disaster, removing all customers present in the system immediately. Retrial queues with negative arrivals (called retrial G-queues) provide a perfect model for various situations, say virus entering a system, production and inventory control, call centers, simple mail transfer protocol (SMPT), etc. Artalejo and Corral [6] initiated the trend of combining repeated customers with negative arrivals. Then Corral and Artalejo [15] also worked in the same area. Since then numerous analysts have scrutinized this field because of its wide area of applications. We have explained the application of retrials G-queue in call centres through a pictorial diagram (Fig. 1).

For detailed applications of recurrent *G*-queues readers may go through Dimitriou [16] who investigated a retrial model under mixed priority and negative units with unreliable servers and multiple vacations. One of the practical examples explained by the author is in context of a computer network, comprising processors that are connected via a central transmission unit. Firstly, a message is sent to this central unit. The messages to be sent are of two kinds, urgent and normal. If the transmission

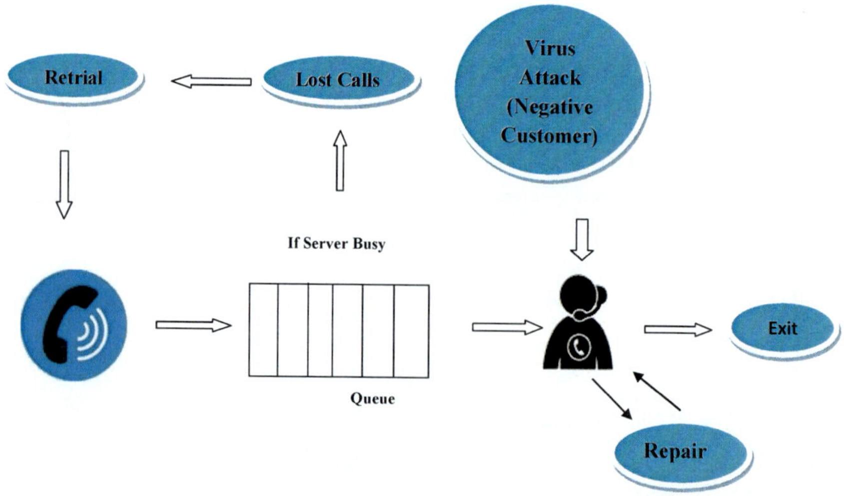


**Fig. 1** Pictorial representation of retrial G-queues in a call center

channel is busy, the urgent messages are stored in a buffer space, while the normal is retransmitted (retrial) later on. On the other hand, if the medium is ready to serve, the central unit sends the message immediately. Clearly urgent messages are delivered immediately and so the medium prioritizes the transmission of urgent messages. Also, such systems are exposed to *virus infection*. When a virus enters the system it can destroy the message under transmission (negative arrival) and forces the system manager to go through a reset of the system (repair from abnormal failure). Moreover, mechanical parts of the system may breakdown, and causes system failure that must be repaired (repair from normal failure).In such a situation the message must be retransmitted. When the system manager finishes all tasks in the buffer or a failure is repaired and no messages are stored in the buffer the medium remains available for a random period. If this time expires the medium is used for secondary jobs as long as urgent messages are stored to be transmitted (multiple vacations).

The remaining paper is organised as follows: Sect. 2 includes all the research papers on the basis of different types of queueing models. Then Sect. 3 classifies the work done till now on the basis of different queueing methodologies. Further, recurrent G-queues under discrete-time environment are discussed in Sect. 4. Also, retrial queues with the disaster are presented in Sect. 5. Following that, we have discussed about inventory systems with recurrent and negative customers in Sect. 6. Finally, conclusions are drawn in Sect. 7.

## 2 Classification on Basis of Queueing Models

To date lot of authors have analyzed recurrent queues as it can be applied very well in telecommunication systems, computer networks, travel and manufacture industries, ATMs, call centers, etc. The incorporation of negative units enhances the practicality of the model. A classification of such queues will make it easy for the upcoming researchers to study this field. Thus, in the following table we have classified them on the criteria of different types of retrial G-queueing models (Table 1).

## 3 Classification on the Basis of Different Techniques

The method applied to study a particular retrial queueing model is very essential to obtain the desired results. Annaswamy [5] and Subramanian [50] have written excellent books on matrix geometric approach and generating function technique respectively. There are enormous techniques which have been used till now. However, we have discussed some common and recently applied methodologies in G-queues with repeated customers as given in Table 2.

**Table 1** Classification of retrial G-queues on the basis of different models

| Models | Authors | Key features |
|---|---|---|
| M/M/c | Ayyappan et al. [9] | Erlang-K service |
| | Do et al. [19] | Working vacations |
| | Farkhadov and Fedorova [20] | M/M/1 model + Calls under heavy load condition |
| M/G/1 | Liu et al. [28] | N-policy + server breakdown + repair |
| | Aissani [2] | Unreliable server |
| | Krishnakumar et al. [25] | Bernoulli feedback |
| | Wu and Yin [57] | Non-exhaustive random vacations + unreliable server |
| | Wu and Lian [55] | Unreliable server + Bernoulli vacation |
| | Berdjoudj and Aissani [12] | Martingale technique |
| | Gao and Wang [21] | Non-persistent customers |
| | Peng et al. [33] | Server breakdown + Delayed repair |
| | Li and Zhang [27] | Working breakdowns |
| | Rajadurai et al. [41] | Multiple working vacation + vacation interruption |
| | Rajadurai [35] | Three types of customers + working vacation |
| | Rajadurai [36] | Unreliable server + working vacation + vacation interruption |
| | Rajadurai et al. [43] | Feedback + working breakdowns |
| $M^X$/G/1 | Yang et al. [58] | Single vacation + breakdown + repair |
| | Divya and Chandrika [17] | Server breakdown |
| | Bhagat and Jain [13] | Preemptive resume |
| | Rajadurai et al. [40] | Orbital search + Bernoulli vacation |
| | Rajadurai et al. [42] | Working vacation + vacation interruption |
| $M^X$/($G_1$,$G_2$)/1 | Rajadurai et al. [38] | Balking + starting failures |
| $M_1^x M_2^x$/($G_1$,$G_2$)/1 | Ayyappan and Thamizhselvi [11] | Working vacation + vacation interruption + priority services + delayed repair |
| M/M/c/c | Nesrine and Natalia [29] | Multiple servers |
| | Nesrine et al. [30] | Exponential abandonments |

**Table 2** Classification of retrial G-queues on the basis of different methods

| Method | Authors | Key features |
|---|---|---|
| Matrix method | Annisimov and Artalejo [4] | Multiple servers |
| | Ayyappan et al. [9] | Erlang-K service + Multiple servers |
| | Wu et al. [56] | Markovian arrival process |
| Embedded Markov chain technique | Krishnakumar et al. [25] | Bernoulli feedback |
| | Gao and Wang [21] | Orbital search + non-persistent customers |
| Supplementary variable technique | Wu and Yin [57] | Unreliable server |
| | Yang et al. [59] | Single vacation + repair |
| | Kirupa and Chandrika [23] | Unreliable server + delayed repair |
| | Peng et al. [33] | Server breakdown + delayed repair |
| | Rajadurai et al. [39] | At most J-vacations |
| | Li and Zhang [27] | Working breakdowns |
| | Radha et al. [34] | Multi-optional services + server breakdown |
| | Rajadurai [35] | Working vacation policy |
| | Rajadurai et al. [43] | Working vacation + vacation interruption |
| | Kirupa and Chandrika [24] | Fluctuating modes of service + Preemptive priority |
| | Nila and Sumitha [31] | Multi-optional service + priority + feedback |
| Direct truncation method | Ayyappan et al. [10] | Coxian phase-type service |
| Adaptive neuro-fuzzy interface system | Singh et al. [48] | State-dependent arrival |
| Maximum entropy principle | Singh et al. [49] | Unreliable server + optional service + delayed repair |

## 4 Discrete-Time Retrial G-Queues

Many computer networks, as well as a communication system, operate via discrete-time systems wherein an event can take place at regular time slots only. For instance, the arrival of customers, service, breakdown, repair, etc. all such activities occur at fixed time epochs. Designing of some of the communication systems such as asynchronous transfer mode (ATM), circuit-switched time-division multiple access (TDMA), etc. is based on the principle of discrete-time retrial phenomenon. While a lot of research is done on continuous-time retrial queues, very little work is done in the field of discrete-time systems. The work done by Yang and Li [59] started the trend of combining repeated attempts with these types of queues. They obtained the queue size distribution of Geo/G/1 discrete-time model with repeated customers. In today's world of high-speed technology, the telecommunication sector involves numerous complications and it is impossible that a virus does not attack the system at all. Thus, studying the concept of negative arrivals under a discrete-time environment is highly recommendable. Wang and Zhang [53] introduced negative arrivals with a non-continuous retrial queue under server breakdown. They calculated the marginal distributions of the orbit size when the server is idle, busy or down. Thereafter, Wang and Zhang [54] incorporated the concept of a server being sent to repair and presented various numerical examples to depict the impact of system parameters on the performance characteristics.

## 5 Retrial G-Queues with Disaster

In today's era, there is a rapid increase in the industry of telecommunication and computer networking. Thus the proper analysis of a retrial queueing model is highly required to control the situation from getting worse and reduce congestion. This can be done by considering the case of negatives units as well since it is next to impossible to create a system without a virus attack. Disaster queues involve those types of negative customers which removes not just the positive customer at the head of the queue but all the customers there in the system. Because of this, a regular busy server becomes defective and unable to serve. This situation leads to a loss for analyzers and system designers and thus there is a need to manage these units perfectly. In previous years, Artalejo and Corral [7] analysed a single-server queueing system with disaster. The necessary performance metrics for a multi-server retrial queue with disaster have been given by Shin [47]. Wang et al. [52] and Chen et al. [14] worked on an M/G/1 recurrent model with disaster. The former included the concept of server failure while the latter involved N-policy vacation. Rajadurai [37] applied a supplementary variable technique for analyzing an M/G/1 retrial model with disaster under working vacations and working breakdown. Recently, Ammar and Rajadurai [3] calculated the performance metrics of an M/G/1 retrial model with disaster under preemptive priority and working breakdown conditions.

## 6 Two Servers (S,S) Inventory Retrial Systems with Negative Customers

Many researchers have studied inventory systems for decades but only recently their attention has turned towards retrial customers in the inventory system. The system starts with S units of inventory on hand. Each arriving customer is served a single unit of the item by any one of the two servers. When the inventory level reaches s, an order is placed for (S-s) units. If the inventory level is zero or both servers are busy, then the arriving customer goes to orbit and becomes a source of repeated calls. The negative arrival plays an important role in these systems as it can control the congestion in the orbit by removing one customer from the orbit. Revathy and Subramanian [45] solved this type of model by applying the direct truncation method. They have also presented very impactful numerical and graphical studies. Rajkumar [44] too worked on (s,S) inventory retrial system with negative arrivals and impatient customers as well. Later on, Laxmi and Soujanya [26] applied the matrix geometric method to solve the retrial inventory model with negative units and server breakdown. They have shown a cost analysis along with some numerical results also.

## 7 Conclusion

In this survey, we have studied different types of retrial queueing models wherein the negative customer may arrive and cause an interruption in the ongoing service. These queues are quite relatable to real-life scenarios as a virus attack is very common in complicated networking systems. Firstly, we have discussed a practical application in detail. Then the prominent research papers are classified on different criteria which will enable the upcoming scholars to study these queues in a systematic manner. Also, recurrent models with discrete-time slots are discussed separately which have a great scope in future years in the field of signal processing. Disaster retrial queues are discussed which are applicable when all units get evacuated due to any type of server failure. Finally, a short study on inventory systems with recurrent and negative customers is presented. We hope that this study will help and motivate the researchers to further analyze such queues for the betterment of the telecommunication and networking sector.

## References


1. Aguir S, Karaesmen F, Aksin OZ, Chauvet F (2004) The impact of retrials on call center performance. OR Spectrum 26:353–376. https://doi.org/10.1007/s00291-004-0165-7
2. Aissani A (2010) An M/G/1 retrial queue with negative arrivals and unreliable server. Lecture Notes in engineering and computer science 1
3. Ammar SI, Rajadurai P (2019) Performance analysis of preemptive priority retrial queueing system with disaster under working breakdown services. Symmetry 11. https://doi.org/10.3390/sym11030419
4. Anisimov VV, Artalejo JR (2001) Analysis of Markov multiserver retrial queues with negative arrivals. Queue Syst 39:157–182. https://doi.org/10.1023/A:1012796517394
5. Annaswamy MGS (2011) Retrial queueing system with priority services matrix geometric approach. Lambert
6. Artalejo JR, Corral AG (1998) Generalized birth and death processes with applications to queues with repeated attempts and negative arrivals. OR Spectrum 20:5–14. https://doi.org/10.1007/BF10545523

7. Artalejo JR, Corral AG (1999) Computation of the limiting distribution in queueing systems with repeated attempts and disasters. Rairo Operat Res 33:371–382. https://doi.org/10.1051/ro:1999113
8. Artalejo JR, Corral AG (2007) Modelling communication systems with phase type service and retrial times. IEEE Commun Lett 11:955–957. https://doi.org/10.1109/LCOMM.2007.070742
9. Ayyappan G, Sekar G, Subramaniam AMG (2010) M/M/1 retrial queueing system with negative arrival under Erlang-K service by matrix geometric method. Appl Math Sci 4:2355–2367
10. Ayyappan G, Sekar G, Subramaniam AMG (2011) Study of influences of negative arrival for single server retrial queueing system with coxian phase type service. QTNA, 53–59
11. Ayyappan G, Thamizhselvi P (2018) Transient analysis of $M_1^x M_2^x/(G_1,G_2)/1$ retrial queueing systems with priority services, working vacation and vacation interruption, emergency vacation, negative arrival and delayed repair. Int J Appl Comput Math 4
12. Berdjoudj B, Aissani D (2014) An analysis of the M/G/1 retrial queue with negative arrivals using a martingale technique. J Math Sci 196:11–14. https://doi.org/10.1007/s10958-103-1068-7
13. Bhagat A, Jain M (2016) N-policy for $M^X/G/1$ unreliable retrial G-queue with preemptive resume and multi services. J Operat Res Soc China 4:437–459. https://doi.org/10.1007/s40305-016-0128-0
14. Chen P, Zhao G, Zhou Y (2014) An M/G/1 retrial queue with subject to disasters and N-policy vacation. Comput Model New Technol 18:509–515. https://doi.org/10.13140/R.G.2.1.4845.4480
15. Corral AG, Artalejo JR (1999) On a single server retrial queue with negative arrivals and request repeated. J Appl Probab 36:907–918. jstor.org/stables/3215450
16. Dimitriou I (2013) A mixed priority retrial queue with negative arrivals, unreliable server and multiple vacations. Appl Math Model 37:1295–1309. https://doi.org/10.1016/j.apm.2012.04.011
17. Divya A, Chandrika KU (2015) An $M^X/G/1$ retrial G-queue with server breakdown. Int J Innov Res Sci Eng Technol 4
18. Do TV (2010) A new computational algorithm for retrial queues to cellular mobile systems with guard channels. Comput Ind Eng 59:865–872. https://doi.org/10.1016/j.cie.2010.08.017
19. Do TV, Papp D, Chakka R, Sztrik J (2014) M/M/1 retrial queue with working vacations and negative customer arrivals. Int J Adv Intel Paradigms 6:52–65. https://doi.org/10.1504/IJAIP.2014.059586
20. Farkhadov M, Fedorova E (2017) Retrial queue M/M/1 with negative calls under heavy load condition. Distributed Comput Commun Netw 406–416. https://doi.org/10.1007/978-3-319-66836-9_34
21. Gao S, Wang J (2014) Performance and reliability analysis of an M/G/1 retrial G-queue with orbital search and non persistent customers. Eur J Oper Res 236:561–572. https://doi.org/10.1016/j.ejor.2014.01.065
22. Hui L, Tao Y (1995) A single server retrial queue with server vacations and a finite number of input sources. Eur J Oper Res 85:149–160. https://doi.org/10.1016/0377-2217(94)E0358-I
23. Kirupa K, Chandrika KU (2014) Batch arrival retrial G-queue and an unreliable server with delayed repair. Int J Innov Res Sci Technol 3:12436–12444
24. Kirupa K, Chandrika KU (2018) Unreliable batch arrival retrial G-queue with fluctuating modes of service, preemptive priority and orbital search. Int J Math Trends Technol (Special Issue):45–53
25. Krishnakumar B, Pavai S, Laxmi SPA (2011) An M/G/1 Bernoulli feedback retrial queueing system with negative customers. Oper Res Int J 13:187–210. https://doi.org/10.1007/s12351-011-0107-5
26. Laxmi PV, Soujanya ML (2015) Retrial inventory system with negative customers and service interruptions. Opsearch 52:212–224. https://doi.org/10.1007/s12597-014-0181-6
27. Li T, Zhang L (2017) An M/G/1 retrial G-queue with general retrial times and working breakdowns. Math Comput Appl 22. https://doi.org/10.3390/mca22010015

28. Liu Z, Wu J, Yang G (2009) An M/G/1 retrial G-queue with preemptive resume and feedback under N-policy subject to the server breakdowns and repairs. Comput Math Appl 58:1792–1807. https://doi.org/10.1016/j.camwa.2009.07.077
29. Nesrine Z, Natalia D (2018) On multiserver retrial queues with negative arrivals. Int J Math Operat Res 13:219–242. https://doi.org/10.1054/IJMOR.2018.094056
30. Nesrine Z, Spiteri P, Djellab N (2019) Numerical solution for the performance characteristics of the M/M/c/k retrial queue with negative customers and exponential abandonments by using extrapolation method. Rairo Operat Res 53:767–786. https://doi.org/10.1051/ro/2017059
31. Nila M, Sumitha D (2020) Analysis of repairable batch arrival retrial queueing system with second multi optional service, orbital search, priority and feedback customers. Int J Adv Sci Technol 29:2873–2882
32. Ohmura H, Takahashi Y (1985) An analysis of repeated call model with a finite number of sources. Electron Commun Jpn 68:112–121. https://doi.org/10.1002/ecja.4410680613
33. Peng Y, Liu Z, Wu J (2014) An M/G/1 retrial G-queue with preemptive resume priority and collisions subject to server breakdowns and delayed repair. J Appl Math Comput 44(1–2):187–213. https://doi.org/10.1007/s12190-013-0688-7
34. Radha J, Indhira K, Chandrasekaran VM (2017) A group arrival retrial G-queue with multi-optional stages of service, orbital search and server breakdown. IOP Conference Series: Material Science and Engineering. 263
35. Rajadurai P (2018) A study on M/G/1 retrial queueing system with three different types of customers under working vacation policy. Int J Math Model Numer Optim 8:393–417. https://doi.org/10.1504/IJMMNO.2018.094550
36. Rajadurai P (2018) A study on M/G/1 retrial G-queue with unreliable server under variant working vacation policy and vacation interruption. Songklanakarin J Sci Technol 40:231–242
37. Rajadurai P (2018) Sensitivity analysis of an M/G/1 retrial queueing system with disaster under working vacations and working breakdowns. Rairo Operat Res 52:35–54. https://doi.org/10.1051/ro/2017091
38. Rajadurai P, Chandrasekaran VM, Saravanaranjan MC (2015) Analysis of $M^X/(G_1, G_2)/1$ feedback retrial G-queue with balking and starting failures under at most J vacations. Appl Appl Math 10:13–39
39. Rajadurai P, Chandrasekaran VM, Saravanaranjan MC (2015) Steady state analysis of batch arrival feedback retrial queue with two phases of service, negative customers, Bernoulli vacation and server breakdown. Int J Math Operat Res 7:519–546. https://doi.org/10.1054/IJMOR.2015.071276
40. Rajadurai P, Chandrasekaran VM, Saravanaranjan MC (2016) Analysis of $M^X/G/1$ unreliable retrial G-queue with orbital search and feedback under Bernoulli vacation schedule. Opsearch 53:197–223. https://doi.org/10.1007/s12597-015-0226-5
41. Rajadurai P, Chandrasekaran VM, Saravanaranjan MC (2017) An M/G/1 retrial G-queue with optional re-service, impatient customers, multiple working vacation and vacation interruption. Int J Operat Res 30. https://doi.org/10.1054/ijor.2017.10006388
42. Rajadurai P, Chandrasekaran VM, Saravanaranjan MC (2018) Analysis of an unreliable retrial G-queue with working vacations and vacation interruption under Bernoulli schedule. Ain Shams Eng J 9:567–580. https://doi.org/10.1016/j.asej.2016.03.008
43. Rajadurai P, Sundaraman M, Narasimhan D (2020) Performance analysis of an M/G/1 retrial G-queue with feedback under working breakdown services. Songklanakarin J Sci Technol 42:236–247
44. Rajkumar M (2013) An (s, S) inventory retrial system with impatient and negative customers. Int J Math Operat Res 6:106–122. https://doi.org/10.1504/IJMOR.2014.057862
45. Revathy HR, Subramanian MG (2013) Two server (s, S) inventory system with positive service time, positive lead time, retrial customers and negative arrivals. Int J Comput Appl 62:0975–8887. https://doi.org/10.5120/10115-4784
46. Roszik J, Kim C, Szbik J (2005) Retrial queues in the performance modelling of cellular mobile networks using MOSEL. Int J Simul Syst Sci Technol 6:38–47

47. Shin YW (2007) Multi-server retrial queue with negative customers and disasters. Queue Syst 55:223–237. https://doi.org/10.1007/s11134-007-9018-9
48. Singh CJ, Jain M, Kaur S, Meena RK (2017) Retrial bulk queue with state dependent arrival and negative customer. In: Proceedings of sixth international conference on soft computing for problem solving. Advances in intelligent systems and computing
49. Singh CJ, Kaur S, Jain M (2020) Unreliable server retrial G-queue with bulk arrival, optional additional service and delayed repair. Int J Oper Res 38:82–111. https://doi.org/10.1054/IJOR.2020.106362
50. Subramaniam AMG (2014) Study of some retrial queueing systems by generating function technique. Lambert
51. Tran-gia P, Mandjes M (1997) Modeling of customer retrial phenomenon in cellular mobile networks. IEEE J Select Areas Commun 15:1406–1414. https://doi.org/10.1199/49.634781
52. Wang J, Liu B, Li J (2008) Transient analysis of an M/G/1 retrial queue subject to disasters and server failures. Eur J Oper Res 189:1118–1132. https://doi.org/10.1016/j.ejor.2007.04.054
53. Wang J, Zhang P (2009) A discrete-time retrial queue with negative customers and unreliable server. Comput Ind Eng 56:1216–1222. https://doi.org/10.1016/j.cie.2008.07.010
54. Wang J, Zhang P (2009) A single server discrete-time retrial G-queue with server breakdowns and repair. Acta Mathematica Applicatae Sinica. 25:675–684. https://doi.org/10.1007/s10255-008-8823-1
55. Wu J, Lian Z (2013) A single server retrial G-queue with priority and unreliable server under Bernoulli vacation schedule. Comput Ind Eng 64:84–93. https://doi.org/10.1016/j.cie.2012.08.015
56. Wu J, Liu Z, Peng Y (2011) Analysis of the finite source MAP/Ph/N retrial G-queue operating in a random environment. Appl Math Model 35:1184–1193. https://doi.org/10.1016/j.apm.2010.08.006
57. Wu J, Yin X (2011) An M/G/1 retrial G-queue with non-exhaustive random vacations and an unreliable server. Comput Math Appl 62:2314–2239. https://doi.org/10.1016/j.camwa.2011.07.018
58. Yang S, Wu J, Liu Z (2013) An $M^X$/G/1 retrial G-queue with single vacation subject to server breakdown and repair. Acta Mathematicae Applicatae Sinica. 29:579–596. https://doi.org/10.1007/s10255-013-0237-z
59. Yang T, Li H (1995) On steady state queue size distribution of the discrete-time Geo/G/1 queue with repeated customers. Queue Syst 21:199–215. https://doi.org/10.1007/BF01158581

# Chapter 19
# Assessment of the Basic Education System of Myanmar Through the Data Envelopment Analysis

**Ankita Panwar, Marlar Tin, and Millie Pant**

## 1 Introduction

In the field of education, evaluation of performance is an essential activity. Performance evaluation helps in identifying the areas that need improvement [1]. The basic education system of Myanmar can be categorized into three groups, Primary school, Lower Secondary middle school, and Upper Secondary high school. After completing 11 years of basic schooling, one can apply for university-level studying.

In this research paper, efficiency is analyzed through the matriculation exam results of high school students in terms of science and mathematics for different states and regions for the year 2018, using Data Envelopment Analysis (DEA). DEA is a multi-criteria decision-making (MCDM) tool and used for measuring the relative efficiency of the decision-making unit (DMUs). Many authors have discussed the implementation of DEA to analyze the education system in the past years. Hu et al. (2009) [2] used DEA to measure the 58 primary schools in Beijing. They selected 6 districts of Beijing and to improve the school efficiency with 10 input and 6 output variables. Agasisti (2013) [3] determined the efficiency of 29 Italian schools using the DEA bootstrapping method. He used 4 input and 2 output variables, and Tobit regression applied for the investigation of the factor affecting school efficiency. Haelermans and Ruggiero (2013) [4] analyzed the allocative and technical efficiency of Dutch secondary schools. They used 4 input and 2 output variables. Singh et al.

A. Panwar (✉) · M. Pant
Applied Science and Engineering, Indian Institute of Technology (IIT), Roorkee, India
e-mail: apanwar@as.iitr.ac.in

M. Pant
e-mail: pant.milli@as.iitr.ac.in

M. Tin
Computer University Dawei, Tanintharyi, Myanmar
e-mail: drmarlartutoungoo@gmail.com



D. Singh et al. (eds.), *Proceedings of International Conference on Scientific and Natural Computing*, Algorithms for Intelligent Systems,
https://doi.org/10.1007/978-981-16-1528-3_19

(2018) [5] determined the relative efficiency of the 126 Indian state education board. They used an integrated DEANN technique to increase the discriminatory power of the DEA model and find the exact rank of the DMUs.

This paper is organized as follows: In the first section, an introduction is given. In Sect. 2, a brief description of Data Envelopment Analysis (DEA). The problem description is defined in Sect. 3, and in Sect. 4 are discussed results and discussion. The conclusion is defined in Sect. 5.

## 2 Data Envelopment Analysis (DEA)

Charnes, Cooper, and Rhodes introduced DEA in 1978 [6]. It is a non-parametric linear programming technique [7] and can be used to assess the relative efficiency of homogeneous DMUs with multiple inputs and multiple outputs [8]. DEA has been popular among researchers as an efficiency measurement tool [8, 9].

The central concept of DEA is based on efficiency measurement, the weighted ratio of the total number of output variables to the total number of input variables [10].

$$\text{Efficiency} = \frac{\text{Weight sum of outputs}}{\text{Weight sum of inputs}} \tag{1}$$

The value of efficiency DMUs lies between 0 and 1.

**Mathematical form**

Suppose we have taken the z DMUs. Let we are choosing one DMU, suppose $z^{\text{th}}$ DMU and maximize its efficiency.

$$\text{maximize} E_z = \frac{\sum_{t=1}^{T} c_{tz} y_{tz}}{\sum_{s=1}^{S} d_{sz} x_{sz}}$$

$$\text{Subject; } 0 \leq \frac{\sum_{t=1}^{T} c_{tz} y_{ty}}{\sum_{s=1}^{S} d_{sz} x b_{sy}} \leq 1 : y = 1, 2, \ldots, Y$$

$$c_{tz}, d_{sz} \geq 0, s = 1, 2, \ldots, S, \text{ and } t = 1, 2, \ldots, T \tag{2}$$

where

| | |
|---|---|
| $E_z$ | efficiency of the $z^{\text{th}}$ DMU |
| $y_{tz} = t\text{th}$ | output of the $z^{\text{th}}$ DMU |
| $c_{tz}$ | weight of the $t^{\text{th}}$ output |
| $x_{sz} = \text{sth}$ | input of the $z^{\text{th}}$ DMU |
| $d_{sz}$ | weight of the $s^{\text{th}}$ input |

**Table 1** The mathematical form of CCR and BCC model

| CCR model [11] | BCC model [11] |
|---|---|
| $maximize E_z = \sum_{t=1}^{T} c_{tz} y_{tz}$<br>$Subject \sum_{s=1}^{S} d_{sz} x_{sz} = 1$<br>$\sum_{t=1}^{T} c_{tz} y_{ty} - \sum_{s=1}^{S} d_{sz} x b_{sy} \leq 0,\ y = 1, 2, \ldots, Y$<br>$c_{tz}, d_{sz} \geq \in, s = 1, 2, \ldots, S, t = 1, 2, \ldots, T$ (3) | $maximize E_z = \sum_{t=1}^{T} c_{tz} y_{tz} + w_{qz}$<br>$Subject \sum_{s=1}^{S} d_{sz} x_{sz} = 1$<br>$\sum_{t=1}^{T} c_{tz} y_{ty} - \sum_{s=1}^{S} d_{sz} x b_{sy} + w_{qz} \leq 0$, y = 1,2,…,Y<br>$c_{tz}, d_{sz} \geq \in, s = 1, 2, \ldots, S, t = 1, 2, \ldots, T$ (4)<br>$w_{qz}$ is unrestricted in sign. |

If the value of any DMU is one, then DMU is said to be relatively efficient, otherwise relatively inefficient [11].

## 2.1 DEA Models Used in the Present Study

### 2.1.1 CCR and BCC Model

This model also called the CCR model in the input orientation model. Charnes et al. proposed in 1978 and Banker et al. proposed the BCC model in 1984. BCC model is the extended model of the CCR model [11]. The mathematical form of the CCR and model is written in Table 1:

### 2.1.2 Scale Efficiency

The ratio of the CCR value into the BCC model represents the value of the scale efficiency (SE) [12]. It measures the impact of scale size on the efficiency of a DMU.

$$\text{Scale efficiency} = \frac{E_{CCR}}{E_{BCC}} \tag{5}$$

where

$E_{CCR}$ = CCR score of the DMU
$E_{BCC}$ = BCC score of the DMU

## 2.2 *Important Terms in the Context of DEA*

### 2.2.1 Decision-Making Unit (DMUs)

DMUs is an entity, which converts inputs into outputs.

### 2.2.2 Peer Groups

A peer group contains more than one efficient DMUs or firms for inefficient DMUs or firms. So we can say that any inefficient DMUs have one or more than one efficient DMU as a peer.

### 2.2.3 Peer Counts

The highest peer count represents the most technically or purest technically efficient DMUs.

### 2.2.4 Increase Returns to Scale (IRS)

A DMU is to operate at IRS if the proportionate increase in all of its input results in a greater than proportionate increase in its outputs. In any firms the sum of its dual weights in the dual model is less than 1, then the firms can be said to operate IRS.

### 2.2.5 Decreasing Returns to Scale (DRS)

A unit is said to DRS if the proportionate increase in all of its input results in a less than proportionate increase in its outputs. In any firm the sum of its dual weights in the dual model is greater than 1, then the firms can be said to operate DRS.

# 3 Problem Descriptions

## 3.1 *Case Study: Efficiency Measurement of the Basic Education System in Myanmar*

The basic Education Sector plays a very vital role in the growth and development of a country. This is particularly relevant for developing countries like that of the Republic of the Union of Myanmar to match up with the standards of the rest of the world. As pointed out by the National Education Strategic Plan (2016–2021), there is a need to improve classroom education.

In Myanmar, there is a total of nine regions, nine states, and one Foreign center. The objective is to determine the relative efficiency of the different regions, states, and Foreign Centre over the academic year 2018. The data set related to the input and output variables is shown in Table 2.

## *3.2 Methodology*

The important steps are defined below:

**Step 1. Selection of DMUs**

The total number of DMUs is 19 with 9 states, 9 regions and 1 Foreign center. States are; Kachin, Kayah, Kayin, Chin, Mon, Rakhine, Shan (North), Shan*(East), and regions are Sagaing, Tanintharyi, Bago (East), Bago (West), Magway, Mandalay, Naypyidaw, Yangon, Ayeyarwady.

**Step 2. Selection of Input and Output variables**

There are two input and six output variables. Table 3 gives a brief description of the input and output variables.

**Step 3. Calculation of relative efficiency**

Once the inputs, outputs, and decision-making units have been finalized according to the above steps, then the relative efficiency of the 19 DMUs is calculated using Eqs. 3–5.

# 4 Results and Discussion

The overall performance of the regions and states of Myanmar with two inputs and six outputs is determined using CCR, BCC, and SE under the output-oriented model. The purpose of the model is the same, to find the relative efficiency of the DMUs.

## *4.1 Overall Performance*

The overall performance of the regions and states through the different models of the DEA. These models are TE (CCR), PTE (BCC), and SE. The results are mentioned in Tables 4 and 5.

**Table 2** Data set for the regions and states

| DMU | Exam centre (I) | Enrolled students (I) | Number of passed students (O) | % of passed students (O) | Distinction in Mathematics (O) | Distinction in Physics (O) | Distinction in Chemistry (O) | Distinction in Biology (O) |
|---|---|---|---|---|---|---|---|---|
| Kachin | 67 | 33212 | 9974 | 30.03 | 355 | 128 | 522 | 1250 |
| Kayah | 17 | 6640 | 1675 | 25.22 | 53 | 14 | 37 | 134 |
| Kayin | 42 | 17593 | 5182 | 29.45 | 118 | 45 | 143 | 518 |
| Chin | 25 | 10168 | 1692 | 16.64 | 20 | 6 | 13 | 137 |
| Sagaing | 207 | 104759 | 36533 | 34.87 | 1816 | 601 | 1799 | 6180 |
| Tanintharyi | 45 | 20237 | 6074 | 30.01 | 241 | 88 | 230 | 646 |
| Bago(East) | 87 | 44313 | 12925 | 29.16 | 472 | 256 | 711 | 2092 |
| Bago(West) | 53 | 25080 | 8379 | 33.40 | 272 | 163 | 446 | 1280 |
| Mon | 56 | 25918 | 9699 | 37.42 | 619 | 305 | 723 | 2257 |
| Magway | 160 | 77221 | 24490 | 31.71 | 903 | 294 | 953 | 3288 |
| Mandalay | 191 | 111398 | 42517 | 38.16 | 4037 | 2460 | 4668 | 12140 |
| Naypyidaw | 41 | 22389 | 7399 | 33.04 | 402 | 257 | 547 | 1369 |
| Rakhine | 121 | 40180 | 8620 | 21.45 | 276 | 168 | 338 | 1016 |
| Yangon | 220 | 111855 | 42006 | 37.55 | 3551 | 2424 | 5752 | 15872 |
| Shan(North) | 79 | 30880 | 9174 | 29.70 | 475 | 208 | 424 | 1431 |
| Shan*(South) | 37 | 16498 | 5342 | 32.37 | 315 | 118 | 396 | 992 |
| Shan*(East) | 16 | 4841 | 936 | 19.33 | 27 | 13 | 24 | 61 |
| Ayeyarwady | 243 | 86518 | 26409 | 30.52 | 676 | 287 | 784 | 2349 |
| Foreign | 20 | 145 | 65 | 44.82 | 7 | 25 | 36 | 100 |

**Table 3** Description of input and output variables

| Input variables | Output variables |
|---|---|
| **1.** Number of exam centre | **1.** Number of passed students |
| **2.** Number of enrolled students | **2.** Passed students % |
| | **3.** Distinction in Mathematics |
| | **4.** Distinction in Chemistry |
| | **5.** Distinction in Physics |
| | **6.** Distinction in Biology |

**Table 4** OTE of the DMUs along with peer counts

| Code | DMUs | OTE (CRS) | Peer counts | Peer groups |
|---|---|---|---|---|
| S1 | Kachin | 0.799 | 0 | R12, R11, F |
| S2 | Kayah | 1.000 | 2 | S2 |
| S3 | Kayin | 0.799 | 0 | R11, F, R12 |
| S4 | Chin | 0.534 | 0 | R12, S2, F |
| R5 | Sagaing | 0.913 | 0 | R11, F |
| R6 | Tanintharyi | 0.824 | 0 | R12, R11, F |
| R7 | Bago(East) | 0.764 | 0 | R11, F |
| R8 | Bago(West) | 0.906 | 0 | R12, R11, F |
| S9 | Mon | 1.000 | 0 | S9 |
| R10 | Magway | 0.831 | 0 | R11, F |
| R11 | Mandalay | 1.000 | 11 | R11 |
| R12 | Naypyidaw | 1.000 | 7 | R12 |
| S13 | Rakhine | 0.561 | 0 | R11, F |
| R14 | Yangon | 1.000 | 0 | R14 |
| S15 | Shan(North) | 0.778 | 0 | R11, F |
| S16 | Shan*(South) | 0.930 | 0 | R12, R11, F |
| S17 | Shan*(East) | 0.739 | 0 | R12, S2, F |
| R18 | Ayeyarwady | 0.799 | 0 | R11, F |
| F | Foreign | 1.000 | 13 | F |
| | Mean | 0.851 | | |

#### 4.1.1 Overall Technical Efficiency (OTE)

The efficiency score of the Myanmar education system is obtained under the CCR model with the help of Eq. 3, and shown in Table 4. Out of the total 19 DMUs taken in this study, the number of efficient DMUs are 6, namely Kayah(S2), Mon(S9), Mandalay(R11), Naypyidaw (R12), Yangon(R14), and Foreign (F). Other remaining 13 DMUs are inefficient. Since the mean score of this model is 0.851 and the value

**Table 5** Efficiency comparison of the DMUs along with peer counts

| Code | DMUs | OTE (CRS) | PTE (VRS) | Peer counts | Peer groups | SE | RTS | Average |
|---|---|---|---|---|---|---|---|---|
| S1 | Kachin | 0.799 | 0.802 | 0 | S9, R12, R11, F | 0.997 | IRS | 0.866 |
| S2 | Kayah | 1.000 | 1.000 | 4 | S2 | 1.000 | – | 1 |
| S3 | Kayin | 0.799 | 0.807 | 0 | S9, S2, R12, F | 0.990 | IRS | 0.865 |
| S4 | Chin | 0.534 | 0.540 | 0 | S9, S2, R12, F | 0.990 | IRS | 0.688 |
| R5 | Sagaing | 0.913 | 0.914 | 0 | R11, F | 1.000 | – | 0.942 |
| R6 | Tanintharyi | 0.824 | 0.830 | 0 | S9, S2, R12, F | 0.993 | IRS | 0.882 |
| R7 | Bago(East) | 0.764 | 0.766 | 0 | R11, F, S9 | 0.997 | IRS | 0.842 |
| R8 | Bago(West) | 0.906 | 0.910 | 0 | S9, R12, R11, F | 0.996 | IRS | 0.937 |
| S9 | Mon | 1.000 | 1.000 | 7 | S9 | 1.000 | – | 1 |
| R10 | Magway | 0.831 | 0.831 | 0 | R11, F | 1.000 | – | 0.887 |
| R11 | Mandalay | 1.000 | 1.000 | 9 | R11 | 1.000 | – | 1 |
| R12 | Naypyidaw | 1.000 | 1.000 | 6 | R12 | 1.000 | – | 1 |
| S13 | Rakhine | 0.561 | 0.562 | 0 | R11, F | 0.999 | DRS | 0.707 |
| R14 | Yangon | 1.000 | 1.000 | 0 | R14 | 1.000 | – | 1 |
| S15 | Shan(North) | 0.778 | 0.778 | 0 | R11, F | 1.000 | – | 0.852 |
| S16 | Shan*(South) | 0.930 | 0.939 | 0 | S9, S2, R12, F, R11 | 0.990 | IRS | 0.953 |
| S17 | Shan*(East) | 0.739 | 1.000 | 0 | S17 | 0.739 | IRS | 0.826 |
| R18 | Ayeyarwady | 0.799 | 0.800 | 0 | F, R11 | 0.999 | DRS | 0.866 |
| F | Foreign | 1.000 | 1.000 | 12 | F | 1.000 | – | 1 |
| | Mean | 0.851 | 0.867 | | | 0.984 | | 0.900 |

of Sagaing (R5), Bago (West) (R8), Shan*(South) (S16) are higher than the average value, and it may be said that the performance of these 3 DMUs is satisfactory.

Since DMU S4 (Chin) (0.534 or 53.4%) is the least efficient compare to other inefficient DMU. So we can say that all criteria or parameters are not relevant for S4, so the performance of the S4 is very poor.

**Peer counts:** Peer counts of five efficient DMUs Kayah(S2), Mon(S9), Mandalay(R11), Yangon(R14), Foreign (F) are 2, 0, 11, 7, 0, 13, respectively. Foreign (F) with the highest peer count as 13, may be considered to be overall technically efficient, while Mon(S9) and Yangon(R14) with peer counts 0 may not be considered as efficient on the basis of peer count efficiency.

### 4.1.2 Pure Technical Efficiency (PTE)

We have calculated the value of the BCC model from Eq. 4, and the result is shown in Table 5. BCC model shows 7 DMUs to be efficient, which are Kayah (S2), Mon (S9), Mandalay (R11), Naypyidaw (R12), Yangon (R14), Shan*(East) (S17), Foreign (F). Since the scores of Sagaing (R5), Bago (West) (R8), and Shan*(South) (S16) is greater than the average value (= 0.867) obtained through BCC performance of these DMUs may be considered to be satisfactory in comparison to other inefficient DMUs.

The inefficiency score of the DMU S13 (Rakhine) is 0.562 (56.2%), which is very less value as compared to other inefficient DMUs. So the performance of S13 is not good for all these criteria. So improvement must be needed.

**Peer counts:** Peer count analysis provides peer counts of Kayah(S2), Mon(S9), Mandalay (R11), Naypyidaw (R12), Yangon (R14), Shan*(East) (S17), Foreign (F) as 4, 7, 9, 6, 0, 0, 12 respectively. Foreign (F) has the highest peer count (= 12), indicating it to be the best performer among efficient units and is called the most pure technical efficiency.

**Returns to Scale:** We observed RTS, S1, S3, S4, R6, R7, R8, S16, and S17 at IRS. Two DMUs R18, S13 are at DRS.

### 4.1.3 Scale Efficiency

SE is the ratio of CCR into the BCC model, calculated from Eq. 5, and represented in Table 5. Through the SE model, a total of 9 DMUs are obtained to be efficient, which are Kayah(S2), Sagaing (R5), Mon (S9), Magway (R10), Mandalay (R11), Naypyidaw (R12), Yangon (R14), Shan (North) (S15), Foreign (F). The average value obtained through SE is 0.984. In this model, only Shan*(East)(S17) is inefficient DMU, and the values are less than the average value. So the performance of these DMUs are not good, and the value of the remaining 9 DMUs are greater than the average value, Kachin (S1), Chin (S4), Tanintharyi (R6), Bago (East) (R7), Bago (West) (R8), Rakhine (S13), Shan*(South) (S16), and Ayeyarwady (R18), so the performance of these DMUs are satisfactory.

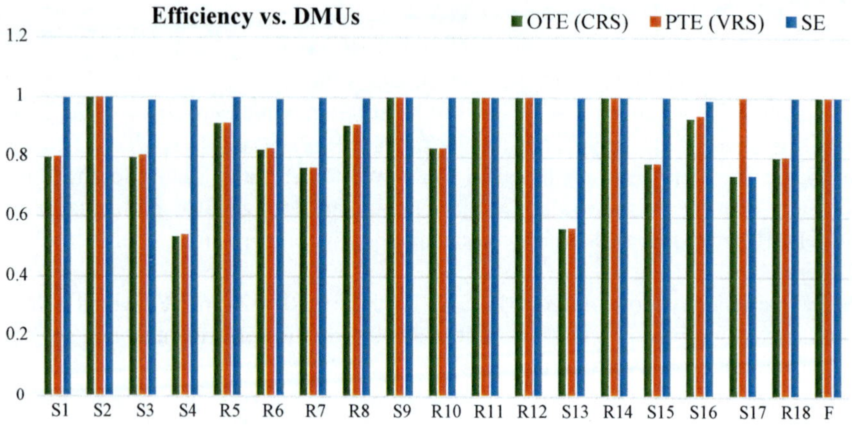


**Fig. 1** Overall efficiency measurement

## *4.2 Performance Assessment of the Region and States Based on the Average Value*

The average value (OTE, PTE, and SE models) of the different states and regions shown in Table 5. Kayah (S2), Mon (S9), Mandalay (R11), Naypyidaw (R12), Yangon (R14), Foreign (F) are efficient, and the efficiency value is 1. Here, State Shan*(South) (S16) has the highest average value, 0.953, that is, S16 is most efficient. Chin (S4) is the least efficient with an efficiency score (0.688).

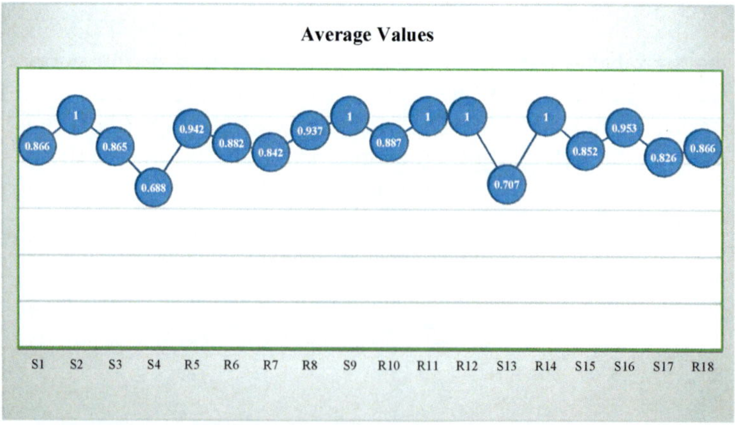


**Fig. 2** Average value of regions and states

## 5 Conclusion

An analysis is done for the Basic education sector of Myanmar on the basis of matriculation exam results of high school students in terms of science and mathematics for different states and regions. DEA is used in this for determining the relative efficiency of DMUs. Some conclusions about the present study can be drawn as follows:

- In terms of the TE and PTE model, the state of Chin is least efficient with scores of 0.534 and 0.540. However, in terms of SE, Shan*(East) is the least efficient (0.739).
- Under the BCC model and CCR model, Foreign Centre (F) is most efficient in terms of peer count.
- Overall performance of the Mandalay (R11) is the best performer in technical efficiency and pure technical efficiency model in terms of regions. In terms of States, Kayah (peer count = 2) is the best performer under the CCR model, but Mon (peer count = 7) is the best performer in the BCC model.
- According to the average value, three Mandalay (R11), Naypyidaw (R12), Yangon (R14) regions are more efficient, and Kayah (S2), Mon (S9) are more efficient states.

**Acknowledgements** The authors acknowledge with thanks the financial support provided by the project as DST-1628-DPT.

## References

1. Ali I, Pant M, Rana US (2018) Relative performance assessment of academic departments of higher educational institute using DEA with sensitivity analysis. Int J Hybrid Intell Syst 14(3):181–191. https://doi.org/10.3233/his-180250
2. Hu Y, Zhang Z, Liang W (2009) Efficiency of primary schools in Beijing, China: An evaluation by data envelopment analysis. Int J Educ Manage 23(1):34–50. https://doi.org/10.1108/09513540910926411
3. Agasisti T (2013) The efficiency of Italian secondary schools and the potential role of competition: A data envelopment analysis using OECD-PISA2006 data. Educ Econ 21(5):520–544
4. Haelermans C, Ruggiero J (2013) Estimating technical and allocative efficiency in the public sector: A nonparametric analysis of Dutch schools. Eur J Oper Res 227(1):174–181. https://doi.org/10.1016/j.ejor.2012.12.004
5. Singh N, Pant M, Goel A (2018) ANN embedded data envelopment analysis approach for measuring the efficiency of state boards in India. Int J Syst Assur Eng Manage 9(5):1092–1106. https://doi.org/10.1007/s13198-018-0743-8
6. Charnes A, Cooper WW, Rhodes E (1978) Measuring the efficiency of decision making units. Eur J Oper Res 2(6):429–444
7. Puri J, Yadav SP (2015) A fully fuzzy approach to DEA and multi-component DEA for measuring fuzzy technical efficiencies in the presence of undesirable outputs. Int J Syst Assur Eng Manage 6(3):268–285. https://doi.org/10.1007/s13198-015-0348-4
8. Aziz NAA, Janor RM, Mahadi R (2013) Comparative departmental efficiency analysis within a university: A DEA approach. Procedia Soc Behav Sci 90(InCULT 2012):540–548

9. Jauhar SK, Pant M, Dutt R (2018) Performance measurement of an Indian higher education institute: a sustainable educational supply chain management perspective. Int J Syst Assur Eng Manage 9(1):180–193
10. Ali I, Pant M, Rana US, Jauhar SK (2017) DEA for measuring the academic performance of a higher educational institute of Uttarakhand, India. Int J Comput Inf Syst Ind Manage Appl 9(January):206–217
11. Cook WD, Seiford LM (2009) Data envelopment analysis (DEA)—thirty years on. Eur J Oper Res 192(1):1–17
12. Mohamed Shahwan T, Mohammed Hassan Y (2013) Efficiency analysis of UAE banks using data envelopment analysis. J Econ Adm Sci 29(1):4–20. https://doi.org/10.1108/10264111311319204

# Chapter 20
# Discrete Cosine Transform with Matrix Technique for a Fractional-order Riesz Differentiator

**Hari Pratap and Amit Ujlayan**

## 1 Introduction

In current years, the digital signal with image processing and the calculus with fractional-order derivative played a vital role in the field of engineering, image processing like signature verification with high resolution or sharpening images, cloud computing, biomedical processing [1–5]. Fractional calculus is used for extracting the information from objects as image with the help of digital filters. In this article, the design of matrix-based Riesz non-integer-order differentiator using discrete cosine transform will be presented. Expression for Riesz kernel for the continuous-time domain as [6]:

$$K(t) = \frac{1}{2\Gamma(-\alpha)\cos(\alpha\pi/2)} \mid t \mid^{-\alpha-1} \tag{1}$$

Fractional derivative for $\alpha$th order of Eq. (1)

$$D^{\alpha}g(t) = \frac{1}{2\Gamma(-\alpha)\cos(\alpha\pi/2)} \int_{-\infty}^{\infty} \mid t-\phi \mid^{-\alpha-1} g(\phi)d\phi \tag{2}$$

Symbol $\Gamma(\cdot)$ denotes gamma function. Conversion of time domain signal to frequency domain signal by using the definition of Fourier transform. So, the relation express in the frequency domain of Eq. (2) is :

$$FT[D^{\alpha}g(t)] = \mid \omega \mid^{\alpha} G(\omega) \tag{3}$$

H. Pratap (✉) · A. Ujlayan
Department of Applied Mathematics(SOVSAS), Gautam Buddha University, Greater Noida, U.P., India
e-mail: pratap.pgdavcollege.hari@gmail.com

A. Ujlayan
e-mail: iitk.amit@gmail.com



D. Singh et al. (eds.), *Proceedings of International Conference on Scientific and Natural Computing*, Algorithms for Intelligent Systems,
https://doi.org/10.1007/978-981-16-1528-3_20

Fourier transform is denoted as $FT(\cdot)$. Ideally definition for matrix Riesz non-integer-order differentiator is

$$R_i(\omega) = | \, \omega \, |^{\alpha} \tag{4}$$

In this paper, matrix filter approach has been used for reducing the complexity of digital system [7]. Section 2, cover the designing approach for matrix-based Riesz non-integer-order differentiator using DCT method [8, 9]. In Section 3, some computative problems are discussed with error graph for different orders for DCT method. Finally, conclusion is discussed.

## 2 Matrix Riesz Non-Integer-Order Differentiator Using DCT Method

Assume the given signals $g(0), g(1), g(2) \cdots g(M-1)$ and $h(0), h(1)h(2) \cdots h(M-1)$ denotes input and output signal sequence, respectively. Both signal sequences can be represented into the vector form as

$$g = [g(0)g(1)g(2) \cdots g(M-1)]^T \tag{5}$$

$$h = [h(0)h(1)h(2) \cdots h(M-1)]^T \tag{6}$$

Matrix size for an input and output signals is $M \times 1$.

When an input signal apply on the matrix system filter, whose transfer function is $R$. Then the generalized expression can be written for an output signal is

$$\mathbf{h} = \mathbf{Rg} \tag{7}$$

The size of transfer matrix $\mathbf{R}$ is $M \times M$

Our task is how to calculate the value of transfer matrix $\mathbf{R}$. $\mathbf{h}$ is an Riesz non-integer-order derivative of an filter input signal $\mathbf{g}$. Exponential signal is an elementary type signals which can be represented by the following vector form as

$$\mathbf{x}(\omega) = (e^{jr\omega})^T = [1e^{j\omega}e^{2j\omega} \cdots e^{(M-1)j\omega}]^T \tag{8}$$

After applying the definition of Riesz non-integer-order derivative on $\mathbf{x}(\omega)$, we obtain

$$\mathbf{y}(\omega) = | \, \omega \, |^{\alpha} \mathbf{x}(\omega) \tag{9}$$

Our main objective is to approximate the output value of the filter $\mathbf{Rx}(\omega)$ to the ideal value of vector $\mathbf{y}(\omega)$.

Basically, there exist several types of DCT [10, 11] of which ***DCT-IV*** method for an input data points is defined as

$$G(k) = \sum_{q=0}^{M-1} \sqrt{\frac{2}{M}} g(q) \cos\left(\frac{(2q+1)(2k+1)\pi}{4M}\right) \tag{10}$$

$$g(p) = \sum_{k=0}^{M-1} \sqrt{\frac{2}{M}} G(k) \cos\left(\frac{(2p+1)(2k+1)\pi}{4M}\right) \tag{11}$$

On applying Riesz FOD (fractional-order differentiator) on eq.(11), we obtain

$$h(p) = D^{\alpha} g(p)$$

$$h(p) = \sum_{k=0}^{M-1} \sqrt{\frac{2}{M}} \left|\frac{(2k+1)\pi}{2M}\right|^{\alpha} G(k) \cos\left(\frac{(2p+1)(2k+1)\pi}{4M}\right) \tag{12}$$

Equating the value of $G(k)$ in eq.(11), we get

$$h(p) = \frac{2}{M} \sum_{k=0}^{M-1} \left|\frac{(2k+1)\pi}{2M}\right|^{\alpha} \sum_{q=0}^{M-1} g(q) \cos\left(\frac{(2q+1)(2k+1)\pi}{4M}\right) cos\left(\frac{(2p+1)(2k+1)\pi}{4M}\right) \tag{13}$$

The Eq. (10) can be written in generalized form as

$$h(p) = \sum_{q=0}^{M-1} r(p, q) g(q) \tag{14}$$

$$r(p, q) = \sum_{k=0}^{M-1} \frac{2}{M} \left|\frac{(2k+1)\pi}{2M}\right|^{\alpha} \cos\left(\frac{(2q+1)(2k+1)\pi}{4M}\right) \cos\left(\frac{(2p+1)(2k+1)\pi}{4M}\right) \tag{15}$$

Average integral square error formula is used to calculate the effectiveness of DCT interpolation method for a fractional-order Riesz differentiator for different orders as

$$E(\omega) = \| \mathbf{Rx}(\omega) - \mathbf{y}(\omega) \|_2^2 / M \tag{16}$$

where $\| \, \|$ represent Euclidean norm function.

$$\|Z\| = \sqrt{\sum_{i=1}^{M} Z_i^2}$$

Definition for **DCT-III** method is

$$G(k) = \sqrt{\frac{2}{M}} \sum_{q=0}^{M-1} c_q g(q) \cos\left(\frac{q(2k+1)\pi}{2M}\right) \tag{17}$$

$$g(p) = \sqrt{\frac{2}{M}} \sum_{k=0}^{M-1} c_k G(k) \cos\left(\frac{p(2k+1)\pi}{2M}\right) \tag{18}$$

$$c_k = \begin{cases} \frac{1}{\sqrt{2}} & k = 0 \\ 1 & \text{otherwise} \end{cases}$$

$$r(p,q) = \frac{2}{M} \sum_{k=0}^{M-1} c_k^2 \left|\frac{(2k+1)\pi}{2M}\right|^{\alpha} \cos\left(\frac{q(2k+1)\pi}{2M}\right) cos\left(\frac{p(2k+1)\pi}{2M}\right) \tag{19}$$

Definition for ***DCT-II*** method is

$$G(k) = \sqrt{\frac{2}{M}} \sum_{q=0}^{M-1} c_q g(q) \cos\left(\frac{(2q+1)k\pi}{2M}\right) \tag{20}$$

$$g(p) = \sqrt{\frac{2}{M}} \sum_{k=0}^{M-1} c_k G(k) \cos\left(\frac{(2p+1)k\pi}{2M}\right) \tag{21}$$

$$c_k = c_q = \begin{cases} \frac{1}{\sqrt{2}} & k = q = M - 1 \\ 1 & \text{otherwise} \end{cases}$$

$$r(p,q) = \frac{2}{M} \sum_{k=0}^{M-1} c_k^2 \left|\frac{k\pi}{M}\right|^{\alpha} \cos\left(\frac{(2q+1)k\pi}{2M}\right) \cos\left(\frac{(2p+1)k\pi}{2M}\right) \tag{22}$$

Definition for ***DCT-I*** method is

$$G(k) = 2 \sum_{q=0}^{M-1} \beta_q g(q) \cos\left(\frac{qk\pi}{M-1}\right) \tag{23}$$

$$g(p) = \frac{1}{M-1} \sum_{k=0}^{M-1} \beta_k G(k) \cos\left(\frac{pk\pi}{M-1}\right) \tag{24}$$

$$\beta_q = \beta_k = \begin{cases} \frac{1}{\sqrt{2}} & k = q = M - 1 \\ 1 & \text{otherwise} \end{cases}$$

$$r(p, q) = \frac{2}{M-1} \sum_{k=0}^{M-1} \beta_k^2 \left| \frac{k\pi}{M-1} \right|^{\alpha} \cos\left(\frac{qk\pi}{M-1}\right) \cos\left(\frac{pk\pi}{M-1}\right) \tag{25}$$

**Example 1** Error results of DCT interpolation methods will be discussed for matrix-based Riesz fractional-order differentiator with effective parameter values $M = 200$, order $\alpha = 0.3$. Figure 1 is shown between the error value $E(\omega)$ with normalized frequency $\omega/\pi$. The normalized frequency value is increasing with decreasing corresponding error value. For an entire frequency range, the DCT-IV method shows a better error size as compared with other DCT interpolation methods.

**Example 2** Error results of DCT interpolation methods will be discussed for matrix-based Riesz fractional-order differentiator with effective parameter values $M = 200$, order $\alpha = 0.5$. Figure 2 is shown between the error value $E(\omega)$ with normalized frequency $\omega/\pi$. The normalized frequency value is increasing with decreasing corresponding error value. For an entire frequency range, the DCT-IV method shows the better error size as compared with other DCT interpolation methods.

**Example 3** Error results of DCT interpolation methods will be discussed for matrix-based Riesz fractional-order differentiator with effective parameter values $M = 200$,

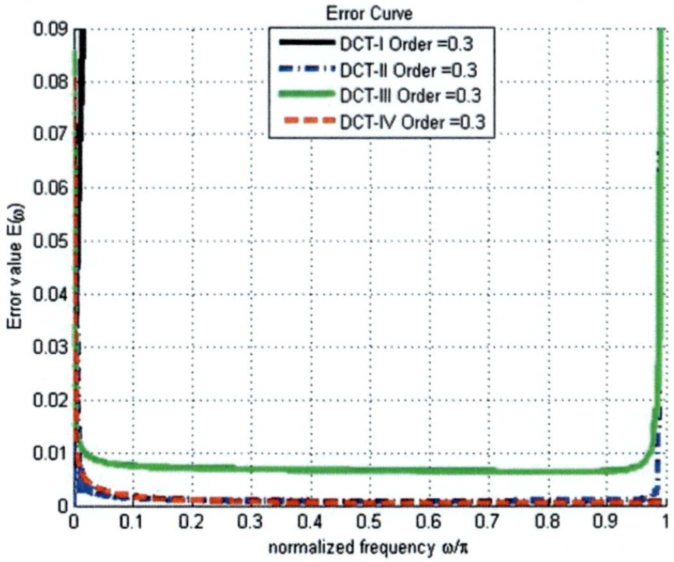


**Fig. 1** Error results for the DCT interpolation for order = 0.3

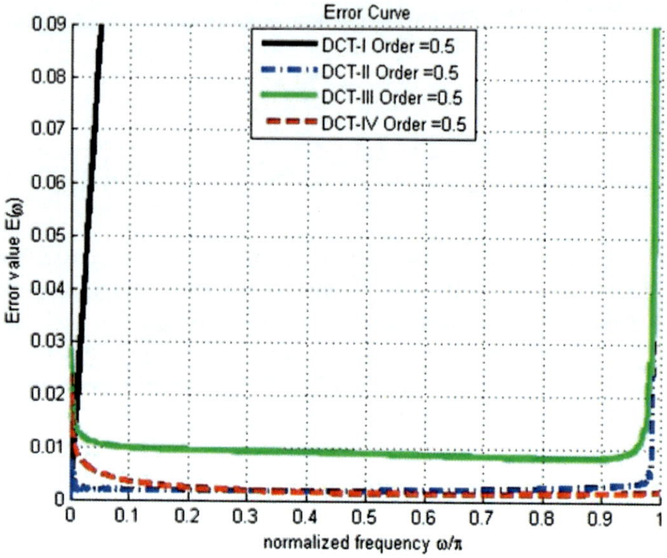


**Fig. 2** Error results for the DCT interpolation for order = 0.5

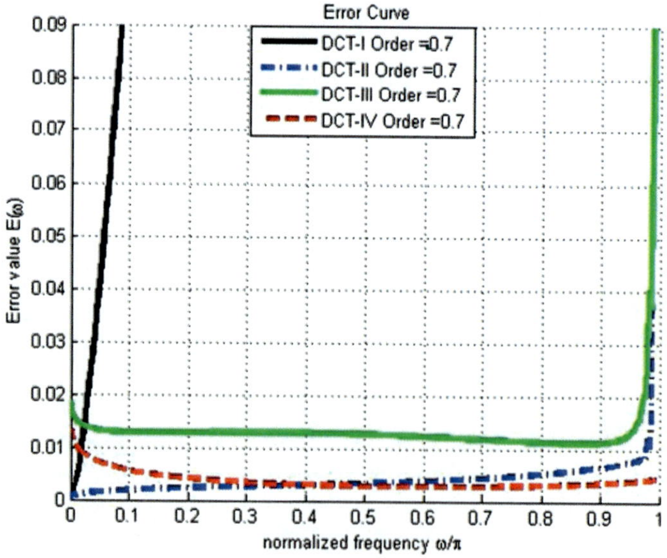


**Fig. 3** Error results for the DCT interpolation for order = 0.7

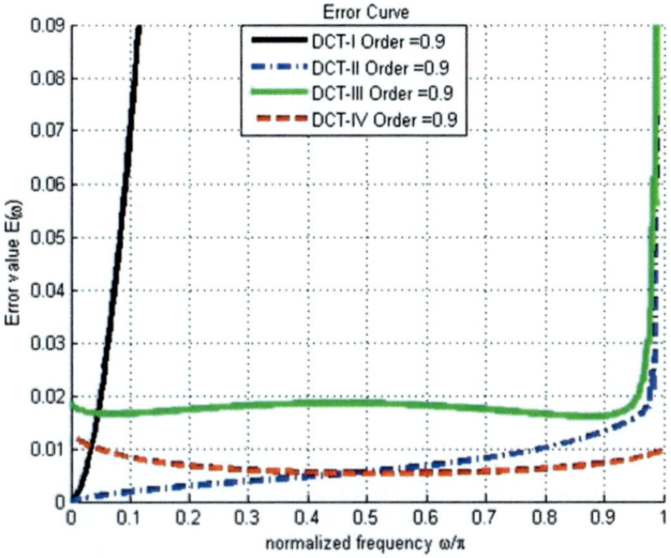


**Fig. 4** Error results for the DCT interpolation for order = 0.9

order $\alpha = 0.7$. Figure 3 is shown between the error value $E(\omega)$ with normalized frequency $\omega/\pi$. The normalized frequency value is increasing with decreasing corresponding error value. For an entire frequency range, the DCT-IV method shows the better error size as compared with other DCT interpolation methods.

**Example 4** Error results of DCT interpolation methods will be discussed for matrix-based Riesz fractional-order differentiator with effective parameter values $M = 200$, order $\alpha = 0.9$. Figure 4 is shown between the error value $E(\omega)$ with normalized frequency $\omega/\pi$. The normalized frequency value is increasing with decreasing corresponding error value. For an entire frequency range, the DCT-IV method shows the better error size as compared with other DCT interpolation methods.

## 3 Conclusion

In this paper, matrix-based Riesz non-integer-order differentiator using DCT interpolation method is presented. Only DCT-IV method shows the better error size value as compared with other DCT interpolation methods. It means DCT-IV method is well suited for the designing of matrix-based Riesz non-integer-order differentiator. In future, we can design matrix-based Riesz FOD using different fractional derivative definitions as Weyl's, Conformable, Riemann–Liouville, etc., and in multidimensional also.

## References


1. Oldham KB, Spanier J (1974) The fractional calculus. Academic Press, New York
2. Miller KS, Ross R (1993) An introduction to the fractional calculus and fractional differential equations. Wiley, New York
3. Podlubny I (2014) Fractional calculus an introduction for physicists, 2nd edn. World Scientific Publishing
4. Ortigueira MD (2011) Fractional calculus for scientists and engineers. Springer
5. Pratt WK (2014) Introduction to digital image processing. CRC Press
6. Tseng C-C, Lee S-L (2014) Design of digital Riesz fractional order differentiator. Signal Process 102:32–45
7. Tseng C-C, Lee S-L (2015) Design of matrix fractional order differentiator. Signal Process 111:73–88
8. Tseng CC (2006) Improved design of digital fractional-order differentiators using fractional sample delay. IEEE Trans Circuits Syst I Reg Papers 53:193–203
9. Singh D, Kumar N, Pratap H, Kushwah H (2018) Discrete cosine transform interpolation approach to design an fractional order hilbert transformer. In: IEEE International conference on advance in computing, communication control and networking, ICACCCN
10. Pratap H, Kumar N (2020) DST interpolation techniques for Reisz based matrix differentiator of non-integer order. In: 2nd international conference on advances in computing, communication control and networking, ICACCCN
11. Naidu VPS (2010) Discrete cosine transform-based image fusion. Defence Sci J 60(1):48–54

# Chapter 21
# Analysis and Design of Memristor Emulator and Its Application in FM Demodulator

**Zeba Mustaqueem and Abdul Quaiyum Ansari**

## 1 Introduction

The memristor device was first discovered by LO Chua as the fourth passive element in electrical circuits [1]. Recently, the physical implementation of the memristor was reported using partially doped $TIO_2$ [2, 3]. The $TIO_2$ memristor fabricated had an I-V characteristic as a pinched hysteresis loop and showed non-volatile behavior. This important feature can potentially find applications in the design of non-volatile memories. Furthermore, it has been reported that memristors also possess key characteristics such as scalability, low power consumption, and compatibility in CMOS technology [4]. However, the exorbitant cost and other technical challenges have limited the production of memristors for practical purposes. It is envisaged that emulators using op-amps and other circuit components can address these technical challenges [2, 5, 8–18].

In this paper, a memristor emulator is proposed. It achieves the performance comparable to the one achieved from the fabricated $TIO_2$ memristor [2] and also shows memristive behaviour for a relatively longer period of time. The basic concept of memristor, the proposed emulator and its mathematical analysis is given in Sect. 2. The performance analysis is presented in Sect. 3 while an improved memristor emulator having memristive behaviour till 3 kHz, which is double to the existing memristors, is reported in Sect. 4. The application of the proposed emulator in the FM detector provided in Sects. 5 and 6 concludes the paper.

Z. Mustaqueem (✉) · A. Q. Ansari
Department of Electrical Engineering, Jamia Millia Islamia, New Delhi, India
e-mail: zeba.ec@gmail.com



D. Singh et al. (eds.), *Proceedings of International Conference on Scientific and Natural Computing*, Algorithms for Intelligent Systems,
https://doi.org/10.1007/978-981-16-1528-3_21

## 1.1 Memristor Device Properties

It is a standard practice to investigate and analyze discrete component-based emulator due to the unavailability of actual memristor models. The possible relationships between the fundamental quantities voltage, current, charge, and flux are given in (1)–(3) [6].

$$dv = Rdi, d\varnothing = Ldi, dq = Cd \tag{1}$$

Equation (1) provides the relation between resistor, capacitor, and inductor. The other expressions relating the charge and flux are called memristive relationship given as [6]:

$$q(t) = \int_{-\infty}^{t} i(\tau)d\tau \tag{2}$$

$$\varnothing(t) = \int_{-\infty}^{t} v(\tau)d\tau \tag{3}$$

A generalized expression to understand memristive behavior is given by:

$$V = R(w_1, w_2, \ldots\ldots . w_k \ldots w_n) \tag{4}$$

where $w_1, w_2 \ldots\ldots$ are state variables.

Equation (4) essentially relates the voltage, current, and the variation in the resistance according to the state of a device at any point of time. It also identifies the charge that passed through the device previously. The constitutive relation for a memristor is f (∅, q) = 0 [6]. The memristor is said to be charge controlled if ∅=∅(q). Then the voltage across the device is given by:

$$V = M(q)I \tag{5}$$

where M (q) is termed as memristance given by:

$$M(q) = \frac{d\varnothing(q)}{dq} \tag{6}$$

Memristor is said to be flux controlled if q = q(∅) of the flux where current across it is given by [6]:

$$I = W(\varnothing) \tag{7}$$

$$W(q) = \frac{dq(\varnothing)}{d\varnothing} \tag{8}$$

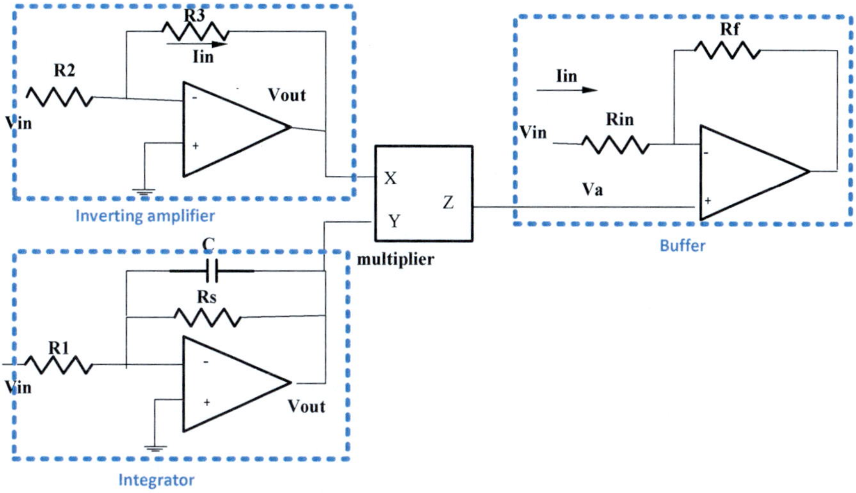


**Fig. 1** Circuit diagram of the proposed emulator (multiplier: AD633, Opamp: UA 741, standard resistors and capacitor)

where W (q) is termed as memconductance.

## 2 Proposed Memristor Emulator

The proposed memristor emulator shown in Fig. 1 aims to emulate the standard characteristics of $TIO_2$ memristor [2]. It consists of a practical integrator, inverting amplifier, multiplier, and buffer. The input is a sinusoidal waveform with different frequencies having a peak to peak voltage of +3 to −3 V and supply voltage of ±15 V. The input is integrated at one end and at the same time, the inverted signal coming from the other end is multiplied by multiplier and finally fed to the buffer. It is apparent that the proposed circuit is extremely simple.

## 3 Mathematical Analysis of Proposed Emulator

A modular analysis is carried out to understand the behavior and operating principle of the proposed emulator. In this context, the current Iin, through the resistance $R_{in}$ is related to the input voltage Vin as:

$$V_{in} = I_{in} R_{in} + V_a \tag{9}$$

Anyhow if (9) is able to make a relation between $V_{in}$ and $V_a$ so that it should be dependent on the current passing through $R_{in}$ and charge (q), then memristance can. be calculated later. Let us assume $V_a$ as input voltage and is equal to the input current multiplied by some variable k, so that the (9) can be substituted with this:

$$V_a = kI_{in} \tag{10}$$

So after the substitution, we get,

$$V_{in} = I_{in}R_{in} + k(I_{in}) \tag{11}$$

And the resistance of the circuit is then

$$R_{in} = (R_{in} + k) \tag{12}$$

Equation (12) shows the parameter 'k', which is a variable and depends on the charge. So, it can be said that resistance is variable with parameter 'k' in the buffer part of Fig. 1. Assuming the operational amplifier is ideal for theoretical analysis. From the integrator of Fig. 1, the output obtained is

$$V_{out} = -\frac{q(t)}{C} \tag{13}$$

where q (t) is the charge stored in the capacitor at any point of time t, C is the value of the capacitor. Value of the $R_s$ is generally chosen to be high so that current does not pass through the resistance and most of the current pass through the capacitor. From the inverting amplifier of Fig. 1, the output obtained is

$$V_{out} = -(R_3)I_{in} \tag{14}$$

The output of the multiplier is obtained by multiplying the output of the integrator and inverting amplifier and driving the output of the two gives (16)

$$V_{om} = V_a = \frac{q(t)}{C} \times (R_3)I_{in} \tag{15}$$

Substituting (15) in (11) gives the final relation between voltage $V_{in}$ and current $I_{in}$ as:

$$V_{in} = I_{in}R_{in} + \frac{q(t)}{C} \times (R_3)I_{in} \tag{16}$$

Equation (16) concludes that, resistance is dependent on charge at any point of time and memristance obtained in the same way explained in standard memristors in (17).

$$M = [R_{in} + \frac{q(t)}{C} \times (R_3)] \quad (17)$$

## 4 Experimental Results

The proposed emulator circuit is implemented on LTspice by using components explained in Fig. 1. Simulation results obtained gives I-V characteristics as Hysteresis loop and report its ability to act as storing element. Observe the following output waveforms at different frequencies when input is taken as a sinusoidal signal with 6 V peak to peak. Output at various frequencies is summarized in Table 1.

Figures 2, 3 and 4 show I-V characteristics of emulator circuit at 400 Hz, 1000 Hz and 1500 Hz respectively and it can be clearly observed that as frequency increases, the hysteresis loop shrinks with a decrease in the current range. The horizontal axis stays from +3 V to −3 V in each plot shown whereas the vertical axis varies according to frequency. So I-V characteristics showed a proper hysteresis loop achieving

**Table 1** Comparison of current range at various frequencies of memristor

| Table column subhead | Frequency in Hz | Current range in uA |
|---|---|---|
| 1 | 100 | −350 to 350 |
| 2 | 200 | −250 to 250 |
| 3 | 300 | −240 to 240 |
| 4 | 500 | −180 to 180 |
| 5 | 1000 | −160 to 160 |

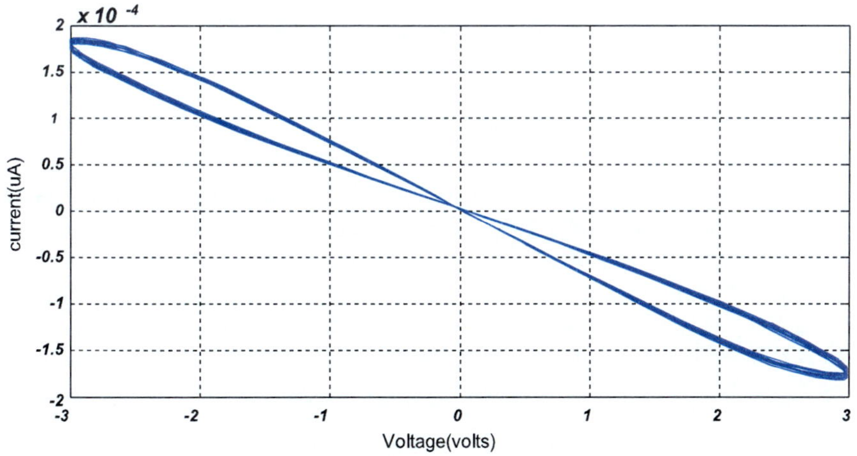


**Fig. 2** I-V characteristics when plotted at frequency 400 Hz

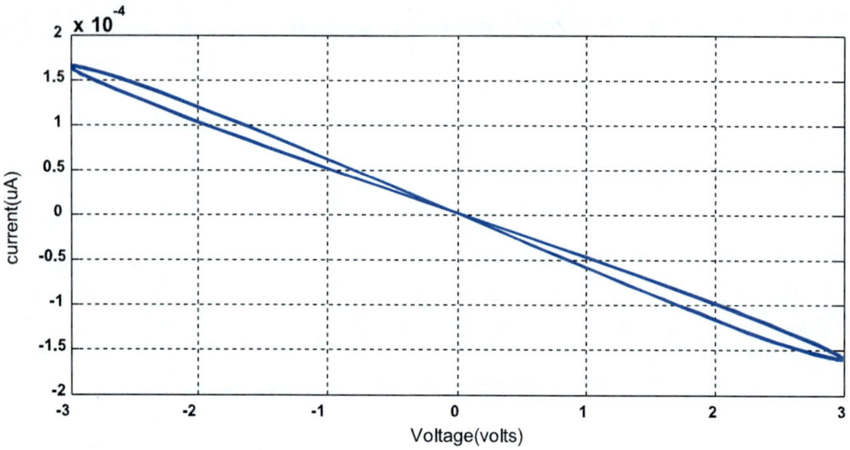


**Fig. 3** I-V characteristics when plotted at a frequency 1000 Hz

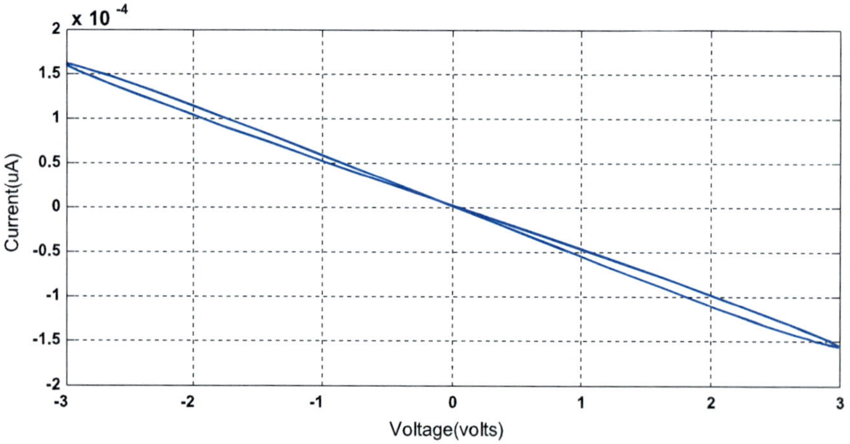


**Fig. 4** I-V characteristics when plotted at a frequency 1500 Hz

its fingerprints. This proposed emulator circuit shows memristive behaviour up-to 1.5 kHz and after this hysteresis loop eventually becomes linear and act as a resistor. The physical implementation of the proposed emulator circuit performed in the lab is shown in Fig. 5. The proposed set up is built using a breadboard and results are obtained on DSO (digital storage oscilloscope) which is shown in Fig. 6a and b clearly tells the imperfection in its characteristics.

Fig. 5 Hardware configuration of the proposed emulator on the breadboard

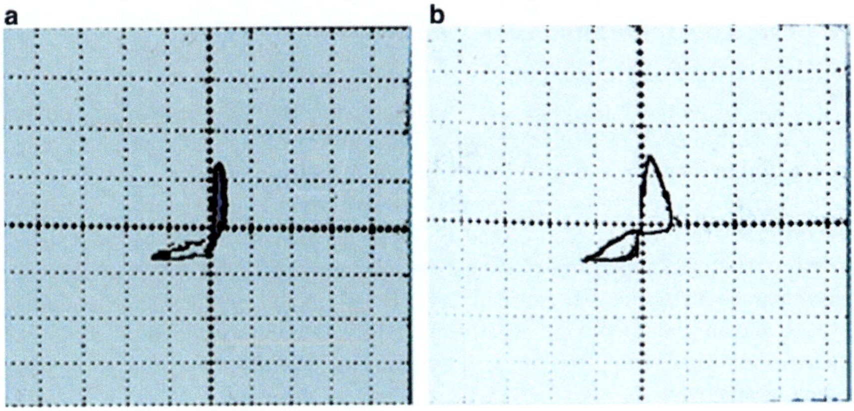

Fig. 6 **a** Hysteresis obtained at around frequency 1.5 kHz. **b** Hysteresis obtained 200 Hz

## 5 Improved Memristor Emulator

Improved memristor emulator basically aims to rectify the distortion obtained previously. This emulator circuit is shown in Fig. 7 consists of summer, integrator, multiplier and a non-inverting amplifier fed its output back to summer. Values of the resistor and the capacitor may change at various frequencies. In this circuit, summer and integrator are connected back to back and the output of both is fed to the multiplier and finally fed back to the initial part to change the state of the circuit. The supply voltage is +15 and −15 V and the input voltage is varied from 0.6 to 3 V. The

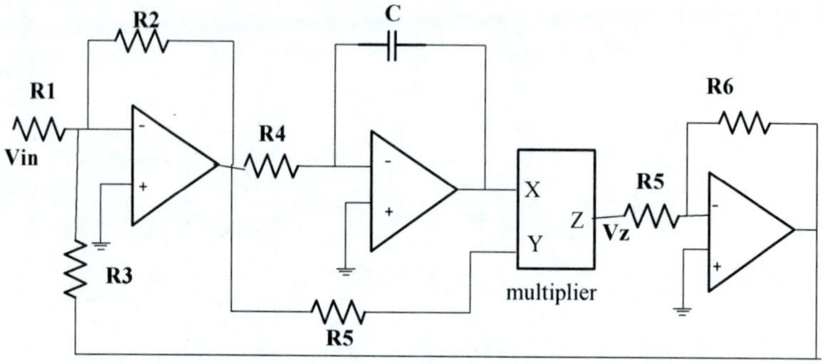


**Fig. 7** Circuit diagram of the improved proposed emulator

operational amplifier gives offset voltage which is prevented using compensation circuit at the time of the simulation.

## *5.1 Circuit Explanation and Analysis*

In this circuit, when input positively increases, the summing amplifier's output increases with the opposite sign charging capacitor of the integrator and accumulates it. Being summer and integrator's output negative, multiplier gives positive result increases current following a parabolic curve which is the second-order signal with quadratic relation between $I_{in}$ and $V_z$ of Fig. 8. Similarly, when the input signal decreases, summer output is negative but the integrator keeps the same sign since its sign cannot change quickly because of larger R and C value giving multiplier output negative and current falls in the negative path. In the same way, the third and fourth quadratic curves are completed making a hysteresis loop.

Multipliers output is:

$$V_z = V_{sum} V_{int} = (\frac{-1}{R_4.C_1}) \int V_{sum}^2 d \tag{18}$$

So after solving and substituting the equations, the relation between $I_{in}$ and $V_z$ is calculated as:

$$I_{in} = \left(\frac{V_{in}}{R_1}\right) + (m).V_z^2 + \left(A.V_{in}^2\right) + (B.V_{in}.V_z) \tag{19}$$

where 'm' is mem conductance, A and B are constants. Equation (19) shows the quadratic equation.

Results obtained on DSO (Digital Signal Oscilloscope) is shown in Fig. 8 at various frequencies. It is finally concluded that this circuit successfully reported its

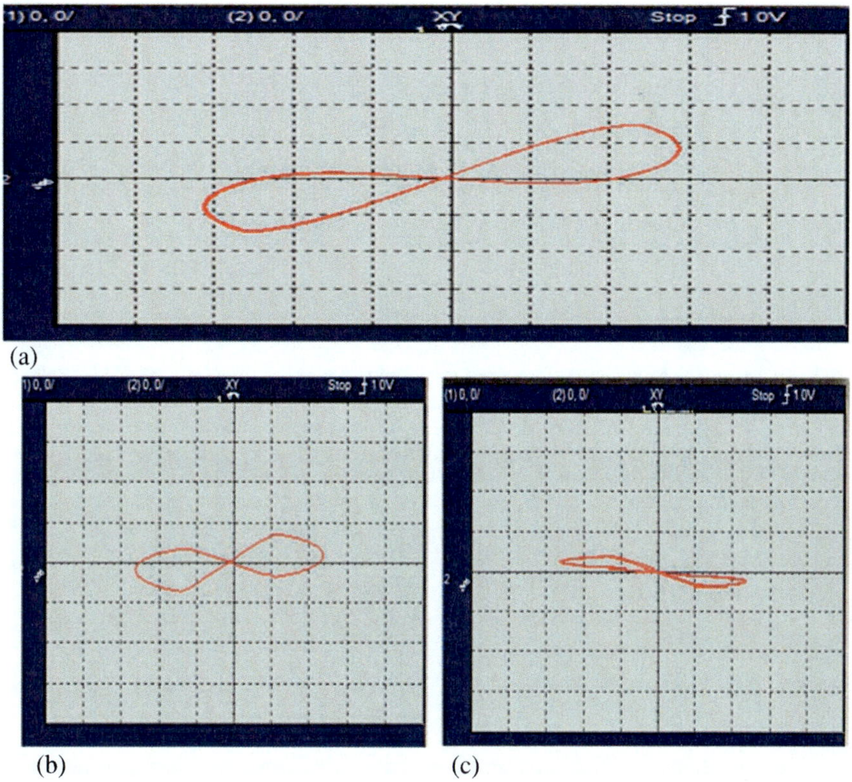


**Fig. 8** I-V characteristics of improved emulator analysed on a breadboard at frequency **a** 100 Hz, **b** 800 Hz and **b** 3 kHz

memristive behaviour till 3 kHz which double of proposed and its hardware results are improved. Both the proposed emulator and improved emulator circuit is tested in various applications and also its compatibility.

## 6 Application

FM demodulator circuit is shown in Fig. 9 is a simple experimental verification to test and check the functionality of the proposed emulator circuit [5]. In frequency to voltage conversion of this circuit, the proposed emulator is used instead of using a floating memristor as explained in [5]. Components values are also taken as in [5]. Figure 10 consists of frequency to voltage converter, envelop detector and low pass filter. Let $R_m$ is the resistance of the memristor and is frequency dependent. The gain of the inverting amplifier is variable and also depends on the input frequency. So when the FM signal is fed as input then it is supposed to get the output of AM signal.

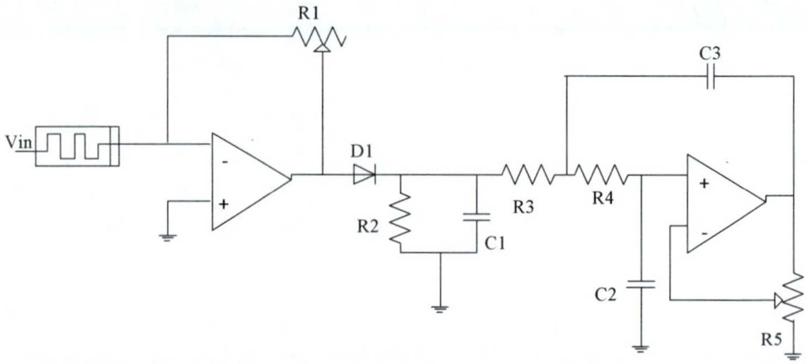


**Fig. 9** FM demodulator using the proposed circuit model consists of envelop detector of diode D1(standard resistor and capacitor) and a low pass filter [5]

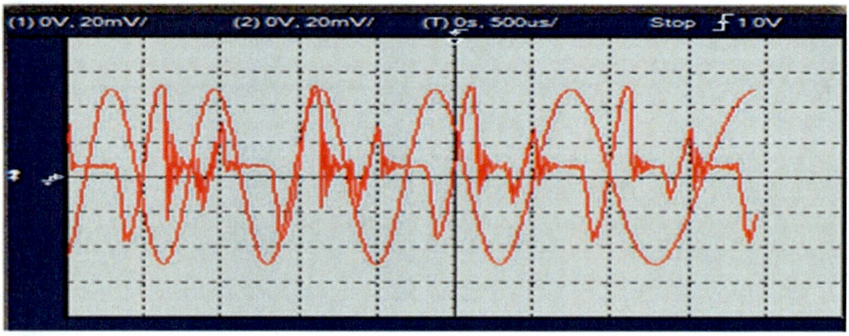


(a)

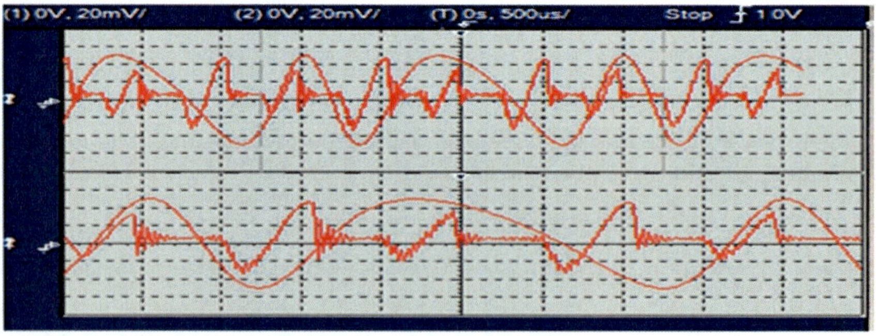


(b)

**Fig. 10** Waveform obtained at a carrier frequency equal to 2 kHz, a modulating frequency of 100 Hz and frequency deviation of 900 Hz with horizontal axis indicating the time and vertical axis indicating voltage were **a** conversion of FM to AM and **b** Delayed version of FM to AM conversion

**Table 2** Waveform in Fig. 10 is obtained with parameters are

| Fig(a) | Fig(b) |
|---|---|
| Rise time-350us | Rise time = 410us |
| Carrier Frequency = 2 kHz | Carrier frequency = 2 kHz |
| Modulating frequency = 100 Hz | Modulating frequency = 150 Hz |
| Frequency Deviation = 900 Hz | Frequency deviation = 950 Hz |

To check the operation, the output is observed at different points of the amplifier, envelop detector and low pass filter [5].

This application is basically focused on FM to AM conversion which is performed physically giving output results in Fig. 10 where FM signal is applied at the input of the circuit and distorted AM signal is obtained after conversion, indicating its functionality though the results are not perfect but appreciable. Conventional methods can be overtaken and usage of capacitors and inductors of larger value for tuning an LC circuit can be avoided. Parameters of waveforms obtained in Fig. 10 are given in Table 2.

## 7 Conclusion

In this paper, a proposed memristor emulator is been represented which behaves and replicates the real memristor. It is successfully reporting memristive behaviour up to 3 kHz and its architecture is quite simpler and easier as compared to other emulators in literature so far. Software results obtained on LTspice and physical implementation on breadboard giving improved result as obtained from spice simulations in the modified emulator. To verify its functionality, both emulator circuits are used in application like FM to AM conversion circuit in FM demodulator making advantage of not using capacitors and inductors of larger value on tuning an LC circuit. It is expected to be used as memory in future if features like speed, endurance, scalability and characteristics are fulfilled. In future work, we will try to investigate the parameters like rising time, fall time, delay and power in designing logic circuits.

## References

1. Chua LO (1971) Memristor-the missing circuit element. Circuit Theory, IEEE Transactions on 18(5):507–519
2. Kim H, Sah MP, Yang C, Cho S, Chua LO (2012) Memristor emulator for memristor circuit applications. Circ Syst I: Regul Papers, IEEE Transactions on 59(10):2422–2431

3. Kavehei O, Iqbal A, Kim YS, Eshraghian K, Al-Sarawi SF, Derek A (2010) The fourth element: characteristics, modelling and electromagnetic theory of the memristor. In Proceedings of the Royal Society of London A: Mathematical, Physical and Engineering Sciences, p. rspa20090553. The Royal Society
4. Kvatinsky S, Friedman EG, Kolodny A, Weiser UC (2013) The desired memristor for circuit designers. Circuits and Systems Magazine, IEEE 13(2):17–22
5. Abuelma'atti, Muhammad Taher, Zainulabideed Jamal Khalifa (2016) A new floating memristor emulator and its application in frequency-to-voltage conversion. Anal Int Circ Signal Proc 86(1):141–147
6. Di Ventra M, Yuriy VP, Leon OC (2009) Circuit elements with memory: memristors, memcapacitors, and meminductors. Proceedings of the IEEE 97(10):1717–1724
7. Adhikari, Shyam Prasad, Maheshwar Pd Sah, Hyongsuk Kim, Leon O. Chua (2013) Three fingerprints of memristor. Circ Syst I: Regul Papers, IEEE Trans 60(11):3008–3021
8. Valsa J, Biolek D, Biolek Z (2011) An analogue model of the memristor. Int J Numer Model Electron Networks Devices Fields 24(4):400–408
9. Preethi E, Mohamed Abbas A, Prabhu Kumar S, Arun Raghesh JT (2016) Design, analysis and simulation of memristor emulator based anti-aliasing filter for biomedical applications. Indian J Sci Technol 9(5)
10. Sánchez-López C, Jorge Mendoza-Lopez MA (2014) Carrasco-Aguilar, and Carlos Muñiz-Montero. A floating analog memristor emulator circuit. Circ Syst II: Exp Briefs, IEEE Trans 61(5):309–313
11. Abuelma'atti MT, Zainulabideen JK (2015) A continuous-level memristor emulator and its application in a multivibrator circuit. AEU-Int J Elect Commun 69(4):771–775
12. Yeşil A, Yunus B, Fırat K (2014) A new DDCC based memristor emulator circuit and its applications.Microelect J 45(3):282–287
13. Alharbi AG, Mohammed EF, Masud HC (2015) Memristor emulator based on practical current controlled model. In Circuits and Systems (MWSCAS), 2015 IEEE 58th International Midwest Symposium on, pp. 1–4. IEEE
14. Alharbi AG, Zainulabideen JK, Mohammed EF, Masud HC (2015) Memristor emulator based on single CCII. In Microelectronics (ICM), 2015 27th International Conference on, pp. 174–177. IEEE
15. Babacan Y, Kaçar F (2017) Floating memristor emulator with subthreshold region. Analog Integr Circ Sig Process 90(2):471–475
16. Babacan Y, Kaçar F (2017) Memristor emulator with spike-timing-dependent-plasticity. AEU-Int J Elect Commun 73:16–22
17. Yesil A, Gül F, Babacan Y (2018) Emulator circuits and resistive switching parameters of memristor. Memristor Memristive Neural Netw IntechOpen 1(1):41–61
18. Hassanein AM, Abdulaziz HE, Lobna AS, Ahmed HM, Ahmed GR (2018) Incremental grounded voltage controlled memristor emulator. In 2018 30th International Conference on Microelectronics (ICM), pp. 156–159. IEEE

# Chapter 22
# Optimization of Culture Conditions for EPS Production in *Lactobacillus rhamnosus* MTCC 5462 Through Taguchi Design Methodology

**Archana Bhati, Anil Kumar Baghel, and Barkha Singhal**

## 1 Introduction

Exopolysaccharides are the long-chain and high molecular mass polymers that are produced mainly by Bacteria and Microalgae. These exopolysaccharides are mainly polymers of sugar derivatives. These sugars are mainly consisting of glucose, galactose, mannose, N-acetyl glucosamine, N-acetyl galactosamine and rhamnose. EPS play important role in the protection of microbes from adverse conditions such as desiccation, nutrient starvation, toxic compounds and osmotic stress. The maximal synthesis occurs in the late log phase or early stationary phase of bacterial growth. EPS also help in the adhesion of bacteria to a solid surface, gastrointestinal tract, and formation of biofilm [1]. Some of the bacterial exopolysaccharides confer an immunomodulatory role in human GIT [2]. These exopolysaccharides found in the microbes in two forms (a) Cell bound exopolysaccharides (c-EPS) which are bound covalently to the cell surface and released exopolysaccharides (r-EPS) that are freely secreted in the surrounding medium. Apart from this classification EPS can also be classified into two categories namely as homopolysaccharides (*e.g.* Levan, Dextran, Cellulose, Pullulan, Curdla) and heteropolysaccharides (*e.g.* Xanthan, Alginate, Fucopol, Galactopol) that are joined together by a single linkage (1–2 or 1–4 linkage) or combination of linkages(1–2 and 1–4). In recent years, the wider application pf EPS as anticancer, anti-biofilm and antioxidant agents in the pharmaceutical industry as well as bio-thickening agents as stabilizing, emulsifying and jelling properties in the food industry have been envisaged. Therefore, the production and purification of EPS from microbial sources is a thrust area of research. Currently,

A. Bhati · A. K. Baghel · B. Singhal (✉)
School of Biotechnology, Gautam Buddha University, Greater Noida, India
e-mail: barkha@gbu.ac.in

A. K. Baghel
Research officer, Panacea Biotec. Ltd., Lalru, Punjab, India

 
D. Singh et al. (eds.), *Proceedings of International Conference on Scientific and Natural Computing*, Algorithms for Intelligent Systems,
https://doi.org/10.1007/978-981-16-1528-3_22

various bacterial species have been identified for the production of EPS but achieving a status of GRAS (Generally Regarded As Safe) various *Lactobacillus* species such as *Lactobacillus rhamnosus*, *L. casei*, *L. plantarum* and *L. hevelticus* are currently being explored for the production of EPS [3–6].

The above discussion revealed that EPS is an important commercial valuable bioproduct thus its production needs to be optimized for achieving its high yields. Among various optimization strategies, media optimization plays a pivotal role in the growth and development followed by product forming capabilities of the cell. The culture medium optimization by "one factor at a time (OFAT)" has been found to be a time and labor-intensive process. Therefore, multiple variables at a time (MVAT) provides a systematic and efficient platform for experimentation for simultaneous study of many factors and their interactions. Therefore, in recent years number of statistical designs such as Plackett-Burman (PB) design, Response surface methodology (RSM) and Taguchi method etc. have been used to rapidly identify the significant parameters influencing productivity. Taguchi method is a robust experimental design that has been used to get an optimum combination set of factors/levels with the lowest cost solution to achieve the product quality requirements. The experimental design proposed by Taguchi involves using orthogonal arrays to organize the parameters affecting the process and the levels at which they should be varied. Thus, in this work EPS production has been optimized using *L. rhamnosus* MTCC 5462 that is an Indian commercialized probiotic.

# 2 Materials and Methods

## *2.1 Growth and Harvesting of Bacterial Cells*

### 2.1.1 Preparation of Primary Culture

The inoculum was prepared by adding a loop of freshly prepared pure culture *L. rhamnosus* MTCC 5462 (procured from IMTECH, Chandigarh) from Petri dish into 25 ml of autoclaved De Man Rogosa and Sharp (MRS) medium was used for cultivation of bacteria.

### 2.1.2 Preparation of Secondary Culture

One milliliters of the primary culture was added to 25 ml of the medium in a 100 ml flask. The flask was incubated in a BOD incubator at 37 °C for 14 h.

## *2.2 Production of EPS*

### 2.2.1 Composition of Synthetic Media for EPS Production (G/L)

Disodium hydrogen phosphate-5.0, Potassium dihydrogen phosphate 6.0, Tri-ammonium citrate 2.0, Magnesium sulfate 1.0, Dextrose- 20.0, Trace elements (g/l) Ferrous sulfate heptahydrate 5.0, Manganese sulfate 2.0, Cobalt chloride 1.0, Zinc chloride 1.0. All the components are autoclaved at 121 °C at 15 psi pressure for 15 min. Dextrose was autoclave separately. Disodium hydrogen phosphate and Ferrous sulfate heptahydrate were filter sterilized.

### 2.2.2 Production of EPS

The selected microorganism *L. rhamnosus* MTCC 5462 was grown in synthetic medium for 16 h at 37 °C. The culture was centrifuged at 9000 rpm for 10 min. The resulting supernatant was transferred to the new falcon tube and an equal volume of chilled absolute ethanol was added to this supernatant. The resulting mixture was stored at 4 °C for 48 h. After 48 h centrifugation was done at 9000 rpm for 10 min and the resulting pellet was dissolved in autoclaved water and stored at −40 °C. The sugar content was determined by the phenol sulfuric acid method.

### 2.2.3 Design of Taguchi Method for the Production of EPS

The initial studies confirmed that glucose as an energy source and tri-ammonium citrate as a nitrogen source plays a pivotal role in the production of EPS. Therefore, the orthogonal design of Taguchi methodology has been implied for the creation of statistical data and results. In this study, Minitab version 16.0 software package has been used for the design of experiments (DOEs).

## 3 Results and Discussion

The optimization of process variables using a statistical approach for EPS production was carried out using Taguchi design with six selected factors after studying the effect of different parameters. The six factors included the following: glucose, tri-ammonium citrate, pH, temperature, inoculum percentage, the incubation time for the efficient EPS producing strains *L. rhamnosus* MTCC 5462. A design model with 27 runs was exhibited and each independent variable was tested at three levels. Each run was performed in triplicate and thus the values of EPS yield were compared with the predicted values of responses which were obtained from using the software

**Table 1** Selected culture conditions as variables and assigned levels for EPS production from *L. rhamnosus* MTCC 5462 by Taguchi design method

| Factors | Level 1 | Level 2 | Level 3 |
|---|---|---|---|
| pH | 7 | 8 | 9 |
| Temperature (°C) | 30 | 37 | 45 |
| Inoculum percentage (%) | 3 | 5 | 10 |
| Incubation time (in hours) | 24 | 48 | 72 |
| Tri- ammonium citrate (g/l) | 1 | 2 | 3 |
| Glucose(g/l) | 10 | 20 | 30 |

Minitab 16.0. The statistical significance was checked by the analysis of variance (ANOVA).

The model of EPS yield obtained from *L. rhamnosus* MTCC 5462 was significant, as is evident from the model F-value and a very low probability value [(Prob > F) = < 0.001]. Table 1 shows the key factors and their levels selected for the optimization process using Taguchi DOE methodology.

After the implementation of 27 experiments based on the orthogonal array the composition of the culture medium and the quantities of the components determine the production of EPS. Table 2 described the variation in EPS production according to the experiments conducted based on the Taguchi design of experiment (DOE) method.

The highest yield has been obtained from experiment no-15 in which 10.82 g/l EPS was obtained. For the assessment of further interaction among various variables analysis of variance (ANOVA) was used to analyze the results of the above-mentioned experiments to determine how much variation was contributed by each factor. Among the factors and their levels studied on the EPS production, glucose and pH showed the strongest influence on the production of EPS. The results are summarized in Table 3. The contribution of each factor has been represented in Fig. 1.

The contribution of individual factors is the key factors for the efficiency of the fermentation process. Thus, by seeing the contribution in the above-mentioned results further the effect of selected variables and their relative influence in EPS production has been evaluated. The results are summarized in Fig. 2. The selected culture conditions mentioned in Table 2 interact with each other to maximize the production of EPS. The best combination that has been obtained from the experimental study can be conferred that from *L. rhamnosus* the optimized composition for EPS products was pH 7.0, Temperature 37°C, Incubation time 72 h, Inoculum percentage 3%, Tri-ammonium citrate 3 g/l and glucose 10 g/l.

**Table 2** Statistical optimization of EPS production through *Lactobacillus rhamnosus* MTCC 5462

| S.No. | pH | Temp. | IC | IP | TAC | Glucose | Response EPS yield (g/l) | Predicted yield (g/l) |
|---|---|---|---|---|---|---|---|---|
| 1 | 6 | 30 | 24 | 3 | 1 | 10 | 1.3 | 0.013 |
| 2 | 6 | 30 | 24 | 3 | 2 | 20 | 8.5 | 8.08 |
| 3 | 6 | 30 | 24 | 3 | 3 | 30 | 1.1 | 1.30 |
| 4 | 6 | 37 | 48 | 5 | 1 | 10 | 3.6 | 3.42 |
| 5 | 6 | 37 | 48 | 5 | 2 | 20 | 3.7 | 3.42 |
| 6 | 6 | 37 | 48 | 5 | 3 | 30 | 3.8 | 3.42 |
| 7 | 6 | 45 | 72 | 10 | 1 | 10 | 0.9 | 1.30 |
| 8 | 6 | 45 | 72 | 10 | 2 | 20 | 6.7 | 5.42 |
| 9 | 6 | 45 | 72 | 10 | 3 | 30 | 8.7 | 9.23 |
| 10 | 7 | 30 | 48 | 10 | 1 | 20 | 0.8 | 0.24 |
| 11 | 7 | 30 | 48 | 10 | 2 | 30 | 6.8 | 6.90 |
| 12 | 7 | 30 | 48 | 10 | 3 | 10 | 7.8 | 7.92 |
| 13 | 7 | 37 | 72 | 3 | 1 | 20 | 7.5 | 6.12 |
| 14 | 7 | 37 | 72 | 3 | 2 | 30 | 8.6 | 9.61 |
| 15 | 7 | 37 | 72 | 3 | 3 | 10 | 10.7 | 10.82 |
| 16 | 7 | 45 | 24 | 5 | 1 | 20 | 1.3 | 0.013 |
| 17 | 7 | 45 | 24 | 5 | 2 | 30 | 4.5 | 4.55 |
| 18 | 7 | 45 | 24 | 5 | 3 | 10 | 1.1 | 1.30 |
| 19 | 8 | 30 | 72 | 5 | 1 | 30 | 3.6 | 3.42 |
| 20 | 8 | 30 | 72 | 5 | 2 | 10 | 5.2 | 5.68 |
| 21 | 8 | 30 | 72 | 5 | 3 | 20 | 6.8 | 6.47 |
| 22 | 8 | 37 | 24 | 10 | 1 | 30 | 4.0 | 5.9 |
| 23 | 8 | 37 | 24 | 10 | 2 | 10 | 1.6 | 1.19 |
| 24 | 8 | 37 | 24 | 10 | 3 | 20 | 4.6 | 5.33 |
| 25 | 8 | 45 | 48 | 3 | 1 | 30 | 6.7 | 6.81 |
| 26 | 8 | 45 | 48 | 3 | 2 | 10 | 2.9 | 3.15 |
| 27 | 8 | 45 | 48 | 3 | 3 | 20 | 2.8 | 3.12 |

# 4 Conclusion

The culture conditions for EPS production have been optimized using recently employed Taguchi design methodology and the optimum culture conditions are as follows (pH 7.0, Temperature 37 °C, Incubation time 72 h, Inoculum percentage 3%, Tri- ammonium citrate 3 g/l and Glucose 10 g/l). The comprehensive studies will be taken for the further validation of the Taguchi optimization at the fermenter scale. Apart from that as EPS production has been done with GRAS organism, therefore, in

**Table 3** Optimum culture conditions and their contributions for EPS production from *Lactobacillus rhamnosus* MTCC 5462

| Factors | Values | Level | Contribution (%) |
|---|---|---|---|
| pH | 7 | 1 | 19 |
| Temperature (°C) | 37 | 37 | 8 |
| Inoculum percentage (%) | 3 | 1 | 4 |
| Incubation time (in hours) | 72 | 3 | 3 |
| Tri-ammonium citrate (g/l) | 3 | 3 | 11 |
| Glucose (g/l) | 10 | 1 | 55 |

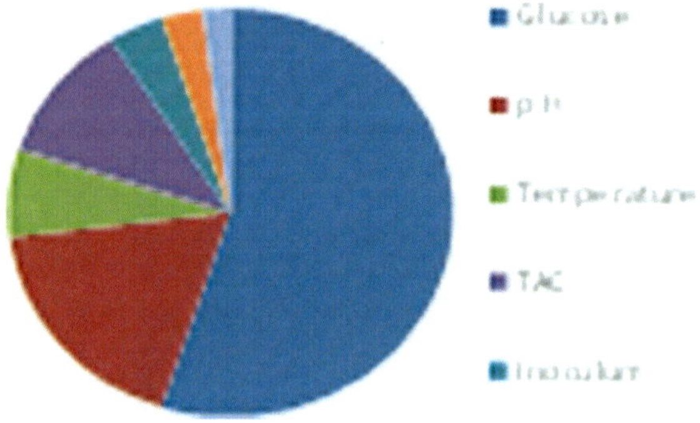


**Fig. 1** Pie chart showing optimum culture condition and their contributions for EPS production from *Lactobacillus rhamnosus MTCC* 5462

the future various bioactive properties like antioxidant, anti-tumor, anti-mutagenic potential of this biomolecule can be characterized for the applicability at an industrial scale.

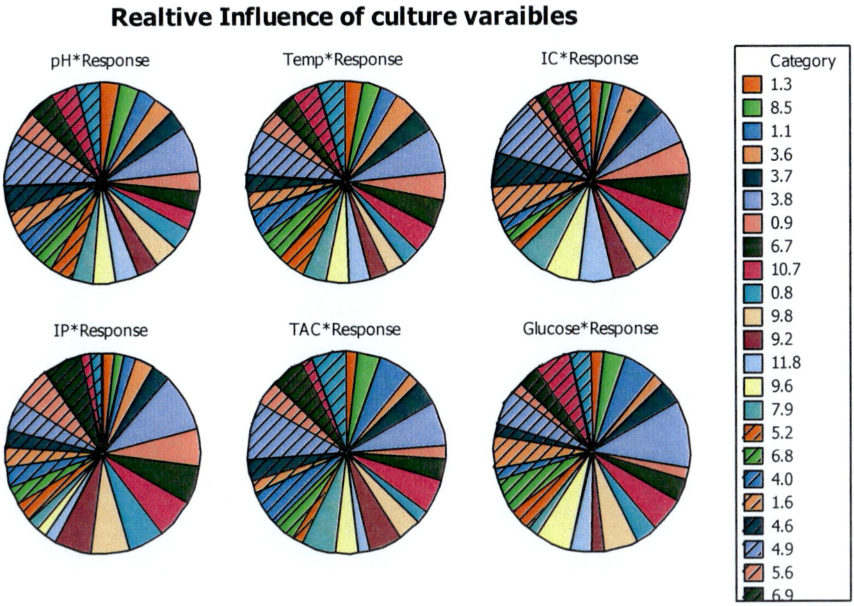


**Fig. 2** Relative influence of culture variables on EPS production by *Lactobacillus rhamnosus* MTCC 5462 (where IP: inoculum percentage, IC: incubation time (hrs), Temp: temperature (°C), Tri-ammonium citrate (g/l), Glucose (g/l)

**Acknowledgements** This research was financially supported by the Gautam Buddha University, Greater Noida. The patented strain used in this study was provided by Anand Agricultural University, Gujrat (India) for the research purpose therefore the authors greatfully acknowledge their support. The authors are highly thankful to the Gautam Buddha University for conferring all infrastructural facilities for the execution of this work.

**Conflicts of Interest** The authors declare no conflict of interest.

## References

1. Dertli E, Mayer MJ, Narbad A (2015) Impact of the exopolysaccharide layer on biofilms, adhesion and resistance to stress in *Lactobacillus johnsonii* FI9785. BMC Microbiol 15(1):8
2. Ruas-Madiedo P, Gueimonde M, Margolles A, de los Reyes-Gavilán CG, Salminen S (2006) Exopolysaccharides produced by probiotic strains modify the adhesion of probiotics and enteropathogens to human intestinal mucus. J Food Protect 69:2011–2015
3. Cerning J, Bouillanne C, Landon M, Desmazeaud M (1992) Isolation and characterization of exopolysaccharides from slime-forming mesophilic lactic acid bacteria. J Dairy Sci 75(692):699
4. Patten DA, Laws AP (2015) Lactobacillus-produced exopolysaccharides and their potential health benefits: a review. Benef Microbes 6:457–471
5. Oleksy M, Klewicka E (2016) Exopolysaccharides produced by Lactobacillus sp.–biosynthesis and applications. Critic Rev Food Sci Nutri 41(13):28–38
6. Živković M, Miljković MS, Ruas-Madiedo P, Markelić MB, Veljović K, Tolinački M, Golić N (2016) EPS-SJ exopolisaccharide produced by the strain Lactobacillus paracasei subsp. paracasei

BGSJ2-8 is involved in adhesion to epithelial intestinal cells and decrease on *E. coli* association to Caco-2 cells. Front Microbiol 7(37)

# Chapter 23
# Assessment of Biochemical Methane Potential in Anaerobic Biodegradation of Industrial Food Waste

**Athar Hussain, Khayati Gaur, and Richa Madan**

## 1 Introduction

The food industry is a complex, global cooperative of diverse businesses that supplies most of the food consumed by the world's population. Only subsistence farmers, those who survive on what they grow and hunter-gatherers can be considered outside of the scope of the modern food industry. The food industry produces large volumes of wastes, both solids and liquid, resulting from the production, preparation and consumption of food. These wastes pose increasing disposal and severe pollution problems and represent a loss of valuable biomass and nutrients. India is the world's second-largest food producer next to China. The food processing industry is a vital link and synergy between the two pillars of our country's economy – industry and agriculture. The food industry is broadly concerned with a number of processing sectors namely fruits and vegetable processing, consumer food processing, meat processing, poultry and eggs processing, fish processing, milk processing, and grains processing.

As per the World Bank (2012) prediction, the generation of municipal solid waste (MSW) in the world will increase from 1.3 billion tonnes per year in 2012 to 2.2 billion tonnes per year by 2025. The European Union (EU) generates approximately

A. Hussain
Associate Professor, Department of Civil Engineering, NSUT WEST CAMPUS (Formerly CBP Government Engineering College Jaffarpur), New Delhi 110073, India

K. Gaur
Postgraduate Student, Department of Civil Engineering, School of Engineering Gautam Buddha University, Noida 201312, UP, India

R. Madan (✉)
PhD Research Scholar, Department of Environment Science, Gurukula Kangri (Deemed to Be) University, Haridwar, Uttarakhand, India
e-mail: richa.madan.92@gmail.com

 
D. Singh et al. (eds.), *Proceedings of International Conference on Scientific and Natural Computing*, Algorithms for Intelligent Systems,
https://doi.org/10.1007/978-981-16-1528-3_23

130 million tonnes per year of organic waste, of which 68% is food waste. A high proportion of the MSW in Indian cities is organic, estimated to range between 40 and 60%, although the MSW generation rates of Indian cities is considerably less, with a total production of the order of 50 million tonnes per year [7]. MSW management has become a global apprehension arising due to escalating human population with extensive resource utilization, rapid urbanization and industrialization and ineffective waste management rules [19]. MSW management is generally allied with landfilling, composting, open dumping, thermal treatment and anaerobic digestion. Among all the waste management techniques, the landfill is the major contributor of greenhouse gases (GHGs) emissions. Furthermore, methane ($CH_4$) emissions from landfills are expected to elevate to 959.4 million tonnes of carbon dioxide equivalent by the year 2030 [4].

Although composting is generally regarded as an environmentally benign waste management option, its potential occupational and public health risks cannot be neglected. In general, the major environmental concerns associated with MSW composting includes generation of bio-aerosol, dust, noise and leachate during various composting operations, as well as their release to the surrounding environment beyond the composting facilities together with potential exposure to heavy metals, organic pollutants, and pathogenic agents in the case of compost application to soil [21]. According to Anyaoku and Baroutian [4] processing of organic fraction of municipal solid waste (OFMSW) with thermochemical conversion technologies such as combustion and gasification may be unsuitable because of the presence of high moisture content of the food waste (FW). Among the treatment technologies available for treating organic solid wastes (OSW), anaerobic digestion (AD) is very apposite because of its limited environmental impacts, cost-efficacy, simplicity and reliability. Therefore, OFMSW may be suitable for processing by anaerobic digestion (AD) technology which can convert it to biogas (a fuel) and digestate (a nitrogen-rich compost material).

An effective maneuver of the AD process needs balanced C/N ratios, suitable pH, substrate and other parameters [14, 22]. Anaerobic digestion is strongly affected by temperature. The optimum temperature of mesophilic digester for biogas production is 35 °C. In the mesophilic range, the activity and growth rate of bacteria decreased by 50% for each 10 °C drop. Fall in biogas production starts, when temperature decreases to 20 °C and the production stops at 10 °C. Increasing the temperature level up to 37 °C leads to the time reduction required for the digestion process. Further increase in temperature decreases the rate of biogas generation. The importance of pH is due to the fact that the methanogenic bacteria are very sensitive to acidic conditions and their growth and methane production are inhibited in an acidic environment. The methane-producing bacteria require a neutral to the slightly alkaline environment, that is, from 6.8 to 8.5. The pH of the anaerobic digestion process is one of the parameters that have a significant effect on the digestion process. The optimum pH range in an anaerobic digester is 6.8 to 7.2. However, the process can tolerate a range of 6.5 up to 8.0.

Furthermore, high moisture content facilitates anaerobic digestion [5]. The carbon to Nitrogen Ratio (C/N) ratio in the organic material plays an important role in AD.

The optimum C/N ratio in anaerobic digestion is ranged from 20 to 30. The low gas production ensues rapid consumption of nitrogen by methanogens due to the high C/N ratio. A lower C/N ratio causes ammonia accumulation and pH values exceeding 8.5, which is toxic to methanogenic bacteria. Optimum C/N ratios of digester materials can be achieved by mixing material of high and low C/N ratio, such as organic solid waste mixed with sewage or animal manure [20]. The anaerobic digestion process involves a wide variety of organisms of which bacteria are the most predominant. These microorganisms digest organic molecules like lipids, carbohydrates and proteins into mainly methane and $CO_2$. In addition, if the wastewater contains sulfate($SO_4^{2-}$), it is used as an electron acceptor by bacteria like desulphovibrio during the oxidation of organic compounds. Denitrifying bacteria oxidize organic substrates and use $NO_3$ as an electron acceptor and liberate $N_2$ in the process. Currently, there are nearly 240 anaerobic digestion facilities around the world operating at a capacity of over 2,500 metric tons of organic waste per year. The total installed capacity of these plants is more than 11 million metric tons per year.

The total anaerobic digestion capacity has increased five-fold worldwide from about two million metric tons in 1990 to over ten million by 2008 [13] Several studies based on the AD process claim the viability and feasibility of the process in the improvement of biogas generation and sustainable handling of waste [8, 17]. Therefore, the present study has been undertaken with an objective to observe methane production potential from biodegradation of industrial food waste under psychrophilic, mesophilic and thermophilic conditions using batch type test study. Furthermore, the study incorporates the calculation of the methane generation rate constant (k) and Biochemical methane potential (BMP) under above-specified conditions.

## 2 Materials and Methods

The food waste used as feedstock in the present study was collected from three food production industries situated in the industrial area of state Delhi and Uttar Pradesh, India. The food waste was collected from Surya Food and Agro limited, A-1 UdyogVihar, Greater Noida, Uttar Pradesh, 201303, City Station Road, Khurja, Bulandshahr, 203131 and Britannia Industries Limited, Keshavpuram, New Delhi, 110034. The characteristics of feedstocks have been determined and are summarized in Table 1. The seed sludge used as inoculum in the study was collected from a real scale 72 MLD upflow anaerobic sludge blanket reactor (UASBR) wastewater treatment plant located in Indirapuram, Ghaziabad, U.P. Prior to charging to the reactor the physicochemical parameters such as total solids (TS), total volatile solids (TVS), chemical oxygen demand (COD), ammonical nitrogen, phosphorus, pH, moisture content of the inoculum were determined and are shown in Table 1.

Collected food samples were screened and further dried at 50 °C for 24 h to reduce the moisture content. Homogenous mixtures of the dried samples were prepared

**Table 1** Characteristics of feedstock(substrate) and inoculum used in the present study

| S. no. | Parameters | Feedstock | | | Inoculum |
|---|---|---|---|---|---|
| | | Biscuits | Rusk | Namkeen | |
| 1. | pH | 7.12 | 7.01 | 7.11 | 7.9 |
| 2. | Total suspended solids (%) | 23.5 | 28.2 | 26 | 16.66 |
| 3. | COD (mg/L) | 79.36 | 21.63 | 71.75 | 1750 |
| 4. | Ammonical nitrogen (mg/L) | 2.62 | 1.36 | 4.75 | 2.8 |
| 5. | Total phosphorous (mg/L) | 3.1 | 1.95 | 2.6 | 4.4 |

through grinding and sieving. Moisture content, pH, COD, TS, ammonical nitrogen and total phosphorus of the feedstocks and inoculums were measured by using the method specified in standard methods [3]. pH values were measured using a digital pH meter (Hatch edition 1). Measurement of COD, ammonical nitrogen and total phosphorus was done through close reflux method using US-VIS Spectrophotometer (HACH make, DR5000). The analytical data quality was guaranteed through the implementation of laboratory quality assurance and quality control methods, including the use of standard operating procedures, calibration with standards, analysis of reagent blanks, recovery of known additions and analysis of replicates. All analyses were carried out in triplicates and the results were expressed as the mean. Stock solutions for the preparation of defined media were prepared according to [9, 12, 16].

To evaluate the impact of the temperature on the anaerobic biodegradability of food waste, batch experiments were performed in three phases in accordance with the methods proposed by [16]. The experiments were carried out in three phases under controlled temperature conditions of psychrophilic range ($15 \pm 2^{\circ}C$) as phase-I, mesophilic range ($30 \pm 2^{\circ}C$) as phase-II and thermophilic range ($45 \pm 2^{\circ}C$) as phase-III. The operating conditions of all phases were carried out at food to inoculum ratio ($\frac{F}{M}$) of 0.75. The serum bottles of working volume 500 mL were initially filled with the measured quantity of food waste, inoculum and nutrients at the same F/M ratio. Thereafter, all the bottles were purged with $N_2$ gas for 5 min in order to maintain the anaerobic conditions then placed in an incubator at a controlled temperature. The blank with nutrients and inoculum were also used in order to calculate the background gas collection. The biogas in the inverted serum bottles was collected by using the water displacement method and $CH_4$ gas generation was measured daily and the values for the same has been reported in STP. The COD reduction or substrate degradation has been calculated in terms of mL of $CH_4$ produced per gram COD consumed or L of $CH_4$ produced per g COD reduction. Biochemical Methane Production runs under anaerobic conditions, most relevant tests for finding biodegradability gives information about biomethanation of the substrate and expressed as mL of $CH_4$ converted to g $CH_4$-COD/g COD destroyed at STP. It measures the maximum amount of methane produced per gram of volatile solids (VS) in batch conditions. The BMP assay can be used as an index of anaerobic biodegradation potential as it provides digestibility and methane production [11].

$$BMP = \frac{\text{Maximum Cumulative Methane (mL)}}{g\text{COD removed}} \quad (1)$$

In the batch test study during the anaerobic biodegradation process, the $CH_4$ production was monitored for a 30-40 days period in all three phases. The cumulative methane generation profile appears to follow an exponential growth curve and data has been used to evaluate the methane generation rate constant k and Biochemical Methane Potential (BMP). The BMP is expressed as $\mu_b$ or ultimate potential to convert substrate COD to $CH_4$–COD. The effect of variation of temperature on k and BMP has also been studied.

## 3 Results and Discussion

The methane production in all the phases was measured for 30–40 days. The blank consisted of seed sludge and media only. Very low methane gas generation was observed in blank as compared to other reactors. The net methane gas production has been calculated by subtracting the value of total methane produced from the blank reactor and is expressed as mL $CH_4$ per gram COD or L $CH_4$ per gram COD. Methane production rate constants (k) were determined as per the method specified by Hussain et al. [1]. The degradation of each sample has been assumed to follow the first order rate of decay. Thus, the production of methane is assumed to follow the equation given below [2].

$$Y = Y_{\max}(1 - e^{-kt}) \quad (2)$$

Where Y is the cumulative methane yield (mL) at the time (t), $Y_{\max}$is the maximum methane yield (mL) at Y-axis, k is the rate constant($d^{-1}$)and t is the time in days. The cumulative methane generation profiles appear to follow logistic exponential growth curves which can be represented by the exponential equation as:

$$-kt = \log\{1 - (Y/Y_{max})\} \quad (3)$$

These equations were employed in the function with non-linear optimization using KY plot software to calculate values of k for maximum biogas yield, k values and, $r^2$ (correlated coefficient) values are summarized in Table 2. Buswell equation has been used in calculating the theoretical $CH_4$ production of the pure substrate, whereas in mixed substrate, the expected $CH_4$ production is calculated based on the following reaction (Eq. 4).

$$CH_4 + 2O_2 \rightarrow CO_2 + 2H_2O \quad (4)$$

One mole of $CH_4$ at standard temperature and pressure (STP:0 °C and one atmospheric pressure) occupies a volume of 22.4 L. In addition, one mole of $CH_4$ (16 g)

**Table 2** Loading concentration of food waste and inoculum in different phases

| Substrate type | F/M | Substrate concentration(mg/L) | Inoculum concentration (mg/L) | Substrate COD (g/L) |
|---|---|---|---|---|
| Namkeen | 0.75 | 19.13 | 0.765 | 7.1 |
| Rusk | 0.75 | 05.76 | 0.23 | 2.1 |
| Biscuits | 0.75 | 21.16 | 0.80 | 7.9 |
| Blank | – | – | 5 | – |

is equivalent to two moles of oxygen (64 g of COD). Thus 64 g COD stabilized to 22.4 L of $CH_4$ and therefore each gram of COD will stabilize the 0.35 L of $CH_4$ at STP.

In the first phase of the study, the serum bottles were kept at a constant temperature of (30 ± 2) °Cat fixed F/M ratio of 0.75 in the incubator. The methane production was observed to be started after 2 days of the study. The cumulative methane production was monitored for 35 days of phase-I of the study. Figure 1 reveals the temporal variation of cumulative $CH_4$ with respect to time in days under mesophilic conditions. The maximum methane yields in reactor 1, 2 and 3 were 60, 75 and 149 mL, respectively. A perusal of the data summarized in Table 3 reveals the fact that the k values using the first-order kinetic model ranges from 0.07 to 0.09 $d^{-1}$ in all the reactors with $R^2$ values 0.96 and 0.98 respectively. It was observed that the maximum $CH_4$ production with a k value of 0.08 $d^{-1}$ was found in the sample of rusk waste. However, the lowest k value of 0.07 $d^{-1}$ was observed in the sample of namkeen waste ensuing the presence of recalcitrant compounds. The BMP value expressed as $\mu_b$ in terms of $CH_4$-COD/g $COD_{fed,}$ that is, COD fed conversion to $CH_4$–COD is another term for the experimental efficiency [6, 15]. The maximum $\mu_b$ value of 0.58 g $CH_4$–COD/g COD was found in the reactor containing rusk waste.

All the experiments of phase-II of the study were carried out under a thermophilic temperature range of (45 ± 2) °C at F/M ratio of 0.75. In this phase, it was observed that the methane production initiated earlier at an accelerated rate as compared to the mesophilic phase because at higher temperature biodegradation process is rapid and fast. The obtained result under temporal variation of cumulative $CH_4$with respect to time in days of the phase-II is depicted through Fig. 2. The maximum methane yield in the waste samples of namkeen, rusk and biscuits were found to be 30, 120, 140 mL, respectively. A perusal of the data summarized in Table 3 indicates that the maximum $CH_4$ production and the maximum k value of 0.1 $d^{-1}$ was observed in a sample containing rusk waste. However, the lowest k value of 0.08 $d^{-1}$ was observed in the reactor containing namkeen waste. Similarly, the maximum BMP value in terms of $\mu_b$ of 0.94 g$CH_4$-COD/g COD was observed in a reactor containing rusk waste.

The phase-III of the study was carried out under psychrophilic condition at (15 ± 2) °C under F/M ratio of 0.75. It was observed that the methane production started slowly as compared to the mesophilic and thermophilic phase because at a lower temperature the biodegradation process retards and takes a long time to degrade the

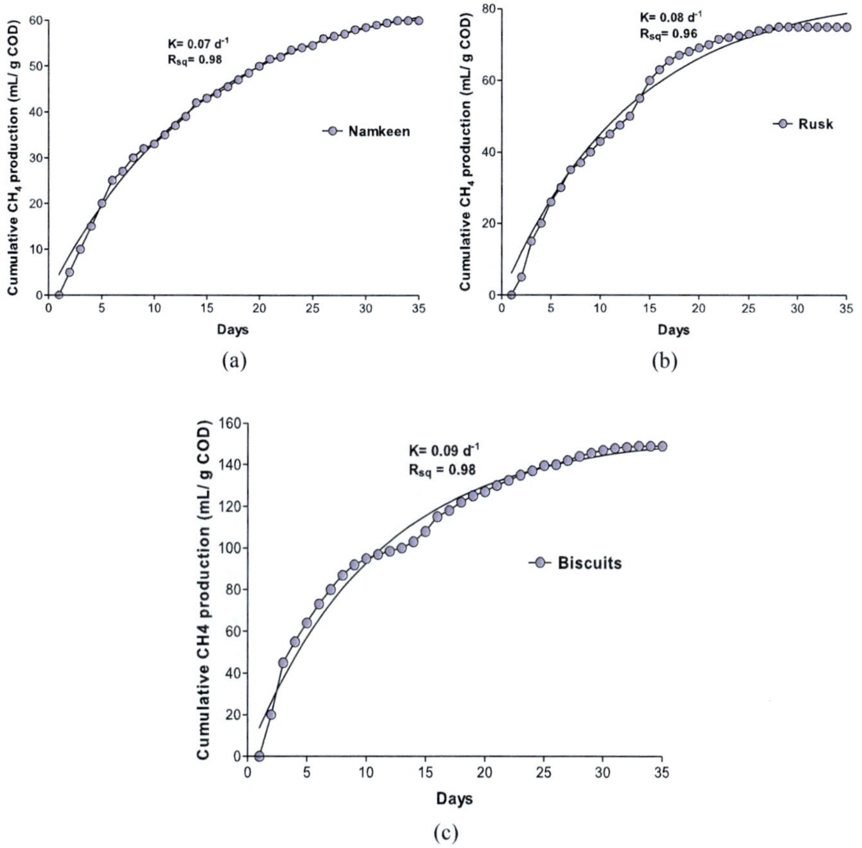


**Fig. 1** **a–c** Temporal variation of cumulative $CH_4$ at F/M ratio at (30 ± 2) °C with substrate as **a** namkeen waste **b** rusk waste **c** biscuits waste

matter. The temporal variation of cumulative $CH_4$ production with respect to time in days under psychrophilic condition is shown through Fig. 3. The maximum $CH_4$ production and the maximum k value of 0.11 $d^{-1}$ was found in reactor 2 containing rusk waste. However, the lowest value of k was observed in the reactor consisting of namkeen waste. The BMP results evaluated in terms of $\mu_b$ is summarized in Table 3. The maximum $\mu_b$ value of 0.66 g $CH_4$–COD/g COD was observed in the reactor containing rusk waste (see Table 3).

## Digestor Quality and Characterization

Nitrogen is the foremost requirement of microorganisms in anaerobic digestion because it is required for protein synthesis. Nitrogenous compounds in organic waste which contain proteins get converted into ammonia by anaerobic digestion. A higher amount of ammonia may hinder the biological process and during methanogenesis, it restricts the methane production by bacteria, but a desirable or lower amount

**Table 3** Methane generation rate constant (k) and BMP with controlled F/M ratio under Mesophilic, Thermophilic and Psychrophilic conditions

| | | Phase (I): Mesophilic temperature (30 ± 2) °C | | | | | | Phase (II): Thermophilic temperature (45 ± 2) °C | | | | | | Phase (III): Psychrophilic temperature (15 ± 2) °C | | | | | |
|---|---|---|---|---|---|---|---|---|---|---|---|---|---|---|---|---|---|---|---|
| Waste type | Waste concentration (mL) | $Y_{max.}$ (mL) | k ($d^{-1}$) | $R^2$ | $CH_4$-COD (g/L) ($a_1$) | $COD_{fed}$ (g/L) ($a_2$) | BMP($\mu_b$) ($a_1/a_2$) | Y max (mL) | k($d^{-1}$) | $R^2$ | $CH_4$-COD (g/L) ($a_1$) | $COD_{fed}$ (g/L) ($a_2$) | BMP ($\mu_b$) ($a_1/a_2$) | $Y_{max}$ (mL) | k ($d^{-1}$) | $R^2$ | $CH_4$-COD (g/L) ($a_1$) | $COD_{fed}$ (g/L) ($a_2$) | BMP ($\mu_b$) ($a_1/a_2$) |
| Namkeen | 100 | 60 | 0.07 | 0.98 | 0.17 | 0.71 | 0.23 | 30 | 0.08 | 0.97 | 0.08 | 0.71 | 0.11 | 65 | 0.08 | 0.95 | 0.18 | 0.71 | 0.25 |
| Rusk | 100 | 75 | 0.08 | 0.96 | 0.21 | 0.36 | 0.58 | 120 | 0.1 | 0.97 | 0.34 | 0.36 | 0.94 | 85 | 0.11 | 0.94 | 0.24 | 0.36 | 0.66 |
| Biscuits | 100 | 149 | 0.09 | 0.98 | 0.42 | 0.79 | 0.53 | 140 | 0.11 | 0.96 | 0.4 | 0.79 | 0.50 | 110 | 0.09 | 0.97 | 0.31 | 0.79 | 0.39 |

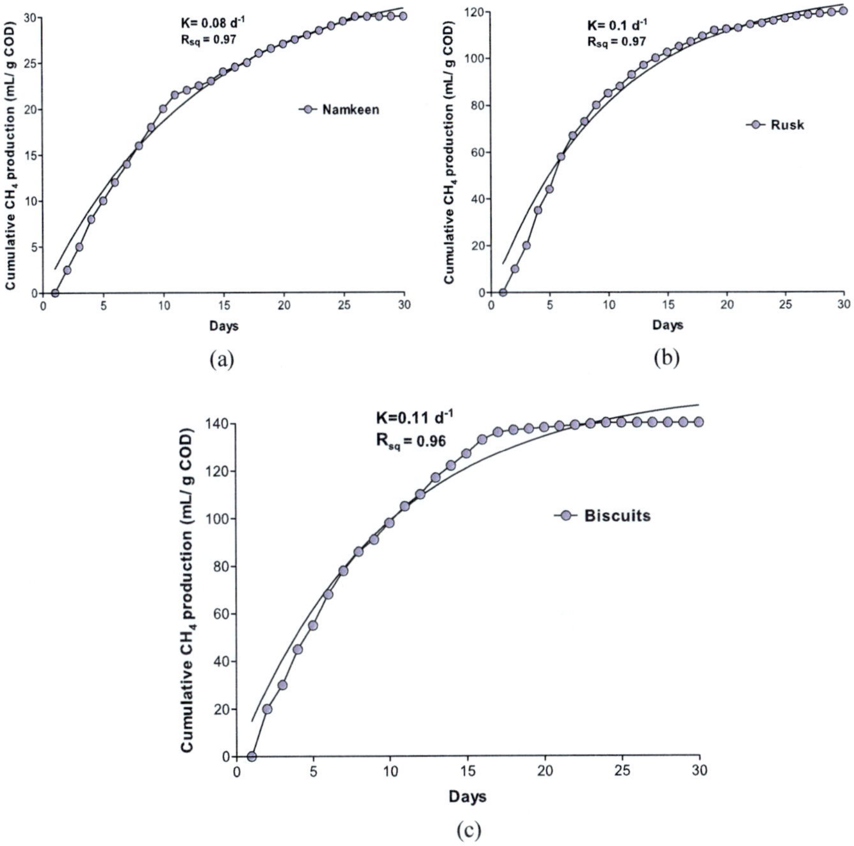


**Fig. 2** **a–c** Temporal variation of cumulative $CH_4$ at F/M ratio at (45 ± 2) °Cwith substrate as **a** namkeen waste **b** rusk waste **c** biscuits waste

of ammonia does not affect the biogas production. It is essential for the growth of the bacterial cell. A nutrient ratio of elements C:N:P:S at 600:15:5:3 is considered sufficient for bio-methanation otherwise greater amount of ammonia restricts the methanogenesis process up to 50%. From Fig. 4 it is observed that under the mesophilic and thermophilic condition the highest ammonical nitrogen removal of 84.73 and 80.91% in biscuits waste as compared to other wastes. Also, a decrement of 74.76 and 74.73% of ammonical nitrogen has been observed in rusk and namkeen wastes under thermophilic condition. The experimental analysis illustrates a small decrement of 19.11%, 12.21% and 11.57% in ammonical nitrogen concentration under the psychrophilic condition for rusk, biscuits and namkeen wastes, respectively.

Phosphorous is needed for both microbial growth and maintaining an optimal pH value. It is also an important constituent of the energy-yielding system of bacterial populations [18]. Like nitrogen, the elevated concentrations of phosphorous inhibits

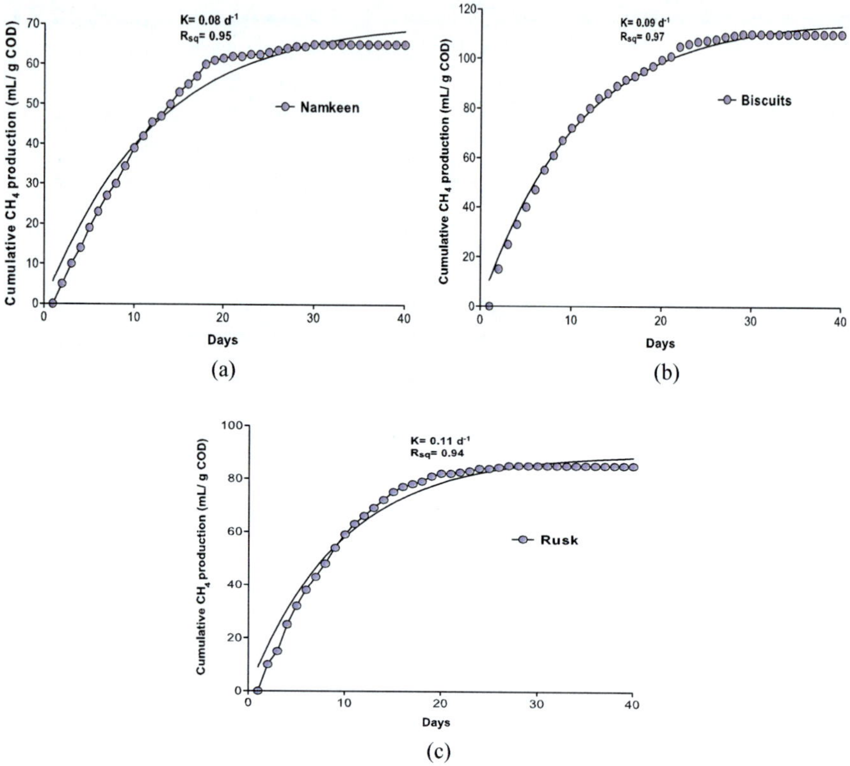


**Fig. 3** **a–c** Temporal variation of cumulative $CH_4$ at F/M ratio (0.75) at (15 ± 2) °C with substrate as **a** namkeen waste **b** rusk waste **c** biscuits waste

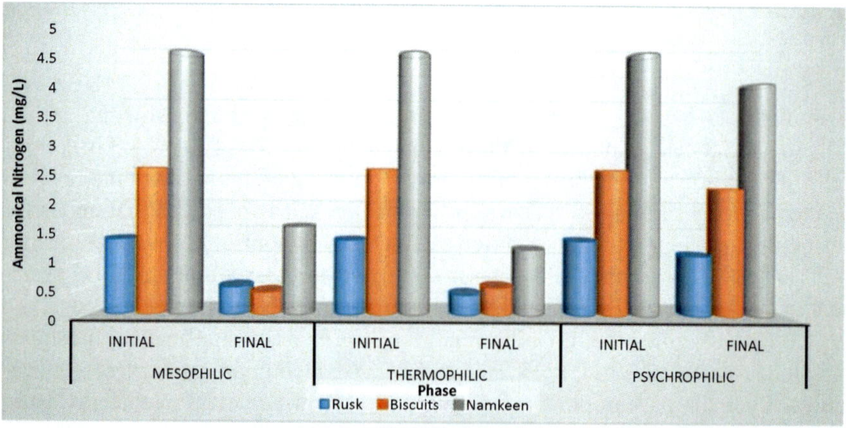


**Fig. 4** Variation of ammonical nitrogen in batch reactors under mesophilic, thermophilic and psychrophilic conditions

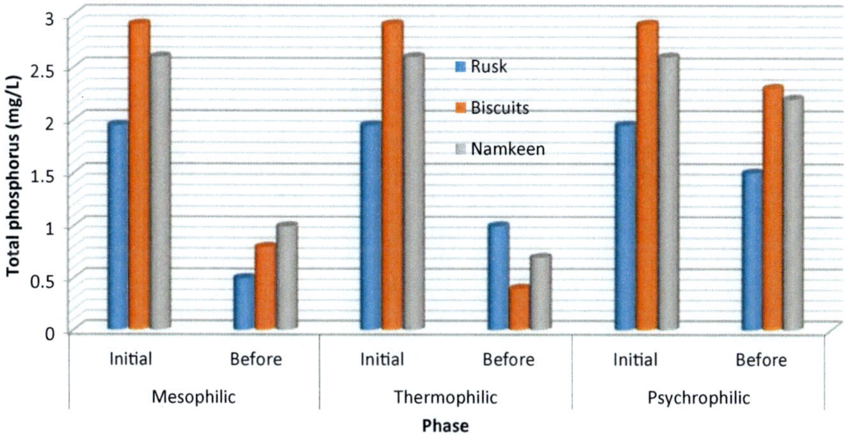


**Fig. 5** Variation of total phosphorus in batch reactors under mesophilic, thermophilicand psychrophilic conditions

the biological process and methanogenesis but if present in a desirable amount then it doesn't hinder anaerobic digestion and it also is useful for the growth and maintenance of cell as nitrogen. The data depicted through Fig. 5 unveils maximum phosphorus removal of 74.35% in rusk waste. However, the phosphorus removal of 72.50% and 61.53% has been observed in biscuit and namkeen waste under mesophilic condition. Furthermore, under thermophilic condition, maximum phosphorus removal of 73.07% and 48.71% has been observed in namkeen waste and rusk waste respectively. Correspondingly, Fig. 5 illustrates lower phosphorus removal of 23.07%, 20.96% and 15.38% under psychrophilic conditions in rusk, biscuits and namkeen wastes respectively.

The COD removal has also been evaluated in the present study and the data is shown through Fig. 6. Experimentation analysis ensures that the amount of methane gas formed is influenced by COD reduction. It is observed that under mesophilic condition COD removal of 20%, 30% and50% has been observed in the biodegradation of rusk, biscuits and namkeen wastes respectively leading to methane conversion of 85%, 48% and 21%, respectively. Similarly, under thermophilic condition, COD removal of 22%, 40% and 70% have been detected in the anaerobic degradation of rusk, biscuits and namkeen wastes, respectively, leading to methane conversion up to 80%, 43% and 11%, respectively. Correspondingly under psychrophilic condition, COD removal of 20%, 25% and 55% have been observed in biodegradation of rusk, biscuits and namkeen wastes, respectively, leading to methane conversion of 70%, 35% and 22%, respectively.

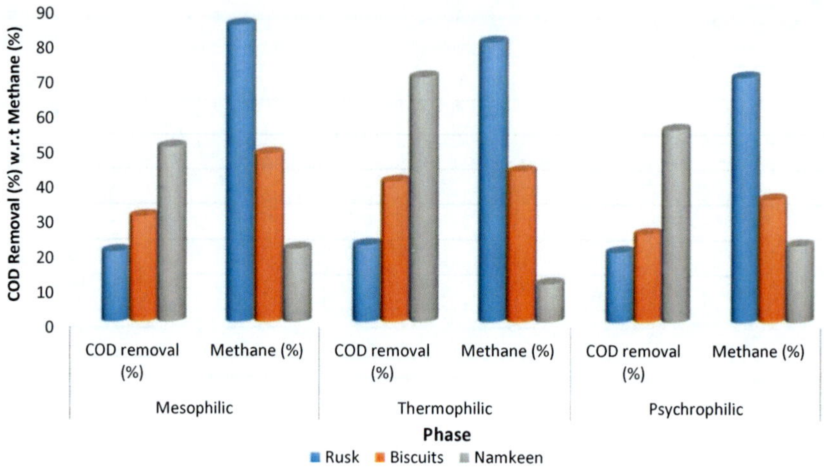


**Fig. 6** COD removal (%) and methane conversion from $COD_{fed}$ (%) at fixed F/M ratio of 0.75 under mesophilic, thermophilic, psychrophilic conditions

## 4 Conclusion

From the present study, it can be inferred that under mesophilic (30 ± 2) °C conditions the gas production ceases after 35 days at a fixed F/M ratio of 0.75. The highest methane production of 0.21L$CH_4$ per g $COD_{fed}$ with methane generation rate constant in terms of k value of 0.08 $d^{-1}$ has been observed in the reactor containing rusk waste. However, under thermophilic (45 ± 2) °C condition the methane gas production ceases after 30 days at a fixed 0.75 F/M ratio. The results indicate that the highest biogas production of 0.34 L $CH_4$ per g $COD_{fed}$ with a rate constant k value of 0.1 $d^{-1}$ has been found in the reactor containing rusk waste. Furthermore, under psychrophilic (15 ± 2) °C conditions the methane gas production ceased after 40 days at a fixed 0.75 F/M ratio. The highest methane production of 0.24 L $CH_4$ per g $COD_{fed}$ with a rate constant k value of 0.11 $d^{-1}$ has been observed in a reactor containing rusk waste. Correspondingly, analysis of digestate revealed that the highest decrement in ammonical nitrogen and total phosphorus were found in a reactor containing rusk waste under mesophilic condition.

## References

1. Hussain A, Kumar P, & Mehrotra I (2008) Treatment of phenolic wastewater in UASB reactor: effect of nitrogen and phosphorous. Bioresour Technol 99(17):8497–8503
2. Gunaseelan VN (2004) Biochemical methane potential of fruits and vegetable solid waste feedstocks. Biomass Bioenergy 26(4):389–399

3. American Public Health Association, American Water Works Association, Water Pollution Control Federation, & Water Environment Federation (2005) Standard methods for the examination of water and wastewater, vol 2. American Public Health Association, Washington, DC
4. Anyaoku CC, Baroutian S (2018) Decentralized anaerobic digestion systems for increased utilization of biogas from municipal solid waste. Renew Sustain Energy Rev 90:982–991
5. Arsova L (2010) Anaerobic digestion of food waste: Current status, problems and an alternative product. Department of earth and Environmental Engineering foundation of Engineering and Applied Science Columbia University
6. Cioabla AE, Ionel I, Dumitrel GA, Popescu F (2012) Comparative study on factors affecting anaerobic digestion of agricultural vegetal residues. Biotechnol Biofuels 5(1):1–9
7. Clarke WP (2018) The uptake of anaerobic digestion for the organic fraction of municipal solid waste–push versus pull factors. Biores Technol 249:1040–1043
8. Gikas P, Zhu B, Batistatos NI, Zhang R (2018) Evaluation of the rotary drum reactor process as pretreatment technology of municipal solid waste for thermophilic anaerobic digestion and biogas production. J Environ Manage 216:96–104
9. Heimersson S, Svanström M, Cederberg C, Peters G (2017) Improved life cycle modelling of benefits from sewage sludge anaerobic digestion and land application. Resour Conserv Recycl 122:126–134
10. Hoornweg D, & Bhada-Tata P (2012) What a waste: a global review of solid waste management
11. Hussain A, Dubey SK (2017) Specific methanogenic activity test for anaerobic degradation of influents. Appl Water Sci 7(2):535–542
12. Hussain A (2007) N and P in anaerobic biotransformation: degradation of phenolic wastewater. A thesis of Department of Civil Engineering. I.I.T. Roorkee, India
13. IEA (2008) IEA bioenergy task 37 plant list 2008. Available http://www.ieabiogas.net/anlagelisten/Plantlist_08.pdf
14. Möller K, Müller T (2012) Effects of anaerobic digestion on digestate nutrient availability and crop growth: a review. Eng Life Sci 12(3):242–257
15. Negi S, Dhar H, Hussain A, Kumar S (2018) Biomethanation potential for co-digestion of municipal solid waste and rice straw: a batch study. Biores Technol 254:139–144
16. Owen WF, Stuckey DC, Healy JB Jr, Young LY, McCarty PL (1979) Bioassay for monitoring biochemical methane potential and anaerobic toxicity. Water Res 13(6):485–492
17. Siddique MNI, Wahid ZA (2018) Achievements and perspectives of anaerobic co-digestion: a review. J Clean Prod 194:359–371
18. Stafford DA, Hawkes DL, Horton R (1980) Methane production from waste organic matter. CRC Press, Florida
19. Tyagi VK, Fdez-Güelfo LA, Zhou Y, Álvarez-Gallego CJ, Garcia LR, Ng WJ (2018) Anaerobic co-digestion of organic fraction of municipal solid waste (OFMSW): progress and challenges. Renew Sustain Energy Rev 93:380–399
20. Vandevivere P, De Baere L, Verstraete W (2003) Types of anaerobic digester for solid wastes. In: Biomethanization of the organic fraction of municipal solid wastes, pp 111–140. Iwa Publishing
21. Wei J, Guo Q, Gong Y, Ding L, Yu G (2017) Synergistic effect on co-gasification reactivity of biomass-petroleum coke blended char. Biores Technol 234:33–39
22. Xu F, Li Y, Ge X, Yang L, Li Y (2018) Anaerobic digestion of food waste—challenges and opportunities. Biores Technol 247:1047–1058

# Chapter 24
# Analysing Distance Measures in Topsis: A Python-Based Tool

**Swasti Arya, Mihika Chitranshi, and Yograj Singh**

## 1 Introduction

Decision making is the process of making choices by identifying a problem, gathering information, and assessing alternative options. The process of making a decision has four main steps—formation of the problem statement, setting the criteria, evaluating the alternatives available and determination of the ideal alternative [1]. Even though decision making is extremely intuitive, a Decision Maker (DM) always follows these four steps in the order specified. Be it purchasing a new mobile phone or deciding the location of a new power plant, the first step is to recognise the need to make a decision, followed by setting standards and evaluating different options based on those standards. Subsequently, the alternative which suits the maximum of the criteria is judged the best. Oftentimes one might desire an alternative that satisfies conflicting criteria.

Consider purchasing a smartphone, better the storage capacity, higher the price. A Decision Maker would still like to attach different importance to the two criteria of storage space and price, in order to find an alternative that makes the correct trade-off to secure a smartphone with low price and high storage capacity. Each criterion could have a different unit of measurement. In the example stated above, the storage capacity of a mobile phone is measured in gigabytes whereas cost is measured in currency. Clearly, the Decision Maker will evaluate his choices on multiple criteria, and hence the name, Multi-Criteria Decision Making.

S. Arya · M. Chitranshi (✉) · Y. Singh
Department of Mathematics, Lady Shri Ram College for Women, University of Delhi, Delhi, India
e-mail: mihichi1699@gmail.com

S. Arya
e-mail: swasarya@gmail.com

Y. Singh
e-mail: yograjchauhan26@gmail.com



D. Singh et al. (eds.), *Proceedings of International Conference on Scientific and Natural Computing*, Algorithms for Intelligent Systems,
https://doi.org/10.1007/978-981-16-1528-3_24

Multi-Criteria Decision Making is an application in the field of Operational Research (OR). It can be considered as a natural outgrowth of Operational Research. MCDM and traditional Operational Research techniques usually focus on maximising or minimising a utility function in the presence of constraints [2]. To facilitate a systemised approach to decision making, Hwang and Yoon [3] suggested that MCDM problems can be classified into two main categories.

### *1.1 Multiple Attribute Decision Making (MADM)*

MADM deals with decision-making problems that have a pre-defined, limited number of alternatives. MADM requires both inter- and intra-attribute comparisons. MADM refers to making preference decisions (e.g., evaluation, prioritization, selection) [3]. MADM problems can further be classified into Multiple Attribute Utility Theory (MAUT) and outranking methods [1]. MAUT methods include the determination of the Decision Maker's preferences, represented as a hierarchical structure by using an appropriate utility function. Although, MAUT methods have been criticised for not taking into consideration the interaction among criteria and its subsequent effect on the final decision. On the other hand, outranking methods determine the best alternative by comparing the preference relation among the alternatives [1]. The earlier example of purchasing a smartphone that satisfies certain criteria is a MADM problem, wherein there are a finite number of alternatives to choose from.

### *1.2 Multiple Objective Decision Making (MODM)*

MODM deals with decision-making problems where there is not a set number of alternatives. The set of alternatives is explicitly defined by constraints using multiple objective programming, leading to a large number of potential alternatives. MODM problems involve designing the best alternative given a set of conflicting objectives [1]. A significant issue related to MODM methods is the increase in a number of dimensions beyond handling capacity, which in turn increases computational costs manifold. For example, automobile manufacturers wish to design a car that maximizes riding comfort and fuel economy and minimizes production cost. The alternatives are created by the design process resulting in a large number of alternatives.

To understand the organisation of the paper: Sect. 2 describes the TOPSIS method and distance measures used, Sect. 3 comprises the literature, Sect. 4 explains the steps of the TOPSIS method, Sect. 5 shows the application and implementation of the method and the last section concludes the study along with highlighting the future scope of research in the area.

# 2 Definitions/Methods

There are a number of statistical and non-statistical techniques developed to aid the process of decision making such as AHP, ELECTRE, TOPSIS, PROMETHEE and Grey theory. This study focuses on Techniques for Order Preference by Similarity to an Ideal Solution abbreviated as TOPSIS.

## 2.1 *Techniques for Order Preference by Similarity to an Ideal Solution (TOPSIS)*

Among the several MCDM methods, this paper focuses on Techniques for Order Preference by Similarity to an Ideal Solution or conveniently synthesized as TOPSIS. The method was proposed by Chen and Hwang (1992), with reference to Hwang and Yoon [3]. TOPSIS is a MADM method which means it chooses the best option from a finite number of alternatives. The main idea behind the method is to choose an alternative that has the shortest distance from the positive ideal solution and the longest distance from the negative ideal solution [4–6]. This is achieved by evaluating the distance of the alternative to the ideal solutions. The one with the shortest distance from the positive ideal solution and the longest distance from the negative ideal solution is ranked the best option. For the purpose of finding the distance to the Positive Ideal Solution or PIS and the remoteness to the Negative Ideal Solution or NIS, a Relative Closeness Coefficient or Similarity Index is determined and the alternative with the highest value of Similarity Index is chosen. There are several methods to calculate distance such as Euclidean, Hamming, Soergel, Canberra, Gower and Lorentzian, with Euclidean being the most popular [7, 8]. This paper uses a modified TOPSIS approach with ten different distance measures to calculate the separation of the alternative to the ideal solutions. TOPSIS is widely used in various multi-attribute decision-making problems such as supply chain logistics, marketing management, environmental management or chemical engineering [9].

## 2.2 *Distance Measures*

There are a number of different families of distance functions, three of which have been included in this paper. The computation of distance and similarity measure requires vectors *a* and *b*, the distance between which is the resulting output of the distance functions [10].

The $L_p$ Minkowski family comprises those distance measures that are derived by the generalized formula, $(\Sigma_i |x_i - y_i|^p)^{1/p}$, where the function differs based on the value of p.

The $L_1$ family comprises distance measures that relate to the absolute difference $L_1$ distance $\Sigma_i|x_i - y_i|$.

The $\chi^2$ family or the $L_2$ family comprises distance measures related to the $L_2$ (Euclidean) norm $\Sigma_i(x_i - y_i)^2$.

## 3 Review of Literature

TOPSIS is a unique method with an 'approach to identify an alternative which is closest to the ideal solution and farthest to the Negative Ideal Solution in a multi-dimensional computing space' [11]. It has several advantages that make it stand out as a decision-making tool. It has a simple process, is easy to use and programmable. The number of steps remains the same regardless of the number of attributes [12]. There have been several form revisions in the method based on normalization techniques, distance measures, the dependence between alternatives etc. Chen-Tung Chen in his work [13] proposed the vertex method to find the distance between PIS and NIS [12]. Several hybrid approaches with AHP and TOPSIS, VIKOR and TOPSIS and other methods have been found in the literature for providing solutions for MCDM problems [12]. A disadvantage is that its use of Euclidean Distance does not consider the correlation of attributes. It is difficult to weigh attributes and keep the consistency of judgment, especially with additional attributes [14, 15]. Developed by Hwang and Yoon in 1981 [3] TOPSIS has been used in various MADM problems such as supply chain logistics, marketing management, environmental management or chemical engineering [9]. Other areas include supply chain logistics, marketing management, environmental management or chemical engineering, human resources management, and water resources management [16]. TOPSIS is preferred over other approaches of MCDM because of (i) its suitability for a large number of attributes and alternatives; (ii) requirement of limited subjective inputs; (iii) its logical and programmable behaviour; and (iv) comparative consistency in the alternative ranking [9, 17].

## 4 Steps of the TOPSIS Method

1. TOPSIS assumes that there are $m$ alternatives, $n$ attributes/criteria and the score of each option with respect to each criterion is known.
2. Let $J$ be the set of beneficial criteria (more is better) and $J'$ be the set of non-beneficial criteria (less is better)
3. Let $x_{ij}$ be the score of option $i$ with respect to criterion $j$. We have a matrix $X = (x_{ij}) = m \times n$ matrix.
4. Construct the normalized decision matrix using the formula:

$$r_{ij} = \frac{x_{ij}}{\sqrt{\sum x_{ij}^2}} for\ i = 1, 2, 3, \ldots, m;\ j = 1, 2, 3, \ldots, n$$

5. Given that the set of weights for each criterion $w_j$ for $j = 1, 2, 3, \ldots, n$. Multiply each column of the normalized decision matrix by its associated weight $w_j$. This generates the weighted normalised decision matrix, each element of which is represented as

$$v_{ij} = w_{ij} r_{ij}$$

6. Calculate the ideal best and the ideal worst value for each criterion. For a beneficial criterion, the best value is the maximum value in the criteria whereas the worst value is the minimum value in the criteria. For a non-beneficial criterion, the best value is the minimum value in the criteria whereas the worst value is the maximum value in the criteria. For a simpler understanding, refer to Table 4.
7. For calculating separation measure from each alternative, we use the formulae of the distance measures as given in Tables 1, 2 and 3.
8. Distance from the Ideal Best Solution ($d_i^+$) and Ideal Worst Solution ($d_i^-$) can be calculated with the help of formulas given in Tables 1, 2 and 3. To calculate the distance from the Ideal Best Solution ($d_i^+$), replace $b_i$ with $V_j^+$. To calculate the distance from the Ideal Worst Solution ($d_i^-$), replace $b_i$ with $V_j^-$.
9. Calculate the relative closeness $P_i$ by the formula:

**Table 1** Lp Minkowski family

| | | |
|---|---|---|
| 1a | Euclidean | $d(a_i, b_i) = \left(\sum (a_i - b_i)^2\right)^{\frac{1}{2}}$ |
| 1b | City block | $d(a_i, b_i) = \sum \lvert a_i - b_i \rvert$ |
| 1c | Minkowski | $d(a_i, b_i) = \left(\sum \lvert a_i - b_i \rvert^p\right)^{1/p}$ |

**Table 2** $L_1$ family

| | | |
|---|---|---|
| 2a | Sørensen | $d(a_i, b_i) = \frac{\sum \lvert a_i - b_i \rvert}{\sum (a_i - b_i)}$ |
| 2b | Gower | $d(a_i, b_i) = \frac{1}{n} * \sum \lvert a_2 - b_i \rvert$ |
| 2c | Canberra | $d(a_i, b_i) = \sum \frac{\lvert a_i - b_i \rvert}{a_i + b_i}$ |
| 2d | Lorentzian | $d(a_i, b_i) = \sum \ln(1 + \lvert a_i - b_i \rvert)$ |

**Table 3** $\chi^2$ family

| | | |
|---|---|---|
| 3a | Squared Euclidean | $d(a_i, b_i) = \sum (a_i - b_i)^2$ |
| 3b | Pearson $\chi^2$ | $d(a_i, b_i) = \sum \frac{(a_i - b_i)^2}{b_i}$ |
| 3c | Neyman $\chi^2$ | $d(a_i, b_i) = \sum \frac{(a_i - b_i)^2}{a_i}$ |

**Table 4** Best and worst values of criteria

| | Beneficial criteria | Non-beneficial criteria |
|---|---|---|
| Best value ($V_j^+$) | $(v_{ij})_{max}$ | $(v_{ij})_{min}$ |
| Worst value ($V_j^-$) | $(v_{ij})_{min}$ | $(v_{ij})_{max}$ |

$$P_i = \frac{d_i^-}{d_i^- + d_i^+} 0 < P_i < 1$$

10. The alternatives are ranked according to their relative scores. Higher the value of $P_i$, better the rank. In other words, the one with $P_i$ value closest to 1 is the best alternative as shown in (Table 4).

## 5 Application

The objective of the paper is to implement TOPSIS on the given dataset using ten different distance measures. The inspiration behind the use of various distance measures was derived from Anandan and Uthra [12]. This paper attempts to recreate the same problem statement with data points from 'Data normalisation techniques in decision making: case study with TOPSIS method' [18]. This numerical illustration is based on a project on the autonomous landing of drones with hazard avoidance named Intelligent Landing of Unmanned Aerial Vehicles (ILUV) [19]. The project objective was to provide services for the government by testing the feasibility for the transfer of Intelligent Planetary Site Selection (IPSIS) technology into Unmanned Aerial Vehicles (UAV) to increase safety, security and efficiency. The aim of the process is to select a suitable site for target-landing which is based on and hence evaluated with respect to the following set of requirements [20]:

- The site should be visible from the camera during the descent phase so that its estimated characteristics can be continuously updated
- The site should be safe in terms of maximum local slope, light level and terrain roughness
- The site should be reachable with the available fuel and the spacecraft propulsive capabilities

Consequently, three criteria namely $C_1$, $C_2$ and $C_3$, corresponding to illumination, reachability and land texture respectively are considered for the application of MCDM methods. Among these, $C_1$ and $C_2$ are beneficial criteria, which implies that a higher value for these criteria is desirable whereas $C_3$ is a cost criterion where a lower value is desirable. $SC_1$ to $SC_{16}$ represent the available landing locations, here considered to be alternatives.

**Table 5** Decision matrix

| Weights | 0.333 | 0.333 | 0.333 |
|---|---|---|---|
| | $C_1$ | $C_2$ | $C_3$ |
| **$SC_1$** | 138.609 | 0.3349 | 6.4543 |
| **$SC_2$** | 154.7214 | 0.3395 | 11.4244 |
| **$SC_3$** | 158.3081 | 0.3441 | 11.4244 |
| **$SC_4$** | 157.3082 | 0.3487 | 6.8542 |
| **$SC_5$** | 144.5976 | 0.3301 | 11.2616 |
| **$SC_6$** | 138.5982 | 0.3346 | 11.2616 |
| **$SC_7$** | 131.5989 | 0.3391 | 11.1988 |
| **$SC_8$** | 132.5988 | 0.3437 | 11.1988 |
| **$SC_9$** | 144.5976 | 0.3252 | 11.2616 |
| **$SC_{10}$** | 138.5982 | 0.3297 | 11.2616 |
| **$SC_{11}$** | 132.5988 | 0.3342 | 11.1988 |
| **$SC_{12}$** | 135.9513 | 0.3387 | 6.8974 |
| **$SC_{13}$** | 119.7141 | 0.3204 | 11.2616 |
| **$SC_{14}$** | 112.7148 | 0.3248 | 11.1988 |
| **$SC_{15}$** | 112.7148 | 0.3292 | 11.1988 |
| **$SC_{16}$** | 128.952 | 0.3337 | 6.8974 |

### *5.1 Implementation Using Python Programming*

The input to the TOPSIS method is in the form of a decision matrix. The multi-criteria decision matrix is set up such that each row represents an alternative and each column denotes a different criterion. The corresponding values are set as elements of the matrix that may or may not be of the same unit of measurement and order of magnitude. Here, we note that a higher value of $C_1$ and $C_2$ is desirable and a lower value for $C_3$. Each of the three criteria have been given equal importance (0.333) in this illustration. Subsequently, we obtain a $16 \times 3$ decision matrix as shown in Table 5.

Table 5 makes the dataset which is taken as input by the Python code. The dataset was saved as a comma-separated (.csv) file, and labelled as < Springer_dataset_TOPSIS > in order to import it to a Python notebook as seen in Fig. 1. Before which two Python packages namely, pandas and NumPy were imported.

### *5.2 Implementing Different Distance Measures*

Figure 1 shows a snippet of the Python programming code that was used to input the dataset, set the criteria weights, normalise the decision matrix and finally calculate

```
[ ] import numpy as np
    import pandas as pd

[ ] df=pd.read_csv('Springer_dataset_TOPSIS.csv')
    df.set_index('Attributes',inplace=True)
    weights=(0.333, 0.333, 0.333)
    df1=np.square(df)
    totals=np.sqrt(df1.sum(axis=0))

    normailised_df=df/totals
    weighted_normalised_df=normailised_df*weights
    wt_nrm_df=weighted_normalised_df.copy(deep=True)

    weighted_normalised_df.loc['Max']=weighted_normalised_df.max(axis=0)
    weighted_normalised_df.loc['Min']=weighted_normalised_df.min(axis=0)
    weighted_normalised_df.loc['Beneficial']=(1,1,0)
    weighted_normalised_df.loc['Non-Beneficial']=(0,0,1)

[ ] weighted_normalised_df.loc['Positive Ideal']=(weighted_normalised_df.loc['Beneficial']*weighted_normalised_df.loc['Max'])+
    (weighted_normalised_df.loc['Non-Beneficial']*weighted_normalised_df.loc['Min'])
    weighted_normalised_df.loc['Negative Ideal']=(weighted_normalised_df.loc['Beneficial']*weighted_normalised_df.loc['Min'])+
    (weighted_normalised_df.loc['Non-Beneficial']*weighted_normalised_df.loc['Max'])
```

**Fig. 1** Python code for the initial steps of TOPSIS

the weighted normalised decision matrix. Following which, the Beneficial and Non-Beneficial criteria have been demarcated. The last part of the code snippet illustrates the calculation of the Positive Ideal Solution and the Negative Ideal Solution. All the programming steps conform to the formulas detailed earlier in the paper. All variables and data frames have been named keeping in mind the ease of reading and comprehension, even for non-coders. Hence, the naming convention is self-explanatory.

The weighted normalized decision matrix which is calculated according to the aforementioned steps is shown in Table 6.

The set of Positive Ideal Solution (PIS) and Negative Ideal Solution (NIS) hence obtained is depicted in Table 7:

Following this, the distance from the Positive Ideal Solution and the Negative Ideal Solution is calculated according to the distance measure being used. The snippets of code in Python language for each of the distance measures are given below (Tables 8, 9 and 10).

## 5.3 *Ranking Alternatives Using TOPSIS Method*

The Relative Closeness Coefficient $P_i$ is calculated and alternatives are ranked according to the rule that the one with a higher value of $P_i$ gets a better rank. The code for the same is given in Fig. 2.

Radar maps are graphical methods to display multivariate data (here, the different criteria). They use three or more quantities/variables in the form of a two-dimensional charts represented on axes. Following are some radar maps along with tables to depict the values of Relative Closeness Coefficient of the alternatives and deduce a comparative analysis.

**Table 6** Weighted normalized decision matrix

| | $C_1$ | $C_2$ | $C_3$ |
|---|---|---|---|
| $SC_1$ | 0.084196 | 0.083351 | 0.052036 |
| **$SC_2$** | 0.093983 | 0.084495 | 0.092106 |
| **$SC_3$** | 0.096162 | 0.08564 | 0.092106 |
| **$SC_4$** | 0.095555 | 0.086785 | 0.055260 |
| **$SC_5$** | 0.087834 | 0.082156 | 0.090793 |
| **$SC_6$** | 0.084190 | 0.083276 | 0.090793 |
| **$SC_7$** | 0.079938 | 0.084396 | 0.090287 |
| **$SC_8$** | 0.080545 | 0.085541 | 0.090287 |
| **$SC_9$** | 0.087834 | 0.080936 | 0.090793 |
| **$SC_{10}$** | 0.084190 | 0.082056 | 0.090793 |
| **$SC_{11}$** | 0.080545 | 0.083176 | 0.090287 |
| **$SC_{12}$** | 0.082582 | 0.084296 | 0.055608 |
| **$SC_{13}$** | 0.072719 | 0.079742 | 0.090793 |
| **$SC_{14}$** | 0.068467 | 0.080837 | 0.090287 |
| **$SC_{15}$** | 0.068467 | 0.081932 | 0.090287 |
| **$SC_{16}$** | 0.078330 | 0.083052 | 0.055608 |

**Table 7** PIS and NIS of criteria

| | $C_1$ | $C_2$ | $C_3$ |
|---|---|---|---|
| Positive Ideal | 0.096162 | 0.086785 | 0.052036 |
| Negative Ideal | 0.068467 | 0.079742 | 0.092106 |

Table 11 gives the Relative Closeness Coefficient $P_i$ of all the alternatives using the distance measures from the $L_p$ Minkowski family.

To better understand Table 11; Fig. 3 depicts radar maps of all the alternatives using the distance measures from the $L_p$ Minkowski family.

Here, $SC_4$ and $SC_{14}$ have the highest and lowest values of relative closeness respectively according to all the three distance measures.

Table 12 gives the Relative Closeness Coefficient $P_i$ of all the alternatives using the distance measures from the $L_1$ family.

To better understand Table 12; Fig. 4 depicts radar maps of all the alternatives using the distance measures from the $L_1$ family.

Here, $SC_4$ and $SC_{14}$ have the highest and lowest values of Relative Closeness Coefficient respectively according to all the four distance measures.

Table 13 gives the Relative Closeness Coefficient $P_i$ of all the alternatives using the distance measures from the $\chi^2$ family.

To better understand Table 13; Fig. 5 depicts radar maps of all the alternatives using the distance measures from the $\chi^2$ family.

Here, $SC_4$ and $SC_{14}$ have the highest and lowest values of Relative Closeness Coefficient respectively according to all the three distance measures. Table 14

**Table 8** $L_p$ Minkowski family

Euclidean

```
[ ] positive_ideal_df=(np.square(wt_nrm_df-weighted_normalised_df.loc['Positive Ideal']))
    negative_ideal_df=(np.square(wt_nrm_df-weighted_normalised_df.loc['Negative Ideal']))

    positive_ideal_df['Separation Ideal Solution']=np.sqrt(positive_ideal_df.sum(axis=1))
    negative_ideal_df['Separation Ideal Worst']=np.sqrt(negative_ideal_df.sum(axis=1))
```

City Block

```
[ ] positive_ideal_df=abs(wt_nrm_df-weighted_normalised_df.loc['Positive Ideal'])
    negative_ideal_df=abs(wt_nrm_df-weighted_normalised_df.loc['Negative Ideal'])

    positive_ideal_df['Separation Ideal Solution']=(positive_ideal_df.sum(axis=1))
    negative_ideal_df['Separation Ideal Worst']=(negative_ideal_df.sum(axis=1))
```

Minkowski (p=3)

```
[4] positive_ideal_df=np.power(abs(wt_nrm_df-weighted_normalised_df.loc['Positive Ideal']),3)
    negative_ideal_df=np.power(abs(wt_nrm_df-weighted_normalised_df.loc['Negative Ideal']),3)

    positive_ideal_df['Separation Ideal Solution']=np.power(positive_ideal_df.sum(axis=1),1/3)
    negative_ideal_df['Separation Ideal Worst']=np.power(negative_ideal_df.sum(axis=1),1/3)
```

summarises the various rankings of all the alternatives in a concise manner that proves instrumental in doing a comparative analysis.

We observe here that $SC_4$ turns out to be the best and $SC_{14}$ the least feasible alternative according to all the distance measures, the second most feasible being the 1st alternative. The ranks according to all the methods are the same till the 10th alternative. Then, the ranks match at $SC_{13}$ (Rank 14) again and are henceforth the same for all the remaining alternatives. The comparison of the rest of the results justifies that the aforementioned distance measures are suitable for similar multi-criteria problems.

## 6 Conclusion

Decision making is a process that has four main steps—formation of the problem statement, setting the criteria, evaluating the alternatives available and determination of the ideal alternative. Multi-Criteria Decision Making is choosing the best feasible solution to a problem on the basis of several criteria. In other words, MCDM encompasses mathematical theories, approaches and tools that aid people in making decisions according to their preferences or requirements.

**Table 9** $L_1$ family

Sørensen

```
[7] positive_ideal_df=(abs(wt_nrm_df-weighted_normalised_df.loc['Positive Ideal'])).sum(axis=1)
    negative_ideal_df=(abs(wt_nrm_df-weighted_normalised_df.loc['Negative Ideal'])).sum(axis=1)
    positive_ideal_df_den=wt_nrm_df+weighted_normalised_df.loc['Positive Ideal']
    negative_ideal_df_den=wt_nrm_df+weighted_normalised_df.loc['Negative Ideal']

    positive_ideal_df['Separation Ideal Solution']=(positive_ideal_df)/(positive_ideal_df_den).sum(axis=1)
    negative_ideal_df['Separation Ideal Worst']=(negative_ideal_df)/(negative_ideal_df_den).sum(axis=1)
```

Gower

```
[11] positive_ideal_df=(abs(wt_nrm_df-weighted_normalised_df.loc['Positive Ideal']))
     negative_ideal_df=(abs(wt_nrm_df-weighted_normalised_df.loc['Negative Ideal']))

     positive_ideal_df['Separation Ideal Solution']=(positive_ideal_df.sum(axis=1))/3
     negative_ideal_df['Separation Ideal Worst']=(negative_ideal_df.sum(axis=1))/3
```

Canberra

```
[ ] positive_ideal_df=(abs(wt_nrm_df-weighted_normalised_df.loc['Positive Ideal'])/
                      (wt_nrm_df+weighted_normalised_df.loc['Positive Ideal']))
    negative_ideal_df=(abs(wt_nrm_df-weighted_normalised_df.loc['Negative Ideal'])/
                      (wt_nrm_df+weighted_normalised_df.loc['Negative Ideal']))

    positive_ideal_df['Separation Ideal Solution']=(positive_ideal_df.sum(axis=1))
    negative_ideal_df['Separation Ideal Worst']=(negative_ideal_df.sum(axis=1))
```

Lorentzian

```
[ ] positive_ideal_df=(np.log(1+abs(wt_nrm_df-weighted_normalised_df.loc['Positive Ideal'])))
    negative_ideal_df=(np.log(1+abs(wt_nrm_df-weighted_normalised_df.loc['Negative Ideal'])))

    positive_ideal_df['Separation Ideal Solution']=positive_ideal_df.sum(axis=1)
    negative_ideal_df['Separation Ideal Worst']=negative_ideal_df.sum(axis=1)
```

Among the numerous methods developed in this regard, this paper focusses on the Techniques for Order Preference by Similarity to an Ideal Solution commonly abbreviated as TOPSIS. The objective of the paper was to extend the traditional TOPSIS method by using ten different distance measures from three distance families. The families proposed in this approach are $L_p$ Minkowski family, $L_1$ family and the $\chi^2$ family. The numerical example for this is inspired by a study of the autonomous landing of drones with hazard avoidance named Intelligent Landing of Unmanned Aerial Vehicles (ILUV) whose aim was to select a suitable site for target-landing with respect to certain requirements.

This modified proposal has been implemented in Python programming language using minimalistic codes and with complexity and runtime as low as possible. For

**Table 10** $\chi^2$ family

Squared Euclidean

```
[ ] positive_ideal_df=(wt_nrm_df-weighted_normalised_df.loc['Positive Ideal'])
    negative_ideal_df=(wt_nrm_df-weighted_normalised_df.loc['Negative Ideal'])

    positive_ideal_df['Separation Ideal Solution']=np.power(positive_ideal_df,2).sum(axis=1)
    negative_ideal_df['Separation Ideal Worst']=np.power(negative_ideal_df,2).sum(axis=1)
```

Pearson $\chi^2$

```
[ ] positive_ideal_df=np.power(wt_nrm_df-weighted_normalised_df.loc['Positive Ideal'],2)
    negative_ideal_df=np.power(wt_nrm_df-weighted_normalised_df.loc['Negative Ideal'],2)

    positive_ideal_df['Separation Ideal Solution']=(positive_ideal_df/weighted_normalised_df.loc['Positive Ideal']).sum(axis=1)
    negative_ideal_df['Separation Ideal Worst']=(negative_ideal_df/weighted_normalised_df.loc['Negative Ideal']).sum(axis=1)
```

Neyman $\chi^2$

```
[ ] positive_ideal_df=np.power(wt_nrm_df-weighted_normalised_df.loc['Positive Ideal'],2)
    negative_ideal_df=np.power(wt_nrm_df-weighted_normalised_df.loc['Negative Ideal'],2)

    positive_ideal_df['Separation Ideal Solution']=(positive_ideal_df/wt_nrm_df).sum(axis=1)
    negative_ideal_df['Separation Ideal Worst']=(negative_ideal_df/wt_nrm_df).sum(axis=1)
```

```
[ ] negative_ideal_df['Relative Closeness']=negative_ideal_df['Separation Ideal Worst']/
    (negative_ideal_df['Separation Ideal Worst']+positive_ideal_df['Separation Ideal Solution'])
    negative_ideal_df['Relative Closeness']

    negative_ideal_df['Relative Closeness'].sort_values(axis=0, ascending=False)
```

**Fig. 2** Python code to calculate the relative closeness coefficient

an improved and visual understanding, the results of each of the ten measures have been depicted using radar maps. The results of the same were compared with that from the standard approach and they indicate close association thus justifying that the proposed approach can be applied to similar datasets of MCDM problems under conflicting criteria.

The following directions can be derived from this study:

- Further work can be done by using different normalisation techniques for calculating the weighted normalised decision matrix. The results can then be compared for accuracy, efficiency, reliability, sensitivity and complexity. This approach can also be combined with various distance measures.
- Majority of the studies available in the field of multi-criteria decision making is based on selecting the best alternative from a set of alternatives. Improvements can be made by experimenting with the values of the decision matrix and comparing ranks.
- Other MCDA methods can be checked to ensure the accuracy of the results.

**Table 11** Relative closeness coefficient of the $L_p$ Minkowski family

| | Euclidean | City block | Minkowski |
|---|---|---|---|
| **$SC_1$** | 0.776283 | 0.79413 | 0.77217 |
| **$SC_2$** | 0.392372 | 0.40463 | 0.38953 |
| **$SC_3$** | 0.413964 | 0.44906 | 0.40946 |
| **$SC_4$** | 0.933790 | 0.94878 | 0.92738 |
| **$SC_5$** | 0.328905 | 0.30870 | 0.33251 |
| **$SC_6$** | 0.284230 | 0.27496 | 0.28737 |
| **$SC_7$** | 0.231146 | 0.23986 | 0.23038 |
| **$SC_8$** | 0.246483 | 0.26329 | 0.24256 |
| **$SC_9$** | 0.326750 | 0.29240 | 0.33225 |
| **$SC_{10}$** | 0.280813 | 0.25866 | 0.28674 |
| **$SC_{11}$** | 0.234264 | 0.23168 | 0.23753 |
| **$SC_{12}$** | 0.734217 | 0.73744 | 0.73105 |
| **$SC_{13}$** | 0.088479 | 0.07438 | 0.09375 |
| **$SC_{14}$** | 0.042700 | 0.03895 | 0.04359 |
| **$SC_{15}$** | 0.056576 | 0.05359 | 0.05641 |
| **$SC_{16}$** | 0.671506 | 0.66397 | 0.67202 |

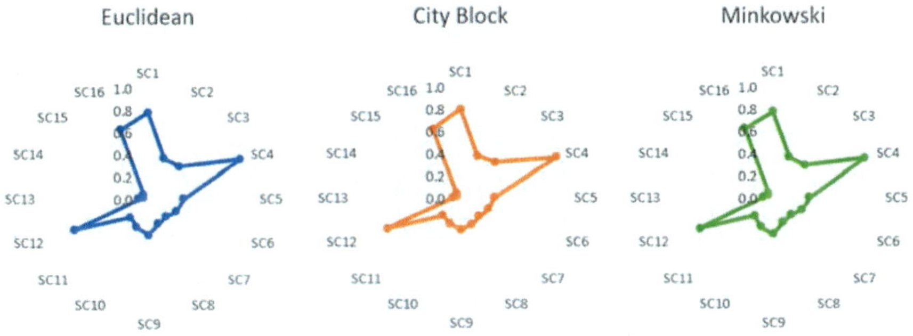


**Fig. 3** Radar maps of the $L_p$ Minkowski family

- Lastly, the approach can be tested with different weight vectors and subsequently analyse the changes in results and conclude the significance of the criteria.

**Table 12** Relative closeness coefficient of the $L_1$ family

| | Sorensen | Gower | Canberra | Lorentzian |
|---|---|---|---|---|
| **$SC_1$** | 0.79222 | 0.79413 | 0.82329 | 0.66146 |
| **$SC_2$** | 0.40211 | 0.40463 | 0.38053 | 0.45272 |
| **$SC_3$** | 0.44648 | 0.44906 | 0.41737 | 0.47538 |
| **$SC_4$** | 0.94824 | 0.94878 | 0.93231 | 0.81043 |
| **$SC_5$** | 0.30642 | 0.30870 | 0.29794 | 0.40140 |
| **$SC_6$** | 0.27282 | 0.27496 | 0.26896 | 0.38215 |
| **$SC_7$** | 0.23789 | 0.23986 | 0.23573 | 0.36092 |
| **$SC_8$** | 0.26120 | 0.26329 | 0.25712 | 0.37539 |
| **$SC_9$** | 0.29019 | 0.29240 | 0.28269 | 0.39205 |
| **$SC_{10}$** | 0.25659 | 0.25866 | 0.25391 | 0.37230 |
| **$SC_{11}$** | 0.22976 | 0.23168 | 0.22857 | 0.35563 |
| **$SC_{12}$** | 0.73519 | 0.73744 | 0.74856 | 0.62528 |
| **$SC_{13}$** | 0.07362 | 0.07438 | 0.07614 | 0.22200 |
| **$SC_{14}$** | 0.03854 | 0.03895 | 0.03432 | 0.16862 |
| **$SC_{15}$** | 0.05303 | 0.05359 | 0.04807 | 0.19336 |
| **$SC_{16}$** | 0.66135 | 0.66397 | 0.68014 | 0.58341 |

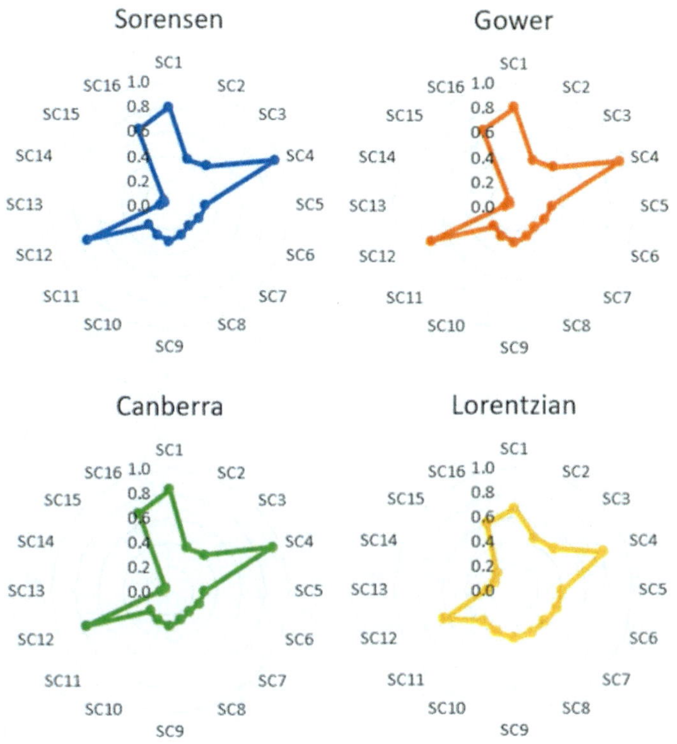


Fig. 4 Radar maps of the $L_1$ family

**Table 13** Relative closeness coefficient of the $\chi^2$ family

| | Squared Euclidean | Pearson | Neyman |
|---|---|---|---|
| $SC_1$ | 0.92332 | 0.92884 | 0.94853 |
| **$SC_2$** | 0.29428 | 0.24027 | 0.29083 |
| **$SC_3$** | 0.33288 | 0.27380 | 0.32453 |
| **$SC_4$** | 0.99500 | 0.99225 | 0.99418 |
| **$SC_5$** | 0.19368 | 0.15732 | 0.19859 |
| **$SC_6$** | 0.13621 | 0.11042 | 0.14442 |
| **$SC_7$** | 0.08289 | 0.06725 | 0.09018 |
| **$SC_8$** | 0.09666 | 0.07782 | 0.10427 |
| **$SC_9$** | 0.19064 | 0.15536 | 0.19520 |
| **$SC_{10}$** | 0.13229 | 0.10773 | 0.14022 |
| **$SC_{11}$** | 0.08558 | 0.06989 | 0.09306 |
| **$SC_{12}$** | 0.88414 | 0.88753 | 0.91299 |
| **$SC_{13}$** | 0.00933 | 0.00798 | 0.01071 |
| **$SC_{14}$** | 0.00199 | 0.00139 | 0.00185 |
| **$SC_{15}$** | 0.00358 | 0.00263 | 0.00343 |
| **$SC_{16}$** | 0.80690 | 0.81186 | 0.85037 |

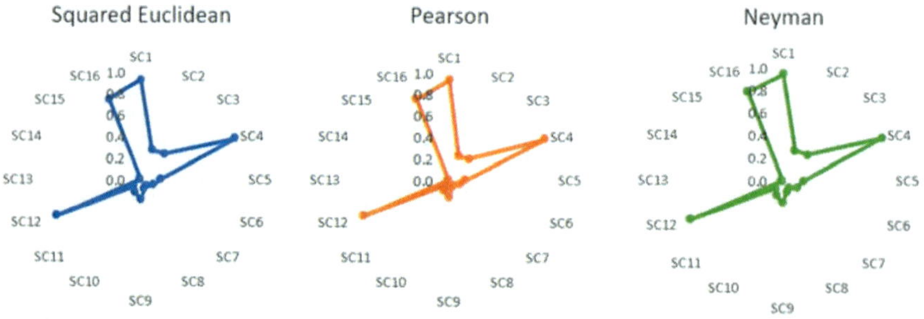


**Fig. 5** Radar maps of the $\chi^2$ family

**Table 14** Ranks of the alternatives obtained using to all the distance measures

| | Euclidean | City block | Minkowski | Sorensen | Gower | Canberra | Lorentzian | Squared Euclidean | Pearson | Neyman |
|---|---|---|---|---|---|---|---|---|---|---|
| $SC_1$ | 2 | 2 | 2 | 2 | 2 | 2 | 2 | 2 | 2 | 2 |
| $SC_2$ | 6 | 6 | 6 | 6 | 6 | 6 | 6 | 6 | 6 | 6 |
| $SC_3$ | 5 | 5 | 5 | 5 | 5 | 5 | 5 | 5 | 5 | 5 |
| $SC_4$ | 1 | 1 | 1 | 1 | 1 | 1 | 1 | 1 | 1 | 1 |
| $SC_5$ | 7 | 7 | 7 | 7 | 7 | 7 | 7 | 7 | 7 | 7 |
| $SC_6$ | 9 | 9 | 9 | 9 | 9 | 9 | 9 | 9 | 9 | 9 |
| $SC_7$ | 13 | 12 | 13 | 12 | 12 | 12 | 12 | 13 | 13 | 13 |
| $SC_8$ | 11 | 10 | 11 | 10 | 10 | 10 | 10 | 11 | 11 | 11 |
| $SC_9$ | 8 | 8 | 8 | 8 | 8 | 8 | 8 | 8 | 8 | 8 |
| $SC_{10}$ | 10 | 11 | 10 | 11 | 11 | 11 | 11 | 10 | 10 | 10 |
| $SC_{11}$ | 12 | 13 | 12 | 13 | 13 | 13 | 13 | 12 | 12 | 12 |
| $SC_{12}$ | 3 | 3 | 3 | 3 | 3 | 3 | 3 | 3 | 3 | 3 |
| $SC_{13}$ | 14 | 14 | 14 | 14 | 14 | 14 | 14 | 14 | 14 | 14 |
| $SC_{14}$ | 16 | 16 | 16 | 16 | 16 | 16 | 16 | 16 | 16 | 16 |
| $SC_{15}$ | 15 | 15 | 15 | 15 | 15 | 15 | 15 | 15 | 15 | 15 |
| $SC_{16}$ | 4 | 4 | 4 | 4 | 4 | 4 | 4 | 4 | 4 | 4 |

## References


1. Tzeng G-H, Huang, J-J.(2011) Multiple attribute decision making: methods and applications. CRC press
2. Zavadskas EK, Turskis Z, Kildienė S (2014) State of art surveys of overviews on MCDM/MADM methods. Technol Econ Dev Econ 20.1:165–179
3. Yoon KP, Ching-Lai H (1995) Multiple attribute decision making: an introduction, vol 104. Sage
4. Qin X-S et al (2008) A MCDM-based expert system for climate-change impact assessment and adaptation planning—a case study for the Georgia Basin, Canada. Expert Syst Appl 34.3:2166
5. Gumus AT (2009) Evaluation of hazardous waste transportation firms by using a two step fuzzy-AHP and TOPSIS methodology. Expert Syst Appl 36(2):4067–4074
6. Shirkhorshidi AS, Aghabozorgi S, Ying Wah T (2015) A comparison study on similarity and dissimilarity measures in clustering continuous data. PloS One 10.12:e0144059
7. Lai Y-J, Liu T-Y, Hwang C-L (1994) Topsis for MODM. Eur J Oper Res 76(3):486–500
8. Chen S-J, Hwang C-L (1992) Fuzzy multiple attribute decision making methods. Fuzzy multiple attribute decision making. Springer, Berlin, Heidelberg, pp 289–486
9. Yadav V et al (2019) PyTOPS: a python based tool for TOPSIS. SoftwareX 9:217–222
10. Tschopp MA, Hernandez-Rivera E (2017) Quantifying similarity and distance measures for vector-based datasets: histograms, signals, and probability distribution functions. No. ARL-TN-0810. US Army Research Laboratory Aberdeen Proving Ground United States
11. Saaty Roseanna W (1987) The analytic hierarchy process—what it is and how it is used. Math Model 9(3-5):161–176
12. Anandan V, Uthra G (2017) Extension of TOPSIS using L1 family of distance measures. Adv Fuzzy Math 12(4):897–908
13. Chen Chen-Tung (2000) Extensions of the TOPSIS for group decision-making under fuzzy environment. Fuzzy Sets Syst 114(1):1–9
14. Palczewski K, Sałabun W (2019) The fuzzy TOPSIS applications in the last decade. Proc Comput Sci 159:2294–2303
15. Salih MM et al (2019) Survey on fuzzy TOPSIS state-of-the-art between 2007 and 2017. Comput Operat Res 104:207–227
16. Papathanasiou J, Ploskas N (2018) Topsis. Multiple Criteria Decision Aid. Springer, Cham, pp 1–30
17. Jahanshahloo GR, Hosseinzadeh Lotfi F, Izadikhah M (2006) An algorithmic method to extend TOPSIS for decision-making problems with interval data. Appl Math Comput 175.2:1375–1384
18. Vafaei Nazanin, Ribeiro Rita A, Camarinha-Matos Luis M (2018) Data normalisation techniques in decision making: case study with TOPSIS method. Int J Inf Decis Sci 10(1):19–38
19. http://www.ca3-uninova.org/project_iluv
20. Bourdarias C et al. (2010) Optimized and flexible multi-criteria decision making for hazard avoidance. In: Proceedings of the 33rd annual AAS rocky mountain guidance and control conference

# Chapter 25
# Removal of Chromium from Synthetic Wastewater Using Synthesized Low-Cost Adsorbents

**Athar Hussain, Deepesh Tiwari, and Zafar Heider**

## 1 Introduction

The presence of heavy metals in the environment imparts toxicity in various forms as they enter the system by regular and anthropogenic methods. Such contaminating sources include natural weathering of the earth's crust, mining, soil disintegration, mechanical release, metropolitan spill over, sewage effluents, irrigation or infectious prevention specialists applied to plants, air contamination aftermath [24, 40]. Although a few people are fundamentally involved in these toxins because of their working environment, for the vast majority the principle course of presentation to these poisonous components is through the eating routine (food and water). The pollution chain of substantial metals quite often follows a cyclic request: industry, climate, soil, water, nourishment and human. Despite the fact that harmfulness and the subsequent danger to human soundness of any impurity are, obviously, an element of concentration, it is notable that persistent presentation to substantial metals and metalloids at generally low levels can cause unfavorable impacts [20, 29].

In this way, there has been expanding concern, majorly in the developed world, about introductions, admissions and assimilation of heavy metals by people. Populaces are progressively requesting a cleaner environment, and decreases in the

A. Hussain (✉)
Department of Civil Engineering, NSUT WEST CAMPUS (Formerly CBP Government Engineering College Jaffarpur), New Delhi 110073, India
e-mail: athariitr@gmail.com

D. Tiwari
Environmental Engineering Department, Delhi Technological University, Bawana Road, Shahbad Daulatpur, Delhi 110042, India
e-mail: deepeshtiwari_2k18en015@dtu.ac.in

Z. Heider
Department of Civil Engineering, School of Engineering, Gautam Buddha University, Great Noida 201312, UP, India

 
D. Singh et al. (eds.), *Proceedings of International Conference on Scientific and Natural Computing*, Algorithms for Intelligent Systems,
https://doi.org/10.1007/978-981-16-1528-3_25

measures of foreign substances contacting individuals because of the rise in human activities. A functional ramification of this pattern, in the developed nations, has been the burden of new and more prohibitive guidelines [20]. The major heavy metals including Lead (Pb), cadmium (Cd), mercury (Hg), and arsenic (As) are broadly scattered in the environment. These components have no gainful impacts on people, and there is no known homeostasis instrument for them. They are commonly considered the most harmful to people and creatures the unfriendly human well-being impacts related with the presentation to them, even at low concentrations, are different and incorporate, yet are not restricted to, neurotoxic and cancer-causing activities [13, 18].

Various studies by different investigators have been carried out to explore the possibilities of treatment measures for the expulsion of heavy metals from wastewater. Physico-synthetic evacuation cycles, for example, adsorption on new adsorbents, particle trade, film filtration, electrodialysis, invert assimilation, ultra-filtration and photocatalysis are some of the useful methods for heavy metal removal [7, 10]. It has been reported that the cycles of natural medicines microorganisms assume a part of settling solids in the arrangement. Activated sludge, trickling filters, stabilization ponds are broadly utilized for treating modern wastewater [19]. Bioadsorption is another organic technique and different ease bioadsorbents prepared from agricultural waste, forest waste, industrial waste and algae can be utilized for the greatest expulsion of substantial metals from wastewater. Bioadsorption strategies are eco-friendly best answers for eliminating heavy metals from wastewater instead of physical-compound techniques [33, 35]. Enacted seepage, streaming channels, adjustment lakes are also broadly utilized for treating modern wastewater [19]. However, synthetic strategies are the most appropriate medicines for poisonous inorganic mixes delivered from different ventures which cannot be eliminated from any natural and actual procedures [17].

The environment contaminated by heavy metals demands sufficient protection and restoration. The policies for environmental protection and public health, at various levels, are planned on the basis of available data of present concentration of certain natural phenomena, mainly comprised in our food chain [21]. The contaminated sites should be managed in a cost-effective manner to provide assurance to the public and the ecosystem [42]. Some of the best technologies for remediation of heavy metal contaminated sites are immobilization, phytoremediation and soil washing [14], but these are generally practised in developed nations only, in spite of being economical and having other field applications [39]. For the adsorption of heavy metals, activated carbon adsorption is an efficient and useful, but expensive technology [6]. There are certain limitations in most of the approaches, therefore, most of the focus is devoted to investigate a low-cost adsorbent with high efficiency of heavy metal sorption, and certain agricultural waste material fits the requirements [41]. These have a high sorption potential to use as heavy metal adsorbents [2]. Studies are done to find heavy metal adsorption efficiency and other characteristics for multiple agricultural and forest waste products like peanut shells [38], wheat bran [11], corncobs [28], rice husk [22], sawdust [8], and many others. Lata et al. 2014 [27] and Sahmoune and Yeddou [34] carried out a study and analyzed the efficiency of rice husk and sawdust as an adsorbent in removing the heavy metals from wastewater with the use of graphical representation.

Considering the above-mentioned facts in view, a study was being planned with objectives in order to remove the chromium concentration in synthetic wastewater using rice husk and saw dust as adsorbent synthesized from agricultural waste. The study on the variation of removal efficiency under the effects of various parameters like adsorbent dose, initial concentration of metal, pH value and contact time, and further analysis is done using adsorption isotherms.

## 2 Methodology

Saw dust and rice husk are the two solid waste materials used as adsorbate for adsorption of heavy metals after proper processing and treatment. The sawdust collected from the local wood shop was used for adsorption of chromium as Cr (VI) after grinding and then passing through a sieve size of 300 μm. Thereafter it was being oven-dried at 80 °C for 6 h for effective color removal and thereafter stored in plastic bags at room temperature. Also, rice husk was collected from an agricultural production process industry. The collected rice husk was dried under the sun and impurities were separated manually. In order to remove the colored compounds, the adsorbate was boiled for 1 h and the filtered residual obtained was dried at 800 °C in hot air oven for 4 h. The obtained sample was ground and sieved through a 300 μm size sieve and stored in the airtight plastic container. The particle size and moisture content were also analyzed for the prepared sample of adsorbents.

Synthetic wastewater containing chromium was being prepared by diluting 99% of pure potassium dichromate reagent in deionized water to obtain a Cr (VI) stock solution concentration of 1000 mg/L. The prepared solution was being utilized for carrying out the adsorption experiments and the adsorption efficiency under different variations has been studied. In each experimental study, 1–5 g of the prepared rice husk and saw dust were used as adsorbents. The samples having solution containing Cr metal ions varying concentration of 10–40 mg/L along with adsorbent was were placed in 250 mL stopper conical flask. The solution was placed in a rotary shaker at 150 rpm for about 100 min at a constant temperature to attain the equilibrium and thereafter filtered using Whatman filter paper to separate the adsorbent from it. The residual concentration of chromium metal ions was determined by analyzing the filtrate using the atomic absorption spectrophotometer (LAB INDIA model AA 8000) and the removal efficiency percentage of metals ions was calculated by using the expression (Eq. 1):

$$R(\%) = \frac{C_o - C_e}{C_0} x100 \tag{1}$$

where

$C_o$ Initial concentration of metal ions in the solution
$C_e$ Final concentration of metal ions in the solution [20]

By utilizing the above equation the efficiency of adsorbents can be assessed in terms of removal of heavy metals concentration with respect to the initial concentration of heavy metals and also with respect to change in adsorbent dose, pH value of sample solution and the contact time of sample solution with these adsorbents. However, under equilibrium, the adsorption rate equals the desorption rate and the remaining amount of adsorbate in the sample solution is measured. The data to plot the Langmuir's and Fredulich's equation isotherms is derived from Table 2, showing the variation in removal percentage with respect to initial concentration ($C_0$) and final concentration($C_e$). The final concentration of adsorbent at equilibrium can be further calculated from equation (Eq. 2):

$$Q_e = (C_o - C_e)x\frac{V}{m} \tag{2}$$

where

$Q_e$ adsorbent concentration after equilibrium (mg/g)
$C_0$ Initial concentration of solution (mg/L)
$C_e$ Equilibrium concentration of solution (mg/L)
V Volume of solution (L)
m Mass of adsorbent (g) [37]

The variables obtained for two isotherms from the observation data of $Q_e$ and $C_e$ are plotted with their trend-lines of data, and the slope and intercept of lines are found using a statistical library of linregress in python. These slopes and intercepts are used to calculate all coefficients of Langmuir and Freundlich isotherm equations. Correlation between actual values and those lying on the line is obtained. However, the Freundlich and Langmuir equations are utilized to describe the experimental data pertaining to adsorption and the data from obtained results may also have a linear correlation [25]. Therefore the sorption theories related to Freundlich and Langmuir isotherms have been used and using the obtained results, the raw data has been processed and compared with the standard adsorption parameters.

## 3 Results and Discussion

In order to carry out the experimental studies, the batch adsorption technique has been used to analyze the effect of various parameters on the removal efficiency of heavy metal from synthetic wastewater. However, prior to the experimental study, the characteristics of each sample was being determined and results for the same has been reported. The particle size of the prepared adsorbent sample through sieving is kept below 300 μm. The moisture content of the rice husk and sawdust adsorbent was found to be 8.9% and 7.9%, respectively. The observations are recorded for the batch adsorption studies to find the removal efficiency of Cr(VI) under the effect of variation in different parameters. It was observed from graph plots that variations in removal efficiency, with respect to change in all the parameters studied, by rice husk and sawdust are highly correlated. The obtained data from the present study has been finally utilized and fitted in the adsorption isotherm models.

### 3.1 *Effect of Adsorbent Dose*

The adsorbent dose is a parameter that is used to determine the removal capacity of adsorbent at a constant initial concentration of heavy metal in the sample [9]. The effect of variation in adsorbent dose on adsorption efficiency of chromium was studied by varying the adsorbent dose from 1 to 4 g. The pH value of all the samples was kept at 5 with a contact time of 100 min under 150 rpm. However, 4 samples have been taken under each dose of adsorbents and it has been observed that a higher adsorbent dose in the experimental study provides the higher accessibility of exchangeable sites for the ions. Therefore the removal efficiency of the chromium is enhanced on increasing adsorbent doses. In the present study, no significant adsorption has been observed at an adsorbent dose greater than 3 g. The maximum chromium removal efficiency of 90.7% and 82% have been observed using rice husk and saw dust as adsorbent, respectively, with the adsorbent dosage of 3 g at an initial chromium concentration of 10 mg/L. This result summarized in Table 1 indicates that after a specific portion of adsorbent, the equilibrium conditions occurred and consequently the quantity of free particles and particles bound to the adsorbent in the arrangement stays constant even with the further addition of the adsorbent doses. A perusal of the data from Fig. 1 indicates that the quantity of adsorbent dose under which equilibrium is reached increases as the initial concentration of Cr increases in solution. However, the correlation factor of 0.84 indicates the variation in removal capacity by two adsorbents to be highly correlated.

**Table 1** Effect of variation in the adsorbent dose and chromium removal efficiency at a constant pH value of 5

| S. no. | Adsorbent dose (g) | Initial Cr concentration (mg/L) | Cr removal (%) using adsorbent | | | |
|---|---|---|---|---|---|---|
| | | | Rice husk | | Sawdust | |
| | | | Mean | Std. Dev. | Mean | Std. Dev. |
| 1 | 1 | 10–40 | 70.85 | 12.83 | 48.35 | 7.36 |
| 2 | 2 | 10–40 | 80.225 | 8.33 | 56.95 | 8.87 |
| 3 | 3 | 10–40 | 84.775 | 5.07 | 74.55 | 7.63 |
| 4 | 4 | 10–40 | 84.575 | 3.56 | 76.125 | 6.66 |

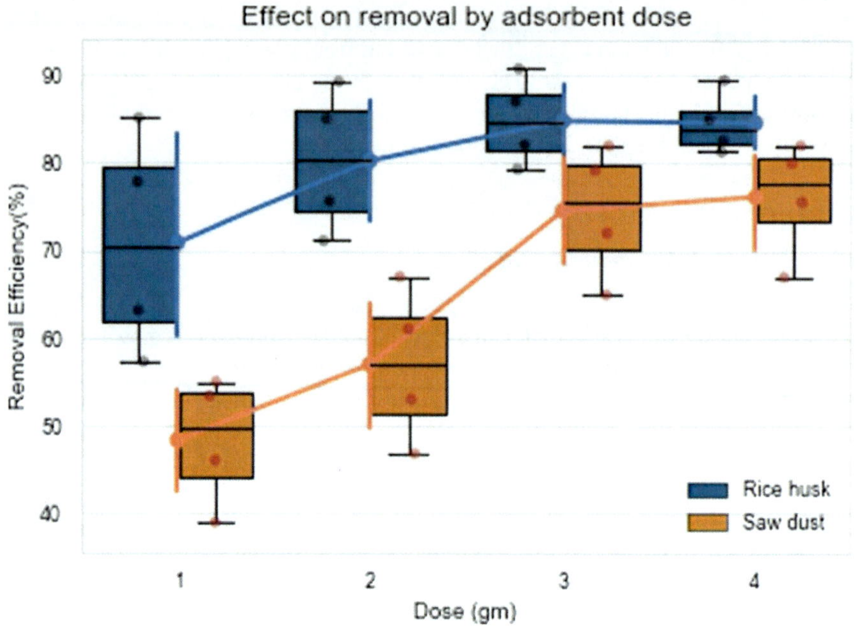


**Fig. 1** Effect of removal efficiency in the variation of adsorbent dose

## 3.2 *Effect of Initial Concentration*

In order to observe the effect of the initial concentration of heavy metals and adsorption rate in a sample the experiments were carried out by varying the Cr concentration from 10 to 70 mg/L with an adsorbent dose of 3 g at a pH value of 5 with a contact time of 100 min. The obtained results are summarized in Table 2. A perusal of the data summarized in Table 2 and Fig. 2 indicates that as the initial concentration of Cr

**Table 2** Effect of Cr removal efficiency at a variation of initial Cr concentration

| S. no. | Initial concentration (mg/L) | Final concentration (mg/L) | Cr removal (%) using adsorbent | |
|---|---|---|---|---|
| | | | Rice husk | Sawdust |
| 1 | 10 | 1.8 | 90.7 | 82 |
| 2 | 20 | 2.1 | 87 | 79 |
| 3 | 30 | 2.8 | 82.1 | 72 |
| 4 | 40 | 3.5 | 79.3 | 65 |
| 5 | 50 | 4.2 | 65.2 | 58 |
| 6 | 60 | 4.9 | 58.9 | 51 |
| 7 | 70 | 5.2 | 55.4 | 48 |

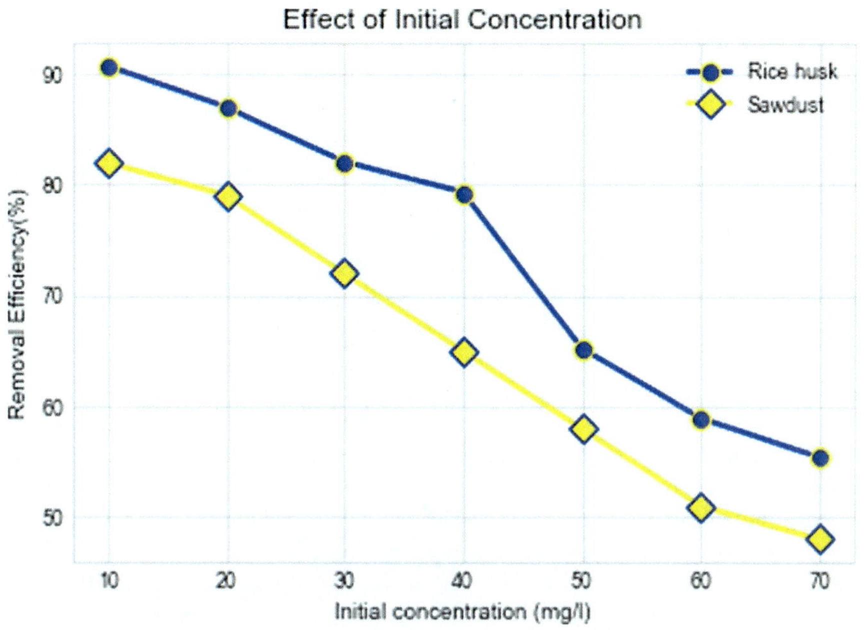


**Fig. 2** Effect of initial Cr concentration on Cr removal efficiency under different adsorbents

increases from 10 to 70 mg/L in a sample a drastic decrease in Cr removal efficiency of 90.7 to 55.4% has been observed when rice husk has been used as adsorbent. However, a decrease in removal efficiency from 82 to 48% Cr ion has been observed when saw dust is used as an adsorbent with Cr concentration in a sample ranging from 10 to 70 mg/L. This lowering efficiency of heavy metals removal at higher concentration may be due to the increased proportion of initial molar concentration of chromium to the vacant sites available.

Also for a given adsorbent dose, the total number of adsorbent sites available was fixed and therefore this limiting amount of adsorbate resulted in a decreased removal efficiency of adsorbate, consequent to an increase in initial Cr concentration. Therefore results from the present study indicates the fact that chromium adsorption depends on its initial concentration and the amount of adsorbent available. However, the correlation in a variation of removal efficiency by the two adsorbents is found to be 0.985 and is an indication of quite close values in terms of removal efficiency of heavy metals in a solution. The study can be well compared with work carried out by Ahmaruzzaman and Gupta, 2011 [3], reporting that the adsorption capacity depends on the initial concentration of heavy metals in the sample. This initial concentration of heavy metals in a solution is proportional to the mass-transfer of adsorbate between the aqueous and solid phases. This parameter includes the factors like availability of specific functional groups at the surface and their ability to bind the metal ions [1].

## 3.3 Effect of PH

pH is another important parameter that controls the adsorption of metal ions in a treatment system as it influences the surface charge of adsorbent, degree of ionization of heavy metal in a solution and dissociation process at the active sites of adsorbents [1, 3]. In the present study, batch experiments have been carried out by varying pH from 2–10 value under room temperature conditions at a rotational speed of 150 rpm with a contact time of 100 min and Cr concentration of 10 mg/L. The obtained results as summarized in Table 3 and Fig. 3 indicates the fact that pH influences the adsorption rate in terms of removal efficiency of the Cr ions in the aqueous solution. The maximum Cr removal of 92.1% has been observed at a pH value of 6 and minimum Cr removal of 75.8% has been found at a pH value of 2 using rice husk as adsorbent. The maximum Cr removal efficiency of 85% has been observed at pH 6 and minimum Cr removal efficiency of 55% has been observed at pH 2 using saw dust as an adsorbent.

Adsorption of metal on adsorbents such as rice husk and sawdust is because the presence of functional groups, such as cellulose, hemicellulose, lignin, tannin, silica, carbohydrates, and others. At low pH, the adsorption efficiency is low as the polar functional groups are protonated and not free for transitions with metal ions. At higher pH, with deprotonation of polar ions, the availability of functional groups for ion transition increases causing a higher adsorption rate [5]. A significant decrease in Cr removal efficiency has been observed as the value of pH decreases or increases from a value of 6. This may be due to the fact that the net positive surface potential of the sorbent decreases with increasing pH, which shows a fall in electrostatic force between adsorbate and adsorbent. Also, as the pH value in a solution containing metallic ions decreases from 6 the heavy metals start precipitating and this ultimately decreases the adsorption potential of the adsorbents. However, the correlation factor of the variations of the two adsorbents in terms of Cr removal efficiency has been found to 0.915 at various pH values.

**Table 3** Effect of variation of pH value on Cr removal efficiency by using the adsorbents

| S. no. | pH | Final concentration (mg/L) | Cr removal (%) using adsorbent | |
|---|---|---|---|---|
| | | | Rice husk | Sawdust |
| 1 | 2 | 2.42 | 75.8 | 55 |
| 2 | 4 | 1.69 | 83.1 | 76 |
| 3 | 6 | 0.79 | 92.1 | 85 |
| 4 | 8 | 1.40 | 86 | 79 |
| 5 | 10 | 1.90 | 81 | 75 |

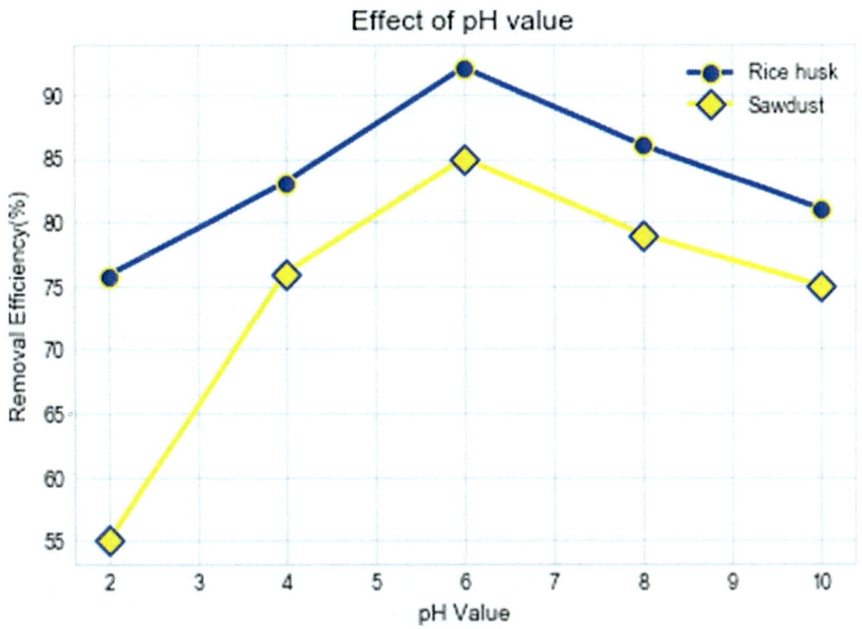


**Fig. 3** Effect of variation in pH values on Cr removal efficiency

## 3.4 *Effect of Contact Time*

The adsorption of heavy metals is also an effect of contact time when ions are to be adsorbed on the solid surface. In the present study, the adsorption of Cr ion has been studied by varying the contact time from 20 to 140 min at Cr concentration of 10 mg/L with a pH value of 6 under 150 rpm. A perusal of the data summarized in Table 4 indicates that the removal efficiency of rice husk increases from 58 to 91% as the contact time in a solution increases from 20 to 100 min. However, from Fig. 4

**Table 4** Effect of Cr removal efficiency in the variation of contact time

| S. no. | Contact time (minutes) | Final concentration (mg/L) | Cr removal (%) using adsorbent | |
|---|---|---|---|---|
| | | | Rice husk | Sawdust |
| 1 | 20 | 4.20 | 58 | 47 |
| 2 | 40 | 3.39 | 66 | 60 |
| 3 | 60 | 2.20 | 78 | 77 |
| 4 | 80 | 1.47 | 85 | 80 |
| 5 | 100 | 0.93 | 91 | 83 |
| 6 | 120 | 0.90 | 91 | 83 |
| 7 | 140 | 0.90 | 91 | 83 |

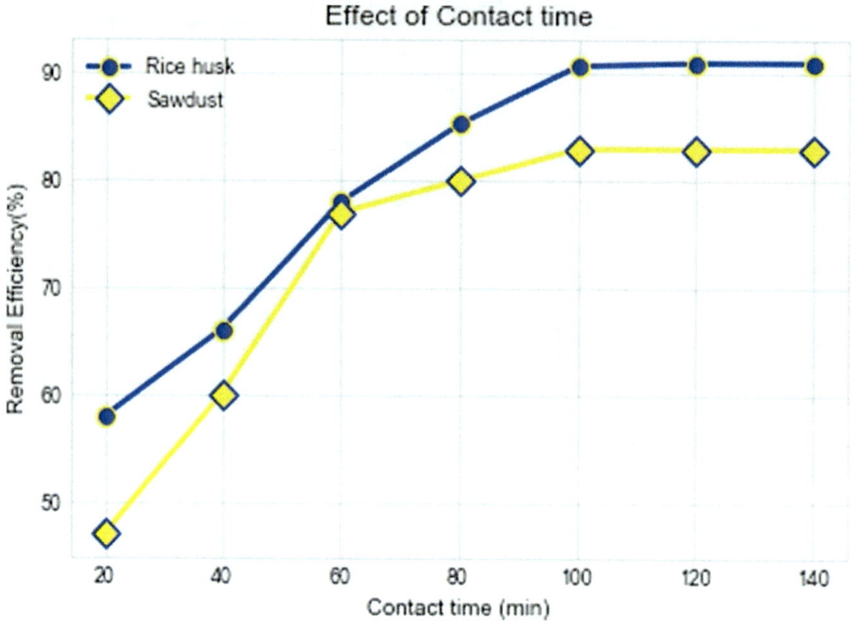


**Fig. 4** Effect of variation of contact time on Cr removal efficiency

it can be inferred that no significant effect in the removal efficiency of Cr ions has been observed as the contact time increases from 100 to 140 min under the same concentration and pH value. However, the Cr removal efficiency has been found to be increased from 47 to 83% using saw dust as absorbent with contact time variation from 20 to 140 min at Cr concentration of 10 mg/L with a pH value of 6. Also, no significant increase in Cr removal efficiency has been perceived as the contact time was varied from 100 to 140 min with no change in Cr concentration and pH value of the solution.

It has been observed that the rate of Cr adsorption using rice husk and saw dust as adsorbents increases up to a contact time of 60 min, then this rate of adsorption decreases up to 100 min and thereafter becomes steady indicating it to be a saturated value with other parameters to be kept constant for variation in contact time. This may be due to the fact that initially higher number of adsorption sites were available and the concentration gradient between metal ions in solution and that present at the surface of the adsorbent is high. After 60 min, only a few number of vacant adsorption sites remains, and the concentration gradient between the two adsorbates also decreases, causing to fall in the rate of adsorption of metal ions [5]. It shows that the adsorption of heavy metal using rice husk as adsorbent is a function of contact time at any initial concentration of metal [12]. Therefore contact time of 100 min by using rice husk and sawdust as adsorbent is the optimum time for the removal of heavy metals [1]. The obtained results are found to be applicable in economical wastewater treatment as equilibrium time is found to be an important parameter.

Adsorption of heavy metal also depends on the contact of functional groups in a solution at the adsorbent surface. Also, the maximum efficiency of heavy metals removal can be obtained when the equilibrium between metal ions and adsorbent in a sample is reached [1]. The variation in Cr removal efficiency by using rice husk and saw dust as an adsorbent with contact time is highly correlated with a correlation factor of 0.98.

### 3.5 Adsorption Isotherms

Further, the effectiveness of rice husk and saw dust as adsorbent materials are compared under diverse operational conditions using adsorption isotherms. Adsorption isotherm describes the quantitative relation of equilibrium between the adsorbate metal concentration in solution and its adsorbed concentration [9]. They can also be used for the purpose to design and optimize an adsorption system. Heavy metal adsorption is usually presented by the classical adsorption isotherms that are Freundlich and Langmuir isotherms [20]. In this study, adsorption isotherms for Langmuir, Freundlich models has been fitted for removal of chromium from aqueous solution using rice husk and saw dust adsorbent.

**Langmuir adsorption isotherm**: Langmuir isotherm is based on an assumption that molecules are adsorbed at constant well-defined active sites which are homogeneously distributed over the adsorbent that means these are identical and bind with the same strength for the same metal. Also, the only single layer of adsorbate is possible. This model determines the maximum capacity of the adsorbent with maximum monolayer cover on its surface [26]. This isotherm model is presented by the following equation (Eq. 3):

$$\frac{1}{Q_e} = \frac{1}{Q_{max}} + \frac{1}{bQ_{max}C_e} \quad (3)$$

where

- $Q_e$ amount adsorbed (mg/g)
- $C_e$ equilibrium concentration of metal ion (mg/L)
- $Q_{max}$ maximum amount of adsorbed/unit mass of adsorbent corresponding to complete coverage of the adsorption sites (mg/g)
- b Langmuir constant representing the affinity (sorption/desorption) of binding sites (L/mg)

The separation factor or equilibrium parameter can be calculated by using following equation (Eq. 4):

$$R_L = \frac{1}{1 + bC_o} \quad (4)$$

where

$R_L$ separation factor or equilibrium parameter (dimensionless)
$C_o$ highest initial concentration of metal ion (mg/g)

In order to determine the b and $Q_{max}$ values a graph between $\frac{1}{Q_e}$ versus $\frac{1}{C_e}$ is plotted and both the values with b as slope and $Q_{max}$ as intercept can be determined. The obtained values of different parameters in the present study using the Langmuir isotherm adsorption model for rice husk and saw dust as adsorbent has been summarized in Table 5. The $Q_m$ and b value of 40.58 mg/g and 0.3072 L/mg has been obtained, respectively, using rice husk as an adsorbent with a correlation coefficient $R^2$ value of 0.99. However, the $Q_m$ and b value of 35.88 mg/g and 0.22 L/mg has been obtained, respectively, using saw dust as an adsorbent with a correlation coefficient $R^2$ value of 0.98 (Fig. 5).

However different studies being carried out under various investigators for the adsorption of heavy metals using rice husk and sawdust as adsorbent using different adsorption isotherms. Pourfadakari et al. [32] in an experimental study on adsorption of chromium Cr(VI) from the aqueous solution used nano-sized cellulose fibers synthesized from rice husk. It has been reported that using Langmuir' s equation isotherm the $Q_{max}$, b and $R^2$ values have been found to be 3.76 mg/g, 0.042 L/mg and 0.998, respectively. Sivakumar [36] in another study observed the $Q_{max}$, b and $R^2$ values to be 78.12 mg/g, 0.06 L/mg and 0.993, respectively for adsorption of Cr(VI) in a tannery industry wastewater using rice husk silica as adsorbent. Mullick et al. [30] has also experiment an experimental study that observed the $Q_{max}$, b and $R^2$

**Table 5** Comparison of Langmuir and Freundlich's adsorbent isotherms computed for the removal of Cr(VI) by rice husk and sawdust in the present study and literature cited

| S. no. | Adsorbent type | Langmuir isotherm values | | | Freundlich isotherm values | | | Literature cited |
|---|---|---|---|---|---|---|---|---|
| | | $Q_{max}$ (mg/g) | $b$ (L/mg) | $R^2$ | $K_f$ (mg/g) | $n$ | $R^2$ | |
| 1 | Rice husk | 3.76 | 0.042 | 0.998 | 1.71 | 3.16 | 0.87 | [32] |
| 2 | Rice husk | 78.12 | 0.06 | 0.993 | 3.89 | 4.46 | 0.962 | [36] |
| 3 | Rice husk | 24.39 | 1.86 | 0.997 | – | – | – | [30] |
| 4 | Treated rice Straw | 1.61 | 1.60 | 0.993 | 2.60 | 4.80 | 0.954 | [9] |
| 5 | Saw dust | 44.05 | 0.22 | 0.999 | 25.48 | 0.10 | 0.923 | [23] |
| 6 | Saw dust | 0.19 | 0.219 | 0.975 | 0.229 | 1.29 | 0.993 | [4] |
| 7 | Saw dust | 3.6 | 2.36 | 0.961 | – | – | – | [8] |
| 8 | Pretreated saw dust | 308.52 | 0.0027 | – | 2.88 | 1.49 | – | [31] |
| 9 | Rice husk | 40.58 | 0.30 | 0.997 | 11.21 | 2.57 | 0.922 | Present study |
| 10 | Saw dust | 35.89 | 0.22 | 0.987 | 9.66 | 2.81 | 0.94 | Present study |

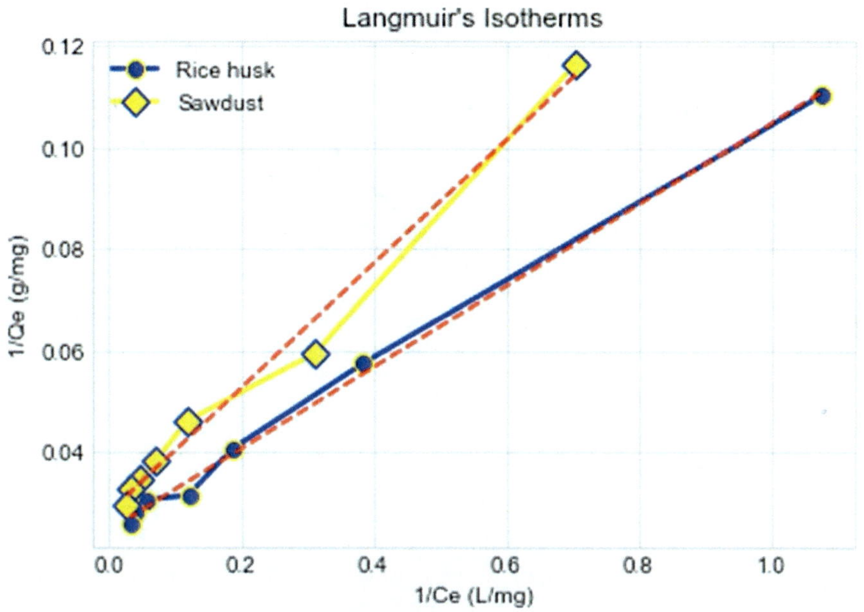


**Fig. 5** Langmuir isotherms model used for assessing the adsorption rate

values to be 24.39 mg/g, 1.86 L/mg and 0.997, respectively, at 298 °K for removal of hexavalent chromium from aqueous solution using rice husk based carbon as adsorbent. Brahmaiah et al. [9] in a study using Langmuir's isotherms obtained the $Q_{max}$, b and $R^2$ values to be 0.28 mg/g, 0.14 L/mg, and 0.898, respectively, for the removal of chromium using untreated rice straw as adsorbent (Fig. 5).

Karthikeyan et al. [23] in an experimental study determined the Cr(VI) adsorption using sawdust activated carbon as adsorbent and assessed the $Q_{max}$, b and $R^2$ values to be 44.05 mg/g, 0.22 L/mg, and 0.999, respectively, at 303 °K. Alibadi et al. [4] in an investigation for the removal of hexavalent chromium using sawdust as adsorbent found the Langmuir's constant $Q_{max}$, b and $R^2$ values to be 0.19 mg/g, 0.21 L/mg and 0.975, respectively. Baral et al. [8] in another study used treated saw dust as an adsorbent for removal of hexavalent chromium and assessed the Langmuir's constants $Q_{max}$, b and $R^2$ values to be 3.6 mg/g, 2.36 L/mg, and 0.961, respectively, at 303 °K. Politi and Sidiras [31] in various sorption experiments for hexavalent chromium removal using sawdust as adsorbent obtained Langmuir's constants $Q_{max}$, b and $R^2$ values as 168.45 mg/g, 0.0042 L/mg and 6.15 in terms of the standard error of estimate (SEE). Same values when sorption performed with pretreated sawdust using 50% diethylene glycol solution at 220 °C for 50 min, found to be 308.52 mg/g, 0.0027 L/mg, and 10.30 as SEE [31].

**Freundlich adsorption isotherm**: Freundlich [16] has given a model that interprets the adsorption over the heterogeneous surface with no restrictions of monomolecular layer formation. Thus it is generally used for heterogeneous surface

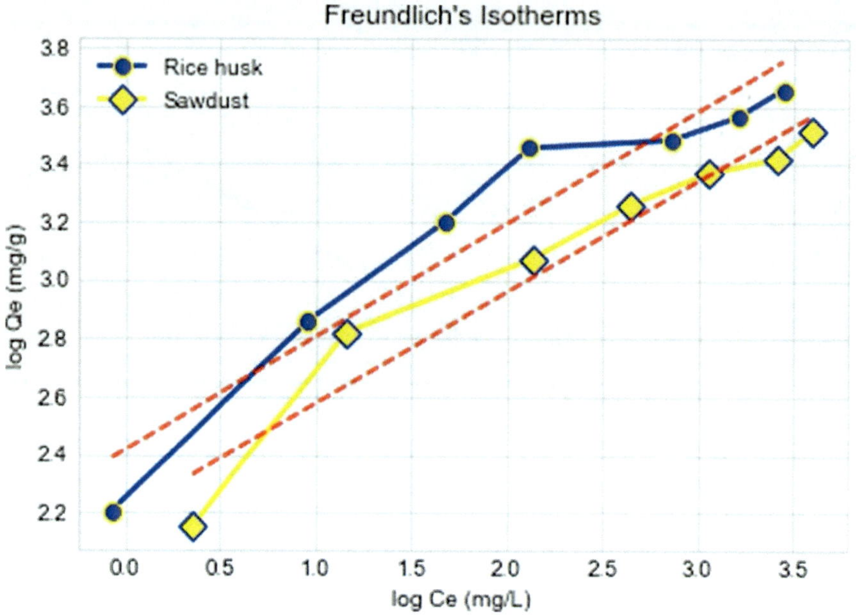


**Fig. 6** Freundlich isotherms model used for adsorption rate determination

adsorbents to present the ratio of the adsorbed concentration of metal on a known mass of adsorbent to the concentration of metal in the sample which differs at different concentrations [16]. Therefore, there is no homogeneous distribution over active sites, and binding strength may differ for the same metal at the same layer of adsorbate due to the heterogeneous distribution of absorbates over the surface of sites. It can be used to model the adsorption of organic and inorganic compounds as well for different adsorbents [15]. The isotherm model is represented by the following mathematical equation (Eq. 5) (Fig. 6):

$$\log Q_e = \log K_f + \frac{1}{n} \log C_e \tag{5}$$

where

$Q_e$ Amount adsorbed (mg/g)
$K_f$ Freundlich constant (L/g) indicating the adsorption coefficient that indicates the relative adsorption capacity of the adsorbent of the bonding energy
n Freundlich equation exponential that represents energetic heterogeneity of the adsorption surface
$C_e$ equilibrium concentration of metal ion (mg/L)

The required parameters in terms of $K_f$ and n values can be obtained by plotting a graph in terms of log $Q_e$ v/s log $C_e$, with 1/n as slope and log $K_f$ as intercept. The computed parameters using the Freundlich isotherm is summarized in Table 5. The

$K_f$ and n value of 11.21 $[(mg/g)(L/mg)]^{1/n}$ and 2.57 has been obtained, respectively, using rice husk as an adsorbent with a correlation coefficient $R^2$ value of 0.92. However, the $K_f$ and n value of 9.66 $[(mg/g)(L/mg)]^{1/n}$ and 2.81 has been obtained, respectively, using saw dust as an adsorbent with a correlation coefficient $R^2$ value of 0.93. The n value greater than 1 indicates good sorption.

Many investigators have performed various studies pertaining to rice husk and sawdust as adsorbents for the removal of heavy metals and assessed the heavy metals removal efficiency using Freundlich adsorption isotherm models. Pourfadakari et al. [32] in an experimental study, has reported that using Freundlich's isotherm model $K_f$, n and $R^2$ values have been reported to be 1.71 $[(mg/g)(L/mg)]^{1/n}$, 3.16 and 0.87, respectively. Sivakumar [36] assessed the Freundlich's constant $K_f$, n and $R^2$ values to be 3.89 $[(mg/g)(L/mg)]^{1/n}$, 4.46 and 0.961, respectively. A study carried out by Brahmaiah et al. [9] reported the Freundlich's constants $K_f$, n and $R^2$ values to be 8.31 $[(mg/g)(L/mg)]^{1/n}$, 1.20, and 0.979, respectively, for the removal of chromium using untreated rice straw as adsorbent.

Karthikeyan et al. [23] in an experimental study observed the removal of Cr (VI) using sawdust activated carbon as adsorbent and observed the $K_f$, n and $R^2$ values to be 25.48 $[(mg/g)(L/mg)]^{1/n}$, 0.10, and 0.923, respectively, at 303 °K. However, a similar study carried out by Alibadi et al. [4] assessed the removal of hexavalent chromium from aqueous solution using sawdust as adsorbent and computed Freundlich's constants $K_f$, n and $R^2$ values to be 0.22$[(mg/g)(L/mg)]^{1/n}$, 1.29 and 0.993, respectively. Politi and Sidiras [31] study computed the Freundlich's constants $K_f$, n and $R^2$ values to be 3.70 $[(mg/g)(L/mg)]^{1/n}$, 1.80 and 8.01 in the removal of hexavalent chromium using untreated sawdust as adsorbent. However, the $K_f$, n and $R^2$ values were reported to be 2.88 $[(mg/g)(L/mg)]^{1/n}$, 1.49, and 11.72, respectively, using pre-treated sawdust as adsorbent [31].

From the present study, it can be compared that the correlation coefficients indicates the fact that Langmuir isotherm provides a better model as compared to the Freundlich model for rice husk as an adsorbent. The adsorption in such a system is based on monolayer sorption on to the surface limiting the finite number of identical sorption sites. However, Freundlich model provides better energetic heterogeneity of the adsorption surface using rice husk and saw dust as adsorbents. Further adsorption studies can be carried out using rice husk and saw dust as an adsorbent for the removal of other heavy metals such as cadmium, mercury, lead, copper, and arsenic. The present study indicates that the adsorbents prepared from agricultural waste can be best utilized for the removal of chromium from the wastewater. This will provide a twofold advantage as the toxicity of heavy metals from wastewater can be removed effectively and the agricultural waste can be utilized as adsorbents thus making the system economical and reducing the burden of agricultural waste on the environment.

# 4 Conclusion

From the present study, it can be concluded that the maximum Cr (VI) removal efficiency of 92% has been observed using 3 g of rice husk as adsorbent at a pH value of 6 with Cr (VI) concentration of 10 mg/L at a contact time of 100 min. However, the maximum Cr (VI) removal efficiency of 82% has been observed using 3 g of saw dust as an adsorbent with the same Cr(VI) concentration and under similar conditions. The correlation coefficient $R^2$ value of 0.997 and 0.987 indicates the Langmuir isotherm to be the best model using rice husk and saw dust as adsorbents, respectively. The $K_f$ and n value of 11.21 $[(mg/g)(L/mg)]^{1/n}$ and 2.57 has been obtained, respectively, using rice husk, and the same value of 9.66 $[(mg/g)(L/mg)]^{1/n}$ and 2.81 using saw dust as an adsorbent with correlation coefficient $R^2$ values of 0.922 and 0.94, respectively, that indicates Freundlich model a better energetic heterogeneity of the adsorption surface using rice husk and saw dust as adsorbents.

# References

1. Abas SNA, Ismail MHS, Kamal ML, Izhar S (2013) Adsorption process of heavy metals by low cost adsorbent: a review. World Appl Sci J 28(11):1518–1530
2. Abdel-Ghani NT, Hefny M, El-Chaghaby GAF (2007) Removal of lead from aqueous solution using low cost abundantly available adsorbents. Int J Environ Sci Technol 4(1):67–73. https://doi.org/10.1007/BF03325963
3. Ahmaruzzaman M, Gupta VK (2011) Rice husk and its ash as low-cost adsorbents in water and wastewater treatment. Indus Eng Chem Res 50(24):13589–13613. https://doi.org/10.1021/ie201477c
4. Alibadi M, Morshedzadeh K, Soheyli H (2006) Removal of hexavalent chromium from aqueous solution by lignocellulosic solid wastes. Int J Environ Sci Technol 3(3):321–325. https://doi.org/10.1007/BF03325940
5. Andrabi SMA (2010) Sawdust of lam tree (Cordia africana) as a low-cost, sustainable and easily available adsorbent for the removal of toxic metals like Pb(II) and Ni(II) from aqueous solution. Eur J Wood Wood Products 69(1):75–83. https://doi.org/10.1007/s00107-009-0398-x
6. Ansari R, Raofie F (2006) Removal of lead ion from aqueous solutions using sawdust coated by polyaniline. E J Chem 3(1):49–59. https://doi.org/10.1155/2006/378619
7. Argun ME, Dursun S, Ozdemir C, Karatas M (2007) Heavy metal adsorption by modified oak sawdust: thermodynamics and kinetics. J Hazard Mater 141(1):77–85. https://doi.org/10.1016/j.jhazmat.2006.06.095
8. Baral SS, Das SN, Rath P (2006) Hexavalent chromium removal from aqueous solution by adsorption on treated sawdust. Biochem Eng J 31(3):216–222. https://doi.org/10.1016/j.bej.2006.08.003
9. Brahmaiah T, Spurthi L, Chandrika K, Chandra LK, Yashas S, Ramanaiah S, Prasad KSS (2015) Removal of heavy metals from waste water using low cost adsorbent. Int J Trend Res Dev 3(1):63–68
10. Buasri A, Chaiyut N, Tapang K, Jaroensin S, Panphrom S (2012) Removal of Cu2 + from aqueous solution by biosorption on rice straw-an agricultural waste biomass. Int J Environ Sci Dev 3(1):10–14
11. Bulut Y, Baysal Z (2006) Removal of Pb(II) from wastewater using wheat bran. J Environ Manage 78(2):107–113. https://doi.org/10.1016/j.jenvman.2005.03.010

12. Chowdhury S, Mishra R, Saha P, Kushwaha P (2011) Adsorption thermodynamics, kinetics and isosteric heat of adsorption of malachite green onto chemically modified rice husk. Desalination 265(1–3):159–168. https://doi.org/10.1016/j.desal.2010.07.047
13. Emsley J (2011) Nature's building blocks: an AZ guide to the elements. Oxford University Press, Oxford
14. Evanko CR, Dzombak DA (1997) Remediation of metals-contaminated soils and groundwater. TE-97-01 Ground-water remediation technologies analysis center, Pittsburgh, USA
15. Febrianto J, Kosasih AN, Sunarso J, Ju Y-H, Indraswati N, Ismadji S (2009) Equilibrium and kinetic studies in adsorption of heavy metals using biosorbent: a summary of recent studies. J Hazard Mater 162(2–3):616–645. https://doi.org/10.1016/j.jhazmat.2008.06.042
16. Freundlich H (1907) About adsorption in solutions. J Phys Chem 57U(1):385–470. https://doi.org/10.1515/zpch-1907-5723
17. Gunatilake SK (2015) Methods of removing heavy metals from industrial waste water. J Multidiscip Eng Sci Studies 1(1):11
18. He Z, Shentu J, Yang X, Baligar VC, Zhang T, Stoffella PJ (2015) Heavy metal contamination of soils: sources, indicators and assessment. J Environ Indicators 9:17–18
19. Hussain A, Ahmed S (2018) Advanced treatment techniques for industrial wastewater. Hershey, IGI Global
20. Hussain A, Maitra J, Khan KA (2017) Development of biochar and chitosan blend for heavy metals uptake from synthetic and industrial wastewater. Appl Water Sci 7(8):4525–4537. https://doi.org/10.1007/s13201-017-0604-7
21. Kabata-Pendias A, Pendias H (2001) Trace metals in soils and plants (2nd edn). CRC Press, Boca Raton, Fla, USA
22. Kalderis D, Bethanis S, Paraskeva P, Diamadopoulos E (2008) Production of activated carbon from bagasse and rice husk by a single-stage chemical activation method at low retention times. Biores Technol 99(15):6809–6816. https://doi.org/10.1016/j.biortech.2008.01.041
23. Karthikeyan T, Rajgopal S, Miranda LR (2005) Chromium(VI) adsorption from aqueous solution by Hevea Brasilinesis sawdust activated carbon. J Hazard Mater 124(1–3):192–199. https://doi.org/10.1016/j.jhazmat.2005.05.003
24. Khan S, Cao Q, Zheng YM, Huang YZ, Zhu YG (2008) Health risks of heavy metals in contaminated soils and food crops irrigated with wastewater in Beijing, China. Environ Pollut 152(3):686–692. https://doi.org/10.1016/j.envpol.2007.06.056
25. Khayyun TS, Mseer AH (2011) Comparison of the experimental results with the Langmuir and Freundlich models for copper removal on limestone adsorbent. Appl Water Sci 9(170):1–8. https://doi.org/10.1007/s13201-019-1061-2
26. Langmuir I (1918) The adsorption of gases on plane surfaces of glass, mica and platinum. J Am Chem Soc 40(9):1361–1403. https://doi.org/10.1021/ja02242a004
27. Lata S, Samadder SR (2014) Removal of heavy metals using rice husk: a review. Int J Environ Res Dev 4(2):165–170
28. Leyva-Ramos R, Bernal-Jacome LA, Acosta-Rodriguez I (2005) Adsorption of cadmium(II) from aqueous solution on natural and oxidized corncob. Separ Purific Technol 45(1):41–49. https://doi.org/10.1016/j.seppur.2005.02.005
29. Mohammed AS, Kapri A, Goel R (2011) Heavy metal pollution: source, impact and remedies. Biomanage Metal Contaminat Soils 1–28. https://doi.org/10.1007/978-94-007-1914-9_1
30. Mullick A, Moulik S, Bhattacharjee S (2017) Removal of hexavalent chromium from aqueous solutions by low-cost rice husk-based activated carbon: kinetic and thermodynamic studies. Ind Chem Eng 60(1):58–71. https://doi.org/10.1080/00194506.2017.1288173
31. Politi D, Sidiras D (2020) Modified spruce sawdust for sorption of hexavalent chromium in batch systems and fixed-bed columns. Molecules 25(21):5156–5178. https://doi.org/10.3390/molecules25215156
32. Pourfadakari S, Jorfi S, Ahmadi M, Takdastan A (2017) Experimental data on adsorption of Cr(VI) from aqueous solution using nanosized cellulose fibers obtained from rice husk. Data Brief 15:887–895. https://doi.org/10.1016/j.dib.2017.10.043

33. Sahmoune MN, Louhab K, Boukhiar A (2010) Advanced biosorbents materials for removal of chromium from water and wastewaters. Environ Progress Sustain Energy 30(3):284–293. https://doi.org/10.1002/ep.10473
34. Sahmoune MN, Yeddou AR (2016) Potential of sawdust materials for the removal of dyes and heavy metals: examination of isotherms and kinetics. Desalinat Water Treat 57(50):24019–24034. https://doi.org/10.1080/19443994.2015.1135824
35. Sen M, Dastidar MG (2010) Chromium removal using various biosorbents. Ir J Environ Health Sci Eng 7(3):182–190
36. Sivakumar D (2015) Hexavalent chromium removal in a tannery industry wastewater using rice husk silica. Glob J Environ Sci Manage 1(1):27–40
37. Vasudevan M, Sakaria PL, Bhatt AS, Mody HM, Bajaj HC (2011) Effect of concentration of aminopropyl groups on the surface of MCM-41 on adsorption of $Cu^{2+}$. Indus Eng Chem Res 50(19):11432–11439. https://doi.org/10.1021/ie201448q
38. Wilson K, Yang H, Seo CW, Marshall WE (2006) Select metal adsorption by activated carbon made from peanut shells. Biores Technol 97(18):2266–2270. https://doi.org/10.1016/j.biortech.2005.10.043
39. Wuana RA, Okieimen FE (2011) Heavy metals in contaminated soils: a review of sources, chemistry, risks and best available strategies for remediation. Int Schol Res Notices Ecol 2011:1–20. https://doi.org/10.5402/2011/402647
40. Zhang M-K, Liu Z-Y, Wang H (2010) Use of single extraction methods to predict bioavailability of heavy metals in polluted soils to rice. Commun Soil Science and Plant Anal 41(7):820–831. https://doi.org/10.1080/00103621003592341
41. Zhang Y, Zheng R, Zhao J, Ma F, Zhang Y, Meng Q (2014) Characterization of $H_3PO_4$-treated rice husk adsorbent and adsorption of copper(II) from aqueous solution. Biomed Res Int 2014:1–8. https://doi.org/10.1155/2014/496878
42. Zhao Q, Kaluarachchi JJ (2002) Risk assessment at hazardous waste-contaminated sites with variability of population characteristics. Environ Int 28(1–2):41–53. https://doi.org/10.1016/s0160-4120(02)00003-x

# Chapter 26
# Some New Results on the Deformable Fractional Calculus Using D'Alambert Approach and Mittag-Leffler Function

**Priyanka Ahuja, Amit Ujlayan, Fahed Zulfeqarr, and Mohit Arya**

## 1 Introduction

Fractional calculus is a conclusion of classical calculus related with the operations of differentiation and integration of non-integer order. The idea of fractional calculus came into existence with the concept of classical ones. The main source can be found in the communication of Leibniz and L'Hospital in 1695 where the question of the sense of the semi-derivative has been lift up [6]. Many types of fractional derivatives have been introduced till now. Most of the definitions of fractional derivative are in integral form [8]. Some famous definition were given by Riemann-Liouville, Hadamard, Caputo, Riesz and Grunwald-Letnikov [9, 13].

In 2014, R. Khalil explained a limit-based definition of fractional derivative named as conformable fractional derivative [5]. His definition does not include zero and negative numbers. So, here we define a new derivative, named as deformable derivative [2, 14]. This derivative is much simpler than conformable fractional derivative. It also overcomes the shortcoming of conformable fractional derivative.

P. Ahuja (✉) · A. Ujlayan · F. Zulfeqarr · M. Arya
Gautam Buddha University, Greater Noida, UP, India
e-mail: pahuja836@gmail.com

A. Ujlayan
e-mail: iitk.amit@gmail.com

F. Zulfeqarr
e-mail: fahedzul@gmail.com

M. Arya
e-mail: mohitarya818@gmail.com



D. Singh et al. (eds.), *Proceedings of International Conference on Scientific and Natural Computing*, Algorithms for Intelligent Systems,
https://doi.org/10.1007/978-981-16-1528-3_26

**Definition 1** Let $g(t)$ be real-valued function defined on interval $(a, b)$. For a given number $\alpha$, $0 \leq \alpha \leq 1$, we define *deformable derivative* by the following limit:

$$\lim_{\epsilon \to 0} \frac{(1+\epsilon\beta)g(t+\epsilon\alpha) - g(t)}{\epsilon}, \quad \text{where } \alpha + \beta = 1. \tag{1}$$

We denote it by $D^\alpha g(t)$, if limit exists.

**Remark 1** It is clear that the definition (1) is compatible with $\alpha = 0, 1$. For, if $\alpha = 0$, $D^0 g(t) = g(t)$ which is the usual convention and if $\alpha = 1$, $Dg(t) = g'(t)$. In the whole article, we assume that $0 < \alpha \leq 1$.

## 2 Preliminary Results

In this section, we established a relation of deformable derivative with function and its derivative. That's why we named it as deformable derivative.

The first theorem shows that *differentiability implies α-differentiability.*

**Theorem 1** *A differentiable g at a point $t \in (a, b)$ is always α-differentiable at that point for any α. And in this case, we have*

$$D^\alpha g(t) = \beta g(t) + \alpha D g(t), \quad \textit{where } \alpha + \beta = 1. \tag{2}$$

***Proof*** By definition, we have

$$\begin{aligned} D^\alpha g(t) &= \lim_{\epsilon \to 0} \frac{(1+\epsilon\beta)\, g(t+\alpha\epsilon) - g(t)}{\epsilon} \\ &= \lim_{\epsilon \to 0} \left( \frac{g(t+\alpha\epsilon) - g(t)}{\epsilon} + \beta g(t+\alpha\epsilon) \right) \\ &= \alpha \cdot Dg(t) + \beta \cdot \lim_{\epsilon \to 0} g(t+\alpha\epsilon). \end{aligned}$$

Since $g$ is differentiable at $t$ so continuous as well. Thus, both the terms exist.

**Definition 2** Suppose $g$ is $n$-times differentiable at $t \in (a, b)$. For any $\alpha \in (n, n+1]$, we define deformable derivative in a very simple way and express it by the following limit:

$$D^\alpha g(t) = \lim_{\epsilon \to 0} \frac{(1+\epsilon\{\beta\}))D^n g\,(t+\epsilon\{\alpha\}) - D^n g(t)}{\epsilon},$$

where $\{\alpha\}$ is the fractional part of $\alpha$ and $\{\alpha\} + \{\beta\} = 1$.

As the result of the above definition, if $g^{(n+1)}$ exists, we have

$$D^\alpha g(t) = \{\beta\} D^n g(t) + \{\alpha\} D^{n+1} g(t).$$

## 3 Basic Properties of Deformable Derivative

This section deals with the discussion of some fundamental properties of deformable derivative like linearity, commutativity and Leibniz rule. Also deformable derivative of some elementary functions are given [3, 10].

**Theorem 2** *The operator $D^\alpha$ holds the following properties:*

*(a) Linearity:* $D^\alpha\,(ag + bh) = aD^\alpha g + bD^\alpha h$.
*(b) Commutativity:* $D^{\alpha_1} \cdot D^{\alpha_2} = D^{\alpha_2} \cdot D^{\alpha_1}$.
*(c) For a constant function c,* $D^\alpha(c) = \beta c$.
*(d)* $D^\alpha(g \cdot h) = (D^\alpha g) \cdot h + \alpha g \cdot Dh$. *Hence, $D^\alpha$ does not follow the Leibniz rule.*

Deformable derivative of the functions, who behave well with respect to differentiation can be obtained from Eq. (2) in Theorem (1). Following preposition lists out the deformable derivatives of some elementary functions.

**Proposition 1** *1.* $D^\alpha(x^r) = \beta x^r + r\alpha x^{r-1}, \quad r \in \mathscr{R}$.
*2.* $D^\alpha(e^x) = (\alpha + \beta)e^x$.
*3.* $D^\alpha(\sin x) = \beta \sin x + \alpha \cos x$.
*4.* $D^\alpha(\log x) = \beta \log x + \frac{\alpha}{x}, x > 0$.

## 4 Fractional Integral

This section introduced fractional integral for deformable derivative as an inverse operator [11, 12]. Some fundamental properties to this fractional integral are discussed. Also, fractional integral of some elementary functions is also given. We are assuming that all the functions considered in this section are continuous.

**Definition 3** Let $g$ be a continuous function defined on $[a, b]$. We define *α-fractional Integral* of $g$, denoted by $I_a^\alpha g$, by the integral

$$I_a^\alpha g(t) = e^{\frac{-\beta}{\alpha}t} \int_a^t e^{\frac{\beta}{\alpha}x} g(x) d_\alpha x, \tag{3}$$

where $\alpha + \beta = 1,\ \alpha \in (0, 1]\, and\ d_\alpha x = \frac{1}{\alpha}dx$.

Next theorem contains some basic properties of this fractional integral.

**Theorem 3** *The operator $I_a^\alpha$ follows the following properties:*

*(i) Linearity:* $I_a^\alpha\,(bg + ch) = bI_a^\alpha g + cI_a^\alpha h$.

*(ii) Commutativity:* $I_a^{\alpha_1} I_a^{\alpha_2} = I_a^{\alpha_2} I_a^{\alpha_1}$, *where* $\alpha_i + \beta_i = 1, i = 1, 2$.

We list out the fractional integral of some elementary functions in the following preposition.

**Proposition 2** *1.* $I_a^\alpha \sin x = \frac{1}{\alpha^2+\beta^2}\left(\beta \sin x - \alpha \cos x + e^{\frac{\beta}{\alpha}(a-x)}\,(\alpha \cos a - \beta \sin a)\right).$

2. $I_a^\alpha e^x = \left(e^x - e^{\frac{(a-\beta x)}{\alpha}}\right).$
3. $I_a^\alpha k = \frac{k}{\beta}\left(1 - e^{\frac{\beta}{\alpha}(a-x)}\right)$*, where k is a constant.*
4. $I_0^\alpha x^n = \frac{1}{\beta}\left(\sum_{k=0}^{n} \frac{(-1)^k\, n!}{(n-k)!}\left(\frac{\alpha}{\beta}\right)^k x^{n-k} + (-1)^{n+1}\, n!\left(\frac{\alpha}{\beta}\right)^n e^{-\frac{\beta}{\alpha}x}\right).$

## 5 Application: D'Alambert Approach

In this section, the solution of deformable fractional differential equation of the type [4, 7],

$$D^\alpha D^\alpha y + P D^\alpha y + Q y = 0, \tag{4}$$

will be discussed, where $P$ and $Q$ are $\alpha$-differentiable functions of $x$.
**Approach:** We start by assuming that (4) has a solution, say, $y_1$. We want to find second solution $y_2$ such that $y_2 = v y_1$, where $v$ is an $\alpha$-differentiable function of $x$.

For this,

$$\begin{aligned} D^\alpha y_2 &= D^\alpha (v y_1) \\ 2\alpha &= (D^\alpha y_1)\, v + \alpha y_1\, (Dv) \end{aligned}$$

and

$$\begin{aligned} D^\alpha D^\alpha y_2 &= D^\alpha \left[(D^\alpha y_1)\, v + \alpha y_1\, (Dv)\right] \\ &= v D^\alpha D^\alpha y_1 + 2\alpha D^\alpha y_1\, (Dv) + \alpha^2 y_1 \left(D^2 v\right). \end{aligned}$$

But $y_2$ is a solution of (4), if and only if

$$\begin{aligned} 0 &= D^\alpha D^\alpha y_2 + P D^\alpha y_2 + Q y_2 \\ &= v D^\alpha D^\alpha y_1 + 2 D^\alpha y_1\, (Dv) + \alpha^2 y_1 \left(D^2 v\right) + P\, (D^\alpha y_1)\, v + P\alpha y_1\, (Dv) + Q v y_1. \end{aligned}$$

Since $y_1$ is assumed to be a solution of (4), we have that

$$D^\alpha D^\alpha y_1 + P D^\alpha y_1 + Q y_1 = 0.$$

With this the above expression becomes

$$2D^{\alpha}y_1(Dv) + \alpha y_1\left(D^2 v\right) + Py_1(Dv) = 0. \tag{5}$$

Using (2), Eq. (5) boils down to

$$\alpha y_1 v'' + (2\beta y_1 + 2\alpha y_1^{'} + Py_1)v' = 0. \tag{6}$$

Now, if we let $v' = w$, then Eq. (6) becomes

$$\alpha\frac{w'}{w} = -\left(2\beta + 2\alpha\frac{y_1^{'}}{y_1} + P\right). \tag{7}$$

The problem becomes find $w$ that satisfies (7).
To do this, we simplify Eq. (7) and get

$$w = A\frac{e^{\left(-2\frac{\beta}{\alpha}x - \int Pd_{\alpha}x\right)}}{y_1^2}, \tag{8}$$

where $A$ is any arbitrary constant and $d_{\alpha}x = \frac{1}{\alpha}dx$. Therefore,

$$v = \alpha\int\left[A\frac{e^{\left(-2\frac{\beta}{\alpha}x - \int Pd_{\alpha}x\right)}}{y_1^2}\right]d_{\alpha}x. \tag{9}$$

Hence, our second solution, $y_2$, for Eq. (4) is given by

$$y_2 = \alpha y_1\int\left[A\frac{e^{\left(-2\frac{\beta}{\alpha}x - \int Pd_{\alpha}x\right)}}{y_1^2}\right]d_{\alpha}x. \tag{10}$$

We state the above approach as follows.

**Theorem 4** *Given a deformable fractional differentiable equation of the type*

$$D^{\alpha}D^{\alpha}y + PD^{\alpha}y + Qy = 0, \tag{11}$$

*where $P$ and $Q$ are $\alpha$-differentiable functions of $x$. Assume $y_1$ is a solution of* (11), *then other solution, $y_2$, is given by*

$$y_2 = \alpha y_1\int\left[A\frac{e^{\left(-2\frac{\beta}{\alpha}x - \int Pd_{\alpha}x\right)}}{y_1^2}\right]d_{\alpha}x, \tag{12}$$

*where $A$ is any arbitrary constant and $d_{\alpha}x = \frac{1}{\alpha}dx$.*

**Example 1** Consider the differential equation

$$D^{\frac{1}{2}}D^{\frac{1}{2}}y - D^{\frac{1}{2}}y + \frac{1}{4}y = 0$$

. Clearly, $y_1 = x$ is a solution of above equation. Here $P = -1, \alpha = \beta = \frac{1}{2}$. Using formula (12), we obtain another solution of the form

$$y_2 = \frac{1}{2}x \int \left[ A \frac{e^{\left(-2x + \int d_{1/2}x\right)}}{x^2} \right] d_{1/2}x.$$

## 6 Application to Fractional Differential Equations Using Mittag-Leffler Function

This section will represent the solution of fractional differential equations using Mittag-Leffler function of one parameter [1].

Mittag-Leffler function in one parameter can be written as

$$E_q(z) = \sum_{k=0}^{\infty} \frac{z^k}{\Gamma(qk+1)}, \tag{13}$$

where $q > 0$.

Taking $q = 1$, Eq. (13) becomes

$$E_1(z) = \sum_{k=0}^{\infty} \frac{z^k}{\Gamma(k+1)} \tag{14}$$

$$= e^z. \tag{15}$$

Equation (15) concludes that Mittag-Leffler function behaves like exponential function at $q = 1$.

**Example 2** We now consider a non-homogeneous linear fractional equation:

$$D^{1/2}y + y = te^{-t}.$$

This can be written as

$$\frac{1}{2}y + \frac{1}{2}Dy + y = te^{-t} \quad \Rightarrow \quad Dy + 3y = 2te^{-t}$$

whose general solution is given by

$$y(t) = Ce^{-3t} + \left(t - \frac{1}{2}\right)e^{-t}.$$

By using (15), we obtain $y(t)$ as

$$CE_1(-3t) + \left(t - \frac{1}{2}\right)E_1(-t),$$

where $C$ is a constant.

## References


1. Agarwal P, Rogosin SV, Trujillo JJ (2015) Certain fractional integral operators and the generalized multi-index mittag-leffler functions. Proc Math Sci 125(3):291–306
2. Ahuja P, Zulfeqarr F, Ujlayan A Deformable fractional derivative and its applications. American Institute of Physics
3. Abdon A, Dumitru B, Ahmed A (2015) New properties of conformable derivative. Open Math 13(1):889–898
4. Iyiola OS, Nwaeze ER (2016) Some new results on the new conformable fractional calculus with application using D'Alambert approach. Progress Fract Differ Appl 2(2):115–122
5. Khalil R, Al Horani M, Yusuf A, Sababhed M (2014) A new definition of fractional derivative. J Comput Appl Math 264:65–70
6. Leibniz GW (1849) Letter from hanover, Germany to Johann G. F. A. L'Hospital, September 30, 1695. Leibniz Mathematische Schriften, Olms-Verlag, Hildeshein, Germany 1962:301–302
7. Miller KS, Ross B (1993) An introduction to the fractional calculus and fractional differential equations. Wiley, New York
8. Oldham KB, Spanier J (1974) The fractional calculus. Academic Press, New York
9. Podlubny I (1999) Fractional differential equations. Academic Press, San Diego CA
10. Tarasov VE (2013) No violation of the leibniz rule. no fractional derivative. Commun Nonline Sci Numer Simul 18:2945–2948
11. Fuat Usta and Mehmet Zeki Sarikaya (2017) Some improvements of conformable fractional integral. Int J Anal Appl 14(2):162–166
12. Yang XJ, Baleanu D, Srivastava HM (2016) Introduction to local fractional derivative and integral operators. Local Fract Integral Transf Appl 24:1–55
13. Zhibao Z, Wei Z, Hongzhe D (2019) A new definition of fractional derivative. Int J Non-Linear Mech 108:1–6
14. Zulfeqarr F, Ujlayan A, Ahuja P A new fractional derivative and its fractional integral with some applications. https://aps.arxiv.org/pdf/1705.00962.pdf

# Author Index





D. Singh et al. (eds.), *Proceedings of International Conference on Scientific and Natural Computing*, Algorithms for Intelligent Systems,
https://doi.org/10.1007/978-981-16-1528-3